Early Modern Studies after the Digital Turn

NEW TECHNOLOGIES IN MEDIEVAL AND RENAISSANCE STUDIES 6

SERIES EDITORS   William R. Bowen and Raymond G. Siemens

MEDIEVAL AND RENAISSANCE
TEXTS AND STUDIES

VOLUME 502

Early Modern Studies after the Digital Turn

Edited by
Laura Estill, Texas A&M University
Diane K. Jakacki, Bucknell University
Michael Ullyot, University of Calgary

Iter Press
Toronto, Ontario
in collaboration with
ACMRS
(Arizona Center for Medieval and Renaissance Studies)
Tempe, Arizona
2016

© Copyright 2016
Iter, Inc. and the Arizona Board of Regents for Arizona State University

Library of Congress Cataloging-in-Publication Data
Names: Estill, Laura, editor. | Jackacki, Diane K., editor. | Ullyot, Michael, 1976– editor.
Title: Early modern studies after the digital turn / edited by Laura Estill, Diane K. Jackacki, Michael Ullyot.
Description: Tempe, Arizona : Arizona Center for Medieval and Renaissance Studies, 2016. | Series: Medieval and renaissance texts and studies ; Volume 502 | Series: New technologies in medieval and renaissance studies ; Volume 6 | Includes bibliographical references and index.
Identifiers: LCCN 2016027567 (print) | LCCN 2016040169 (ebook) | ISBN 9780866985574 (hardcover : alk. paper) | ISBN 9780866987257 (ebook)
Subjects: LCSH: Digital humanities--Research--Methodology. | English literature--Early modern, 1500-1700--Research--Methodology. | English literature--Early modern, 1500-1700--Information resources.
Classification: LCC AZ105 .E27 2016 (print) | LCC AZ105 (ebook) | DDC 001.30285--dc23
LC record available at https://lccn.loc.gov/2016027567

Cover image:
File: A13730.jpg. National Gallery of Art | NGA Images, accessed September 6, 2016, http://www.nga.gov/content/ngaweb/Collection/art-object-page.50722.html.

Portrait details:
Detail of *Portrait of a Merchant*, c. 1530 (oil on panel), Gossaert, Jan (Netherlandish, c. 1478–1532) / National Gallery of Art, Washington, DC. Ailsa Mellon Bruce Fund. Accession number 1967.4.1.

Contents

Acknowledgements vii

Introduction
Laura Estill, Diane K. Jakacki, and Michael Ullyot 1

Books in Space: Adjacency, EEBO-TCP, and Early Modern Dramatists
Michael Witmore and Jonathan Hope 9

Plotting the "Female Wits" Controversy: Gender, Genre, and Printed Plays, 1670–1699
Mattie Burkert 35

A Bird's-Eye View of Early Modern Latin: Distant Reading, Network Analysis, and Style Variation
Maciej Eder 61

Displaying Textual and Translational Variants in a Hypertextual and Multilingual Edition of Shakespeare's Multi-text Plays
Jesús Tronch 89

Re-Modeling the Edition: Creating the Corpus of Folger Digital Texts
Rebecca Niles and Michael Poston 117

Collaborative Curation and Exploration of the EEBO-TCP Corpus
Martin Mueller, Philip R. Burns, and Craig A. Berry 145

"Ill shapen sounds, and false orthography": A Computational Approach to Early English Orthographic Variation
Anupam Basu 167

Linked Open Data and Semantic Web Technologies in *Emblematica Online*
Timothy W. Cole, Myung-Ja K. Han, and Mara R. Wade 201

Mapping Toponyms in Early Modern Plays with the *Map of Early Modern London* and *Internet Shakespeare Editions* Projects: A Case Study in Interoperability
Janelle Jenstad and Diane K. Jakacki 237

Microstoria 2.0: Geo-locating Renaissance Spatial and Architectural History
Fabrizio Nevola 259

Gazing into Imaginary Spaces: Digital Modeling and the Representation of Reality
John N. Wall 283

Cambridge Revisited?: Simulation, Methodology, and Phenomenology in the Study of Theatre History
Jennifer Roberts-Smith, Shawn DeSouza-Coelho, and Paul J. Stoesser . . . 319

Staying Relevant: Marketing Shakespearean Performance through Social Media
Geoffrey Way 345

Contributors 373

Acknowledgements

We would like to thank our contributors and all of those who helped to make this volume possible, including the English Department at Texas A&M. Special thanks to our editorial assistant, Tommy Pfannkoch, and to Quinn Dombrowski, our web consultant.

Introduction

Laura Estill
Texas A&M University
lestill@tamu.edu

Diane K. Jakacki
Bucknell University
dkj004@bucknell.edu

Michael Ullyot
University of Calgary
ullyot@ucalgary.ca

The essays in this volume show the importance of understanding, analyzing, sustaining, and creating digital resources to undertake effective and compelling research in early modern studies. The humanists who digitally manipulate their objects of study demonstrate how digital tools provoke research questions that were unanswerable, even inconceivable, before texts were digitized, data were visualized, realities were augmented, and media were social.

Digital methods and manipulations are not particular to the scholars in this volume, nor even to the wider community of self-described "digital humanists." Digital resources and databases have become naturalized to our scholarly habits because they empower and extend our thinking. Even traditionally-inclined and classically-trained scholars use search tools to explore digital repositories and surrogates such as online facsimiles and editions.

We use new research tools not as ends in themselves, but as "telescopes for the mind." Margaret Masterman coined that phrase in 1962 to "suggest computing's potential to transform our conception of the human world just as in the seventeenth century the optical telescope set in motion a fundamental rethink of our relation to the physical one," as Willard McCarty writes (2012, 113). Where human eyes see stars, telescopes reveal distant galaxies that were always there, even if they were formerly invisible. Michael Witmore describes this emergent knowledge as "something that is arguably true now about a collection of texts [that] can only be known in the future" (2011). As such, our tools are "a means to illumine ignorance by provoking us to reconsider what we think we know," as McCarty paraphrases Roberto Busa (2012, 115). The essays in this book show how we can look to these new horizons in early modern intellectual and cultural history to see what was previously unseen.

This book has two goals: to highlight the emerging research methodologies enabled by digital projects, and to induce readers to imagine and pursue future questions. The chapters that follow reveal how different digital methodologies provide new interpretations of diverse primary texts, from early modern drama to social media posts to emblem books to neo-Latin literature. They bring the research methodologies of the digital humanities to bear on the research objects of the early modern period. Within each of these fields, moreover, we are uniting subfields like emblem studies, theatre history, audience reception, literary studies, geohumanities, and network analysis.

These essays feature scholars who have created digital humanities projects, those who use digital tools for analysis, and those who theorize online spaces. Each approach shows how modes of engaging with early modern culture are enlarged by digital capabilities and tools. The projects discussed in this collection show the breadth and depth of the questions we can ask using new methods and resources.

While these projects focus on early modern objects of study, they signal the effects of the digital turn across literary and cultural studies. They showcase how the use of different digital approaches—from geographical information systems (GIS) to large corpus analysis—can shed new light on a historical period and encourage future scholarly inquiry. These projects invite other scholars to use these methods and tools in their own inquiry. They also help us to understand the bounds, methodologies, and goals of digital projects we use (and undertake). In order to enlarge and enrich our inquiry into intellectual and cultural history, we must theorize and problematize our use of these texts and artifacts.

Our new methods can, for instance, inquire into the validity of settled orthodoxies. Consider the text analysis software DocuScope, used in this collection by Jonathan Hope and Witmore and by Mattie Burkert. Its algorithms are "productively indifferent to linear reading and the powerful directionality of human attention," as Hope and Witmore have written elsewhere (2010, 359). They can reveal dimensions of texts that human readers cannot. For instance, Burkert's "productive estrangement from familiar materials" demonstrates that "women were blamed for a larger theatrical trend [of reviving heroic tragedy] followed by men and women alike." Burkert's close reading of primary sources and DocuScope's distant-reading algorithms enable this re-evaluation of a common opinion. Using established digital tools to explore an early modern corpus can help us see new patterns of genre, language, and influence. Maciej Eder, for instance, uses linguistic analysis to establish webs of influence between classical writers and neo-Latinists. Corpus analysis can help us see change over time, be it in genre (as Burkert proves) or orthography (as Anupam Basu shows).

As N. Katherine Hayles explains, digital media can facilitate reading from the micro- or close-reading level to the macro- or corpus level. One of the more contested orthodoxies is which texts or artifacts we should use to make a broader argument. Proponents of distant reading or macroanalysis such as Franco Moretti, Matthew Jockers, and Ted Underwood encourage scholars to reorient our arguments from exemplary, canonical texts to the continuum of historical text-corpora—most of which consists of neglected, obscure texts. As Hope and Witmore write in this volume: "If we count the right things, we can recover texts, and relationships between texts, currently lost to literary history." So Burkert, for one, leads us to reconsider texts like Edward Cooke's *Love's Triumph*, which is often overshadowed by more anthologized Restoration plays such as William Wycherley's *The Country Wife*. Moving away from arguments based on a few exemplary texts or artifacts—whether or not they are canonical and familiar—can only be a positive inducement to new arguments. As Stephen Ramsay writes about text-analysis tools, "[t]he evidence we seek is not definitive, but suggestive of grander arguments and schemes" (2011, 10).

This volume stands alongside the others in the New Technologies in Medieval and Renaissance Studies series (including, notably, Brent Nelson and Melissa Terras's *Digitizing Medieval and Early Modern Material Culture*, 2012) as a way to reflect on existing digital humanities practices while also illuminating future trends. Other recent important works in the field include Janelle Jenstad, Jennifer Roberts-Smith, and Mark Kaethler's *Shakespeare's Language and Digital Media: Old Words, New Tools* (forthcoming); Christie Carson and Peter Kirwan's *Shakespeare and the Digital World* (2014); and the *Shakespearean International Yearbook*'s 2014 special issue focused on "Digital Shakespeares," edited by Brett D. Hirsch and Hugh Craig. Our volume goes beyond Shakespeare to show how digital projects can reflect non-canonical texts as well as the international and multilingual realities of the early modern period. It is important that further work is published on digital approaches to early modern studies as the field continues to evolve.

Early Modern Studies after the Digital Turn shows us not only how digital tools can help us to understand the past, but also how to theorize and understand these tools themselves, and how the process of developing and engaging with them is in itself a form of humanistic inquiry. The projects in this book cover the range from nascent to mature, from Fabrizio Nevola's *Hidden Florence* project to John Wall's award-winning *Virtual Paul's Cross*.[1] We juxtapose digital projects along this spectrum because there is much insight to be gained by openly discussing research trajectories. As Roberts-Smith, Shawn DeSouza-Coelho, and Paul Stoesser realized in the process of developing SET (the Simulated Environment for Theatre), "the knowledge is in the doing, rather than subsequent to, an outcome of, or an ancillary to the doing; the process is our product and our argument."

There are also essays in this volume that are not focused on the creation and use of digital tools, but think more holistically critical methods like text encoding and analytic reading itself. As Geoff Way demonstrates, critically reading social media interactions sheds light on twenty-first-century ways of approaching the early modern period. Way's essay analyzes social media posts rather than (say) network-scatterplot-type analysis in the style of Basu and Eder. Rebecca Niles and Michael Poston argue that encoding is editing is scholarship. Martin Mueller, Phil Burns, and Craig Berry's argument that we need more accurate digital surrogates for edited texts underscores the data curation that enables our interpretive work. Jenstad and Diane Jakacki, similarly, explore the necessary encoding that underpins scholarly inquiry, and argue that interpretations are embedded in encoding. Furthermore, encoding and digital presentation offer more avenues for display and exploration than print editions, as Jesús Tronch envisions.

Many of the chapters in this book reveal how digital projects can help scholars both access and present early modern international and multilingual culture. Tronch, for instance, considers how digital versioning can offer multi-text editions that move beyond the basic parallel-text codices and awkwardly-formatted editorial conventions. Nevola's *Hidden Florence* allows English and Italian speakers to experience layers of history in one of the most storied cities in Europe. Timothy W. Cole, Myung-Ja K. Han, and Mara R. Wade focus on the metadata apparatus surrounding emblem books, those illustrated and multilingual collections of apothegms that were popular both on the continent and in England during the early modern period. Although,

---

[1] The *Virtual Paul's Cross Project* won the 2014 Digital Humanities award for best data visualization: http://dhawards.org/dhawards2014/results/.

today, different languages and literatures can be siloed in separate departments across a university campus, the reality of the early modern period was one of fluid borders, trade, and multilingualism. While some of the projects described in this volume actively work toward understanding this multicultural and polyglot past, others could serve as building blocks for future comparative or international projects.

Increasingly, digital projects illustrate how methods such as geohumanities need to be opened up to support richer and more varied approaches. Wall reflects on how visual and acoustic recreations of place can inform textual analysis. His consideration of spatial acoustics has revealed John Donne's process in preparing for and presenting his sermons to audiences in Paul's Churchyard. Jenstad and Jakacki's interconnected reading of the *Map of Early Modern London* and the *Internet Shakespeare Editions* demonstrates how teasing out both explicit and textual references to London spaces problematizes the understanding of locational meaning in plays, both for modern and contemporary audiences. Nevola's *Hidden Florence* reveals that the augmented reality embedding history in a contemporary landscape enriches and transforms the experience of Florentine geography. What connects these varied approaches to virtual, augmented, and remediated historical places is how scholars explore ambient sound, storytelling, avatars, and changes in the identity of place over time.

As part of the invitation to apply new research methodologies to increase our understanding of the early modern world, this volume offers a digital component that shares tools and corpora online. On ems.itercommunity.org you will find materials that support the findings in this volume, such as Eder's interactive graphs tracing the influence in Latin literature and video demonstrations from the Simulated Environment for Theatre (SET). You will also find links to ongoing research explorations such as Martin Mueller's blog, Scalable Reading, which documents his continuing work with AnnoLex and MorphAdorner; and Witmore and Hope's blog, Wine Dark Sea, which reflects on quantifying aspects of early modern drama.

The web page for *Early Modern Studies after the Digital Turn* also serves as a way to find the online projects discussed in these pages, including *Folger Digital Texts, Emblematica Online,* the *Map of Early Modern London, Hidden Florence,* and *Virtual Paul's Cross*. Most importantly, when possible, we link to the tools used here. You can, for instance, download the Simulated Environment for Theatre (SET) and use your computer to map out blocking for a stage play. You can explore the texts and corpora (notably, EEBO-TCP, the Early English

Books Online-Text Creation Partnership) that were mined, transformed, and analyzed to offer the results presented here. The online apparatus for this book invites readers to participate in furthering our knowledge about the early modern period by adopting or changing these digital approaches in order to build on the questions raised in these essays.

We introduce new projects and overlooked texts, as well as highlight ongoing developments in early modern studies and digital humanities. Following Kathleen Fitzpatrick's model, we present these publications not as an archive of thoughts, but rather prompts and invitations that will lead to new uses for these tools. Our volume and its online resource enable digital research on the early modern period to serve as records, models, and prompts for future work. With new projects emerging, with the constant upgrading of existing projects, and with the ability to apply new tools to new objects (textual or otherwise), it is imperative for humanists to share their ongoing research.

Each chapter in this volume offers potential avenues for future research. Their provocations, methodologies, and questions invite us to reimagine what it means to study the early modern period in the digital age. This book, then, is a call for increased digital access, new research projects, and continued improvements to existing projects. For scholars to draw the most accurate conclusions, more material needs to be added to our datasets. But as Mueller, Burns, and Berry argue, the problem is equally one of quality as of quantity. Our text corpora must be regularized and curated until they are as reliable as print editions—and as Niles and Poston argue, those texts must also be rigorously edited.

Software can be developed in perpetual beta, where changes are made gradually, over time, without definitive releases. Scholarship on digital humanities (and, indeed, scholarship writ large) is also in perpetual beta: it is not an undertaking that can ever be finished. As we apply new critical lenses to texts, corpora, databases, and other objects, we gain different insights that build on, refine, and redirect earlier questions. The "new technologies in Renaissance studies" from 1990 were not new in 2000, and those from 2016 will not be new in 2026: but future research will be richer if it accounts for its predecessors. The multifaceted research featured in this book assesses where we are today, and envisions the many pathways we can take as we extend our research.

These essays address the digital humanities' core tensions: fast and slow; surficial and nuanced; quantitative and qualitative. Scholars design algorithms and projects to process, aggregate, encode, and regularize historical texts and artifacts in order to position them for new and further interpretations. Every essay in this book is concerned with the human-machine dynamic, as it bears on early modern research objects and methods. The interpretive work in these pages and in the online projects discussed orients us toward the extensible future of early modern scholarship after the digital turn.

WORKS CITED

Carson, Christie, and Peter Kirwan, eds. 2014. *Shakespeare and the Digital World.* Cambridge: Cambridge University Press.

Fitzpatrick, Kathleen. 2011. *Planned Obsolescence: Publishing, Technology, and the Future of the Academy.* New York: New York University Press.

Hayles, N. Katherine. 2012. *How We Think: Digital Media and Contemporary Technogenesis.* Chicago: University of Chicago Press.

Hirsch, Brett D., and Hugh Craig, eds. 2014. *The Shakespearean International Yearbook: Volume 14: Special section, Digital Shakespeares.* Burlington, VT: Ashgate.

Hope, Jonathan, and Michael Witmore. 2010. "The Hundredth Psalm to the Tune of 'Green Sleeves': Digital Approaches to Shakespeare's Language of Genre." *Shakespeare Quarterly* 61 (3): 357–90.

Jenstad, Janelle, Jennifer Roberts-Smith, and Mark Kaethler, eds. Forthcoming. *Shakespeare's Language and Digital Media: Old Words, New Tools.* Abingdon, UK: Routledge.

Jockers, Matthew. 2013. *Macroanalysis: Digital Methods and Literary History.* Champaign: University of Illinois Press.

McCarty, Willard. 2012. "A Telescope for the Mind?" In *Debates in the Digital Humanities,* edited by Matthew K. Gold, 113–23. Minneapolis: University of Minnesota Press. http://dhdebates.gc.cuny.edu/debates/text/37.

Moretti, Franco. 2013. *Distant Reading.* London: Verso.

Nelson, Brent, and Melissa Terras, eds. 2012. *Digitizing Medieval and Early Modern Material Culture.* Toronto: Iter; Tempe: Arizona Center for Medieval and Renaissance Studies.

Ramsay, Stephen. 2011. *Reading Machines: Toward an Algorithmic Criticism.* Urbana: University of Illinois Press.

Underwood, Ted. 2013. *Why Literary Periods Mattered: Historical Contrast and the Prestige of English Studies.* Stanford, CA: Stanford University Press.

Witmore, Michael. 2011. "The Ancestral Text." *Wine Dark Sea.* 9 May. http://winedarksea.org/?p=979.

Books in Space: Adjacency, EEBO-TCP, and Early Modern Dramatists

Michael Witmore
Folger Shakespeare Library
mwitmore@folger.edu

Jonathan Hope
University of Strathclyde
jonathan.r.hope@strath.ac.uk

Imagine being given the keys to your own personal copyright library, containing every book printed in English since 1450 to the present day. You rush to the door and open it, keen to start exploring—but once inside the huge space you find there is a problem. There are miles of shelving, rooms and rooms of books, but there is no catalogue, and there seems to be no principle governing the arrangement of books on the shelves. You rush from floor to floor, scanning titles and authors you do not recognize, desperately looking for a familiar text. You have a Ph.D. in English literature, but none of these books were on the syllabus: there are hundreds of thousands of them, millions. You are lost.

This is the situation we are all about to be in—indeed, are all, to some extent, already in. A series of book digitization and transcription projects—EEBO-TCP, ECCO-TCP, HathiTrust, Google Books—is making almost every English printed book available.[1] Already, anyone can download 25,000 phase 1 TCP texts—with another 40,000 to follow. But "made available" is not the same as "made useable." Such an increase in the amount of available data has all kinds of effects: practical ones in terms of storage and processing; methodological ones in terms of how we manipulate it and measure it; and theoretical ones in terms of how it changes our subject or object of study. In this essay we will explore these effects, and suggest ways in which we can deal with them. We'll

---

[1] For EEBO-TCP (Early English Books Online), see http://www.textcreationpartnership.org/tcp-eebo/; for ECCO-TCP (Eighteenth Century Collections Online), see http://quod.lib.umich.edu/e/ecco/; for HathiTrust, see http://www.hathitrust.org; for Google Books, see https://books.google.com/intl/en/googlebooks/about/history.html—and also the alternative front-ends to the Google, and other, corpora built by Mark Davies: http://corpus.byu.edu. Text Creation Partnership to: "EEBO-TCP (Early English Books Online—Text Creation Partnership)" and "ECCO-TCP (Eighteenth-Century Collections Online—Text Creation Partnership)."

focus on EEBO-TCP for examples, as it is the data set most readers will know about or be familiar with (and it is the one we are most familiar with).

At the time of writing, anyone with an Internet connection can go to the following URL and download a file called TCP.csv:

> https://github.com/textcreationpartnership/Texts/blob/master/TCP.csv

This file lists all of the text files in EEBO-TCP—both those freely available as "Phase 1," and those limited to subscriber access. The csv file should open in any spreadsheet or statistics program as a structured spreadsheet with 61,315 rows (in the version available in April 2015)—corresponding to 61,315 text files, or "books."

Let's read *Hamlet*.

Text files in TCP are "named" numerically rather than with the titles of the books they contain. The TCP.csv file has the TCP number for each volume transcribed by TCP, as well as metadata such as "title," "date," "author," "terms" (roughly, subject or genre), and "pages" (i.e., length). So to read *Hamlet*, we need to find out the relevant TCP number from the csv file. Once we have that, we can download the right file from:

> https://github.com/textcreationpartnership

Let's search the csv file for "*Hamlet*."

If you have access to a networked computer, it might be instructive for you to try to do this: download the csv file, open it in a spreadsheet program, and do a search for the string "hamlet." When we did this with the csv file open in the software we use for data analysis (a commercial statistical package called JMP), we got a series of hits (depending on the software you use, and any updates to the TCP.csv file, you may get different results). The first hits are texts by authors with "hamlet" in their name. For example:

> A12788   Spenser, John, 1559-1614.; Marshall, <u>Hamlett</u>.   1615
> A learned and gracious sermon preached at Paules Crosse by that famous and iudicious diuine, Iohn Spenser ... ; published for the benefit of Christs vineyard, by H.M. Bible. -- O.T. -- Isaiah V, 3-4 -- <u>Sermons</u>.; Sermons, English -- 17th century.   60

> A56269   Puleston, <u>Hamlet</u>, 1632-1662.1661 Monarchiæ Britannicæ singularis protectio, or, A brief historicall essay tending to prove God's especial providence over the Brittish monarchy and more particularly over the family that now enjoys the same / by Hamlett Puleston ...   Monarchy -- Great Britain.   67

Then there's a hit in a title, but it's from a book about "villages and hamlets":

> A08306   Norden, John, 1548-1625?; Keere, Pieter van den, ca. 1571-ca. 1624, engraver.   1593   Speculum Britanniae. The first parte an historicall, & chorographicall discription of Middlesex. Wherin are also alphabeticallie sett downe, the names of the cyties, townes, parishes hamletes, howses of name &c. W.th direction spedelie to finde anie place desired in the mappe & the distance betwene place and place without compasses. Cum priuilegio. By the trauaile and vew of Iohn Norden. Anno 1593; Speculum Britanniae. Part 1   Middlesex (England) -- Description and travel -- Early works to 1800.   140

Then we get a play called *Hamlet*, but it is Davenant's adaptation:

> A59527   D'Avenant, William, Sir, 1606-1668.; Shakespeare, William, 1564-1616. <u>Hamlet</u>.   1676   The tragedy of Hamlet, Prince of Denmark as it is now acted at His Highness the Duke of York's Theatre / by William Shakespeare.   94

Only after a while do we find TCP number A11959, which is the file name for the TCP transcription of the 1603 quarto of *Hamlet*:

> A11959   Shakespeare, William, 1564-1616. 1603   The tragicall historie of <u>Hamlet</u> Prince of Denmarke by William Shake-speare. As it hath beene diuerse times acted by his Highnesse seruants in the cittie of London: as also in the two vniuersities of Cambridge and Oxford, and else-where; Hamlet   66

Great. Super.

Except this is Q1—the so-called "bad quarto," and very different from the text most of us are used to reading and seeing performed. It is nice to know that this important variant text is available in TCP, but we really wanted to read something closer to the "standard" text. We search on, but there is no sign of Q2 *Hamlet* (1604), on which most modern editors base their texts. We have fallen afoul of one of the principles of TCP: the aim is to include one version of each text, not multiple editions, even if they are very different, as is the case with Q1 and Q2 *Hamlet*.

Then a great idea strikes us: surely Shakespeare's first folio is in? And of course, searching for "hamlet" will not find that, as *Hamlet* is not mentioned in the title of the collected volume. So, we remember that the title of the folio is *Mr William Shakespeares Comedies, Histories & Tragedies*, and we do a search on "mr william shakespeare." Nothing. Unable to believe that the first folio is not in TCP, we sort the spreadsheet by "date" and scroll to the section for 1623. There it is:

> A11954   Shakespeare, William, 1564–1616.; Heminge, John, ca. 1556–1630.; Condell, Henry, d. 1627.   1623   Mr. VVilliam Shakespeares comedies, histories, & tragedies Published according to the true originall copies.; Plays   916

Our initial search failed because "William" has been (accurately) transcribed "VVilliam."

So now we know that the first folio has the TCP number A11954. Before we download this file, though, let's reflect on this as a search: we were looking for something we knew existed; something we are very familiar with. Our familiarity with Shakespeare enabled us to know that A59527 was an adaptation, and that A11959, while interesting textually, was not exactly the text we wanted. It also enabled us to guess where to look to find the text, even though the title of the file gives no indication that it is present. This is fine for Shakespeare, but there are over 60,000 texts in TCP—the vast majority by writers of whom we have virtually no knowledge. What are the chances of us finding something by one of them? If we search, and get no results, how will we know if this is a true or a false negative?

Putting that uncomfortable thought aside, let's to GitHub, armed with our TCP number: A11954. In fact, to save time, you can use this URL with any TCP number at the end to take you to the relevant text file:

> https://github.com/textcreationpartnership/A11954

Now we are in business: because this text is in TCP phase 1, we get a link on the text's GitHub page to a nice HTML edition:

> http://tei.it.ox.ac.uk/tcp/Texts-HTML/free/A11/A11954.html

as well as a live download link for an EPUB version, and a link to the page images on JISC Historical Texts (for UK users at subscribing academic institutions only—other users at subscribing institutions may have access to the page images via EEBO). If we had tried this with a phase 2 text, however, we might have found ourselves trying to open an XML-coded file—and if we had been able to do that, we might well have found a text marked frequently with encoding errors. Not only is everything not there; what is there is not perfect.[2]

It is worth spending some time playing with the TCP texts, if only to temper the excitement that has surrounded their release. For any collection of books to be usable, or understandable, we need paths through it—paths that allow us to do at least two very different things. First, we want to be able to find texts we already know about: "I want to read *Hamlet*." And even this apparently simple request, as we have seen, can be tricky. Secondly, and perhaps more importantly, we want to be shown texts we *don't* currently know about,

---

[2] TCP transcribers were told not to spend too much time trying to work out illegible sections of text. As they were working from the same images online subscribers to EEBO get, they were using digital images of microfilms of early modern books. EEBO users are well aware of the many problems with reading these images. Consequently, the TCP files have regular marked gaps where transcribers could not read the image. Martin Mueller is leading AnnoLex (http://annolex.at.northwestern.edu/about/), a project to enable the collaborative correction of errors and gaps.

In addition to the gaps in individual TCP texts, we should remind ourselves that TCP does not contain "the whole" of the early modern print record (although it is tempting, and easy, to fall into such rhetoric). TCP, of course, can contain only surviving printed texts—so we must bear in mind likely survival rates when using it to represent early modern culture (Alan Farmer has a forthcoming essay on survival rates of printed material). We must also remember not only lost manuscript material, but the huge amount of surviving manuscript material not in TCP—although we can also welcome projects like Early Modern Manuscripts Online: http://folgerpedia.folger.edu/Early_Modern_Manuscripts_Online_(EMMO). But even taking these materials out of consideration, TCP does not contain "everything." The aim was to have one copy of each *text*, not a copy of each *book*, so texts are usually included in one edition only. Anupam Basu's graph of TCP text counts against ESTC entries (http://earlyprint.wustl.edu/tooleeboestctexts.html) is a striking visual reminder of the difference this makes. Finally, we should remember that TCP is ongoing: texts are being added constantly. In a recent comparison between drama texts recorded in the Database of Early English Playbooks (http://deep.sas.upenn.edu/index.html), Beth Ralston found a number of dramatic texts not currently included in TCP (the data from this study, funded as part of the Visualizing English Print project, is available from http://winedarksea.org).

but which are relevant to our interests: "Out of these tens of thousands of texts, I want to read texts with a relationship to *Hamlet*."

Typically, in the past, such relational paths through collections of books have been based on humans making high-level, subjective comparisons between texts. The pathway of traditional literary history, for example, was made by clearing away the vast majority of books, leaving a narrative formed out of just a chosen few (Seneca leads to *Hamlet*, which leads to ...). More comprehensively, library science developed a subject-based arrangement, which placed books on similar topics close to each other, allowing the apparent serendipities of open-stack research: effectively projecting and visualizing the results of human-based content analysis in the three-dimensional spaces of the library. Crucially, both literary history and librarianship rely on metadata (title, author, date) and comparison-based sorting (this text is drama, that one is religious prose; this text is "good," that one is not).

How do we deal with this within the new digital collections? Increasing numbers of books have always been a problem for scholars, so perhaps we can learn from the past. Let's move away from GitHub and csv files for a moment, and think about medieval libraries and the impact of Renaissance humanism on the employment rates of carpenters.

Imagine you are standing in a medieval library.[3] There are a couple of things to note. There are not many books, and those that are there are stored in fixed positions, chained to desks. When in use they are placed open on a reading surface. When not in use, the books are generally stored lying horizontally, either on top of the desk, or on a shelf beneath the reading surface. This tells us a lot about the tenor of the medieval intellectual world, and the lack of climate control in medieval libraries. It is not too much of a caricature to say that medieval scholasticism meant that intellectual life was focused on a small number of authoritative texts. The main job of a library was to hold copies of that small number of texts, and replace them with new copies as they wore out with use and the depredations of damp, mould, cold, and heat. Books were stored horizontally in piles of one because (a) there were not very many, so space was not an issue; and (b) the clasps and decorative metalwork affixed to the covers of books would damage other books if they were piled on top of each other.

---

[3] Our discussion of libraries, storage, and cataloguing methods draws on the following sources (complete information can be found in the Works Cited): Balsamo 1990; Campbell and Pryce 2015; Leedham-Green and Webber 2006, especially the essays by Gameson, Sargent, and McKitterick; Norris 1939; Petroski 1999; and Webster 2015.

With the rise of Renaissance humanism, however, things changed. The manuscript hunters scoured Europe and beyond for "lost" works which were added to the store of classics that libraries might be expected to hold. Humanists started to write new books—some commenting on the old ones, and some even introducing new ideas. And of course the invention of the printing press allowed books to be reproduced more cheaply and in greater numbers than had previously been possible.

One consequence of this intellectual revolution was a storage crisis. Libraries had to find space for more books. In the Renaissance, librarians responded in two ways: they employed carpenters, and they rotated books through 90 degrees. The carpenters added shelves to the wooden reading desks of medieval libraries, both above the reading surfaces and in the hollows below them where seated scholars had previously been able to put their knees. At the same time, librarians began to store books standing vertically, rather than horizontally, since this was more space-efficient. The Renaissance can be thought of as a ballet of books, thousands of them spinning gracefully through 90 degrees so that they become upright in space.

People realized pretty early on in the Renaissance that having more and more books was not an unmitigated good. Not only was there nowhere to put your knees when reading, but the stalls carpenters were building above reading desks now blocked out the light from the small, low windows usual in medieval libraries. It is a rather nice paradox that the new books brought into being by the light of the Renaissance quite literally blocked out the light required to read them.

And there was another worry: how could you possibly read them all? The restricted medieval canon had something going for it in terms of removing the stress of the unread. As book production gathered momentum, however, scholars became uneasy in the face of all the new knowledge: how could any one person master it? Very soon, the first attempts at listing books, and significantly organizing and excerpting them, appeared—because, of course, it *is* a good thing to have more books, as long as you can have some kind of meaningful access to them.[4]

---

[4] The pioneer in this field was Conrad Gessner, whose work is frequently cited in several landmark studies of information management in the Renaissance and beyond: see Blair 2010, Krajewski 2011, and Rosenberg 2013. For specific work on Gessner, see Blair 2003; Nelles 2009; Rissoan 2014; and Rosenberg 2003.

Returning to our medieval library for a moment, if you wanted to find a book, you asked the librarian, whose job it was to know where the books were in physical space—on which table or shelf (and if the librarian died suddenly, libraries would become less usable). In the Renaissance, we get the rise of the shelfmark. Those interested in digital humanities would do well to read up on the history of libraries, bibliographies, and cataloguing techniques: this is a large part of what we do, and librarians are pretty good at it. Originally, of course, the shelfmark was exactly that: generally a three-part mark consisting of letters and numbers, which typically identified the press or bookcase ("A"), the shelf ("3"), and the book's position on the shelf ("11"). In our example, then, the book is the eleventh book on the third shelf on bookcase A. The book's position is fixed in physical space. And the position of the book is very likely to be decided by the librarian's assessment of its subject matter: bookcases contain all the books relating to a certain subject area.

This may seem pretty obvious, but in fact its implications are conceptually substantial. Libraries that organize books by subject are effectively three-dimensional search engines, physical instantiations of the Amazon algorithm that generates those "If you liked X you may also like Y" messages.

Scholars of our generation like to gush nostalgically about the serendipities of the open-stack library. Geoffrey Hill did it recently in one of his Oxford poetry lectures in the midst of a brag about never going online, claiming that going to libraries was better because it allowed for the serendipitous discovery of the book you needed, but did not know you wanted.[5] But in fact Hill is wrong—there's nothing serendipitous about these finds, despite the air of self-congratulation that usually accompanies narratives of such "discoveries"—"Wasn't I clever to spot this?" Well, no. The librarians and the catalogue were clever to place similar books proximately in physical space. Otherwise you'd have had to wander the stacks pulling books randomly off shelves.

Those "serendipitous" discoveries are thanks to the invention of relational cataloguing systems like Dewey, which constitute an advance on literal shelfmarks—one again occasioned by the needs of storage. Relational systems number books relative to each other and their subject areas, but have no necessary or fixed relation to the physical space of the library. They are meant to

---

[5] Hill 2013. This lecture was not published, but the audio is available at the link provided in the Works Cited. We are grateful to Mary Erica Zimmer for this reference, and apologetic to Professor Hill for singling him out. We have all made similar claims about open-stack research.

allow books to be shifted around in the library space as new books, shelves, rooms, floors, even buildings, are added. The introduction of such numbers is another significant event in intellectual history, severing the organization of cultural materials from the organization of the buildings they are stored in—it is driven by a practical need (the maximal utilization of space), and perhaps a theoretical development (the multiplication of categories of intellectual culture).

So the physical organization (and cataloguing system) of a library is one way of combating the information overload of the Renaissance, the Enlightenment, and the present. But of course, modern libraries have to deal with exponential rates of book publication, and the biggest new libraries deal with this by an even more radical severing of the subject–location relationship, and even by (actually or potentially) severing the relational links between books noted above.

Book storage hangars like the new Bodleian book depository, or the British Library facility at Boston Spa, store books in largely human-free zones. Books tend to be fetched by robot. They are identified by bar code. This may sound clinical, anti-human. Perhaps it is. But perhaps we should remember that humans are bad for books, with their coughs and sneezes and greasy fingers, and a liking for high temperatures and humid spaces. These new book depositories are atmospherically, environmentally kinder to books than human-ridden open-stack libraries. And there's another payoff. With books identified by bar code in vast hangars patrolled by robots, we can break the final link between subject and physical space. In these spaces, books can sit beside any other books. Geoffrey Hill is never going to be allowed in to wander randomly—and even if he were, he wouldn't fit between the towering shelves, which are not accessible to humans, however slim, and are far too tall for safe browsing. This gives us the potential of "organizing" these books in any number of ways—in the actual physical space of the facility, they may be added, and clustered, by date of cataloguing, or size. But our access to these books, since it can't be physical, is virtual—opening up all the possibilities of modern search engines. And here we have a parallel with the opportunities and methodologies afforded by the digitization of collections such as TCP: by digitizing all of our books, we enable multiple reorganizations—either radical or traditional.

So you can think of the digital humanities as a million different ways to organize the books on a shelf, allowing you to make "serendipitous" discoveries more frequently, and more mind-bendingly, than in any open-stack library.

Indeed, digital tools allow you to have your books on a hundred different shelves simultaneously—rather like the high-dimensional library imagined at the end of the film *Interstellar*. This ability to reorganize enables us to map out multiple pathways through the book collection. If we count the right things, we can recover texts, and relationships between texts, currently lost to literary history: but everything depends on us counting the right thing, and being able to interpret the results.[6]

So let's have a look at some of the things that can happen when you have easily countable texts. You can learn things about texts by counting pretty simple things, although the things you learn tend to be quite simple in themselves. But we will begin with some simple examples because they give us the basic principles. Let's count the frequencies of a few words in Shakespeare's plays, starting with the word "king."[7]

Table 1 arranges Shakespeare's plays by relative frequency of the word "king," from highest to lowest. When we do this, the results are impressive, if predictable: all the histories go to the top. Only one non-history play gets amongst them: *King Lear* (which Shakespeareans will remember is actually called a history in its quarto publication). But this is telling us something really very obvious: there are lots of kings in the history plays, so the word king gets used a lot. Behold the golden new dawn of digital humanities!

---

[6] Here the paradigm shift turns back to literary scholars: computer science and statistics (and corpus linguistics) have an array of well-established techniques for counting and analyzing—but the decision of what to count, and the analysis of the results, can only sensibly be made by literature specialists.

[7] Before we count, let's note that an apparently simple phrase like "We'll count the word *king* in each Shakespeare play" makes a whole host of assumptions and covers up a lot of work and thinking: which plays do we mean by "each Shakespeare play"? Are we including *The Two Noble Kinsmen, Pericles, Sir Thomas More*? Which texts will we use? Paper, electronic—Q1 or folio? Edited? Unedited? Which parts of the text will we count? All of the text, or just those parts spoken on stage (i.e., not speech prefixes or stage directions, or running titles at the top of pages)? How will we define "king"? As just the character string < _king_ > or as the "lemma" *king(n)*, including *kings, king's*, but not including *king (v)*? How will we report our results? As raw totals for each play, or as standardized relative frequencies adjusted for length to allow direct comparison? In this case, we "simply" counted the string <_king_>, using our project's web-based text tagger Ubiqu+Ity (http://vep.cs.wisc.edu/ubiq/)—which automatically reports results in relative frequencies. The edition of Shakespeare used was that of the Folger Digital Texts (http://www.folgerdigitaltexts.org).

Play	Genre
Richard II	History
3 Henry VI	History
2 Henry VI	History
Henry VIII	History
Henry V	History
1 Henry VI	History
Richard III	History
King John	History
1 Henry IV	History
King Lear	Tragedy
2 Henry IV	History
Hamlet	Tragedy
Pericles	Late
Winter's Tale	Late
Macbeth	Tragedy
Tempest	Late
Love's Labour's Lost	Comedy
All's Well	Comedy
Cymbeline	Late
Titus Andronicus	Tragedy
Two Noble Kinsmen	Late
Antony & Cleopatra	Tragedy
Midsummer Night's Dream	Comedy
Measure for Measure	Comedy
Troilus & Cressida	Tragedy
Julius Caesar	Tragedy
Two Gentlemen of Verona	Comedy
Twelfth Night	Comedy
Merchant of Venice	Comedy
Merry Wives	Comedy
Romeo & Juliet	Tragedy
Taming of the Shrew	Comedy
Much Ado about Nothing	Comedy
As You Like It	Comedy
Othello	Tragedy
Coriolanus	Late
Comedy of Errors	Comedy
Timon of Athens	Tragedy

Table 1. Shakespeare's plays arranged in order by relative frequency of the word "king" (highest to lowest)

But note the principle of what we've done here: we have rearranged our books on the shelf, using an unusual method (frequency of one word, rather than alphabetical order of title, or date of writing). This has had the effect of isolating a single group normally identified by other methods, and has picked out a further play with a potentially interesting generic relationship to that group. Strolling round the library of serendipity, we have apparently stumbled on an idea for a scholarly essay (albeit not a very original one).

We have also established a quantitative test for identifying the genre of Shakespeare's plays. Based on this evidence, if someone discovered a previously lost play by Shakespeare, and we wanted to find out if it was a history or not, all we'd need to do would be to count the frequency of the word "king" in it. Above a certain level, we'd be happy saying it was a history; below a certain level, we'd start to think it was something else.

Of course, that's a bit of a daft way to decide if a newly discovered play is a history or not. What we'd really do is get humans to read it and argue about the genre for a bit—and it would probably be a short argument because humans are pretty good at ascribing texts to genres. We have a good idea of what makes a play a history play, and it is not the frequency of the word "king": we ascribe a play to the genre "history" on the basis of relatively high-level features such as its relationship to its source material, and its thematic concerns. We can see, given this definition of a history play, why Shakespeare's history plays have a high frequency of the word "king." This is a statistical fact about history plays, but not, we would suggest, a very interesting fact. It is not very interesting because it is not surprising—it doesn't tell us anything we did not already know, or challenge our assumptions about history plays.

What happens if we count a different word? Let's try "love." Table 2 shows Shakespeare's plays arranged in order of frequency of the word "love." The comedies now come to the top, with an extra added tragedy, *Romeo and Juliet*, hardly a surprise.

Play	Genre
Two Gentlemen of Verona	Comedy
Midsummer Night's Dream	Comedy
Romeo & Juliet	Tragedy
As You Like It	Comedy
Love's Labour's Lost	Comedy
Much Ado about Nothing	Comedy
Twelfth Night	Comedy
Two Noble Kinsmen	Late
Taming of the Shrew	Comedy
Othello	Tragedy
Merchant of Venice	Comedy
All's Well	Comedy
Troilus & Cressida	Tragedy
Richard III	History
Hamlet	Tragedy
Merry Wives	Comedy
King Lear	Tragedy
King John	History
Timon of Athens	Tragedy
Julius Caesar	Tragedy
3 Henry VI	History
Antony & Cleopatra	Tragedy
Henry V	History
Richard II	History
Pericles	Late
Comedy of Errors	Comedy
Measure for Measure	Comedy
1 Henry VI	History
1 Henry IV	History
Macbeth	Tragedy
Titus Andronicus	Tragedy
Cymbeline	Late
Coriolanus	Late
Winter's Tale	Late
Henry VIII	History
2 Henry IV	History
Tempest	Late
2 Henry VI	History

Table 2. Shakespeare's plays arranged in order by relative frequency of the word "love" (highest to lowest)

The separation here isn't as good as with histories, but it is not bad—and note a couple of "interesting" results:

1. two comedies are very low on "love": *Comedy of Errors* and *Measure for Measure*;
2. four of the late plays are right at the bottom of the "love" ladder.

No doubt literary scholars could offer several theories about these findings, and this illustrates one of the benefits of counting things and rearranging books on the shelf. We don't really do it to allow us to identify newly discovered histories or comedies: we do it to suggest new questions about texts we already know. What is it about *Comedy of Errors* and *Measure for Measure* that makes them behave differently in this case? What shifts between the main comic group and these plays? Why are so many of the late plays so low in "love"? One of their typifying features is supposed to be the redemption of parents through the love of lost and rediscovered children, so this is a surprising result.

Note here that we are not only generating questions for humanities scholars; we are observing absence. Humans are pretty good at seeing things that are present in texts—especially things that happen relatively infrequently—but we are not very good at spotting things that aren't there, or are there relatively less frequently. Computers, being undiscriminating, are really good at this.

For our final word, let's pick "might"—the results of our count are in Table 3.[8]

---

[8] In choosing "might," we were inspired by the work of Lynne Magnusson on modal verbs in Shakespeare (for example, Magnusson 2009).

Play	Genre
Twelfth Night	Comedy
Hamlet	Tragedy
All's Well	Comedy
Pericles	Late
Two Noble Kinsmen	Late
2 Henry IV	History
Antony & Cleopatra	Tragedy
Timon of Athens	Tragedy
Measure for Measure	Comedy
Winter's Tale	Late
Tempest	Late
Julius Caesar	Tragedy
As You Like It	Comedy
Cymbeline	Late
Othello	Tragedy
Coriolanus	Late
King John	History
Richard III	History
Henry VIII	History
1 Henry VI	History
Love's Labour's Lost	Comedy
Merry Wives	Comedy
King Lear	Tragedy
2 Henry VI	History
Comedy of Errors	Comedy
Two Gentlemen of Verona	Comedy
Troilus & Cressida	Tragedy
Midsummer Night's Dream	Comedy
3 Henry VI	History
Macbeth	Tragedy
Much Ado about Nothing	Comedy
Henry V	History
Merchant of Venice	Comedy
Titus Andronicus	Tragedy
Richard II	History
1 Henry IV	History
Romeo & Juliet	Tragedy
Taming of the Shrew	Comedy

Table 3. Shakespeare's plays arranged in order by relative frequency of the word "might" (highest to lowest)

Look at how the late plays all come to the top—the only genre completely in the top half of the table. This result makes us want to investigate hypothetical and speculative language across Shakespeare's career—and it makes us wonder if he writes more definitive language early in his career, and gradually becomes less certain, more indefinite, as he develops.[9]

So far, we've been reorganizing our books along one shelf—one dimension—at a time, using just one frequency count. We could do the same reordering for anything we can count: nouns, verbs, murders, marriages, references to the Bible, scenes with more than three speaking characters. But we don't have to use just one dimension all the time. We can look at two frequency counts at once, arranging our books on a two-dimensional plane, as we do in Figure 1.

Here we plot Shakespeare's plays in a two-dimensional space, with the position of each play fixed by a pair of coordinates: the frequency of "king" on the vertical axis, and the frequency of "love" on the horizontal one. Now we can see something of the relationship between the two features. The "L"-shaped distribution of the dots, with a blank space in the top right of the plot, tells us that no play has high frequencies of both words—which might lead us to the republican hypothesis that where there are many kings there is no love—although more prosaically it shows that at high frequencies the words are negatively correlated, while at low frequencies there is no necessary relationship between them (a high value of one word is a reliable predictor of a low value for the other, but a low value of one does not reliably predict the value of the other).

---

[9] We have found some support for this notion in other studies of Shakespeare's texts: see Hope and Witmore 2014.

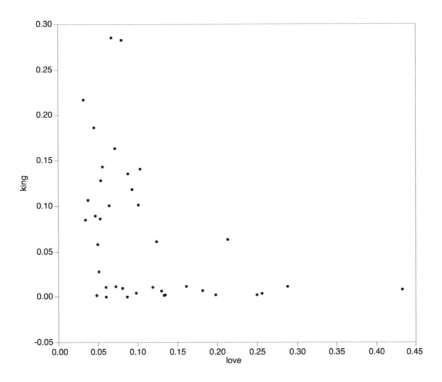

Figure 1. Shakespeare's plays plotted according to the relative frequencies of the words "king" and "love."

And there is no need for us to stop at two dimensions. We can further explore the relationships between our texts by adding a third feature to the plot, such as the frequency of the word "might," and projecting the texts into a three-dimensional space. But the things we've counted here—the frequencies of just three words, two of which are pretty obvious—restrict the new questions we are generating, and they do not capitalize on the other great advantage of computers: their ability to count lots of things across hundreds of texts. It would just about be possible for a human to count every instance of "king," "love," and "might" across all of Shakespeare's plays, but soon we will all have access to tens of thousands of texts. Digital tools allow us to work at scales beyond those that limit human readers, so let's shift up from three words in 38 Shakespeare plays, and start counting over 72 linguistic dimensions in 554 early modern plays. We are currently working on a project funded by Mellon to produce software tools, and methodologies, to allow humanities scholars to access and work with large corpora like EEBO-TCP and

ECCO. We are trying to find ways to allow humanities researchers to make sensible use of all of this data: how can they find what they want, and how can they find new things?

One obvious way to come to grips with an expanding data set is to begin with what you know and work outwards. So we began with Shakespeare, and we are working out to "the whole of" early modern printed drama—about 554 texts, depending, of course, on how you define "drama."[10] Rather than counting individual words, we've been using a piece of linguistic analysis software called DocuScope.[11] DocuScope counts functional language units and sorts them into groups called "Language Action Types" (LATs). Each LAT consists of words and phrases that have the same function—marking first person, for example, or encoding anger, or introducing turns in rapid dialogue. Because LATs are more sophisticated linguistically than individual words, counting them gives us a more complex, nuanced picture of what's going on linguistically in the texts we're comparing. DocuScope was designed to pick up the different things writers do when they try to achieve different things with their texts—and because it was designed using the OED as a base, it is surprisingly good at "reading" early modern English.

---

[10] Our current definition is "plays" printed before 1660 which either were performed, or were intended to be performed, or look as though they could have been performed (note the vagueness of the last category—some of our texts are closet dramas). A corpus of less tightly defined "dramatic" texts including masques, entertainments, and so on, would run to more than 700 texts. Including dialogues (a form often employed in philosophy and instructional texts) in the corpus would push that over the thousand mark—but dialogues were not intended for performance. The point here is that there is no single "right" corpus of early modern drama: our corpus of 554 attempts to be inclusive, but necessarily lacks any plays known only in manuscript, all lost plays, and several plays not yet transcribed by TCP. The needs of any one researcher are likely to differ from ours (we have already spoken to scholars who want to include Peele's Lord Mayor entertainments in "the" corpus)—and one of the aims of Visualizing English Print is to give scholars tools that allow them to construct their own corpus from TCP easily. Doing this kind of work is very good for the soul: it makes you define your object of study very precisely!

[11] For DocuScope, see https://www.cmu.edu/hss/english/research/docuscope.html. The language theory underpinning DocuScope, and the categories it sets up, are detailed in Kaufer et al. 2004. A number of studies illustrating its use in the classroom, and authorship work, are listed at http://wiki.mla.org/index.php/Docuscope.

Now, we just moved from one, to two, to three-dimensional analysis, by adding an extra feature at each stage—an extra frequency score. Effectively, we arranged our books on a line, a flat plane, and then in a cube, allowing us to see the relationships between the books, and in the case of two- and three-dimensional representations, between the features themselves. When we get beyond three dimensions, our puny human brains seize up: we can't imagine adding a fourth axis to the three-dimensional graph (or not without a lot of difficulty). But mathematicians have long known how to describe spaces with more than three dimensions, and once you regard these spaces as mathematical objects, there is no limit to the number of dimensions you can project data points, or books, into.

So we can continue adding dimensions to our virtual library right up to the number of features DocuScope counts—which happens to be 72.[12] The resulting spreadsheet has 72 values for each of the 554 plays: 72 coordinates fixing them precisely, if unimaginably, in 72-dimensional space. All well and good—but how can we look at this space to see where the books are on the shelves, which ones are serendipitously next to each other? The answer is that, having built up a space too complex for us to see, we use statistics to simplify it—effectively to throw away information we hope isn't important, allowing us to visualize the space in a form we can read. Now, there are lots of ways of doing this: statistics is an art, not a science. The method we have been using, not without qualms and unease, is principal component analysis (PCA)—a well-known method for looking at relationships between features in complex data sets.[13]

To cut a long and complex story short (which is, effectively, what PCA itself attempts to do), PCA collapses the axes making up our 72-dimensional space into a series of super-axes, each one of which attempts to summarize some of the variation, or space, of our original space. If we take the two super-axes (called "principal components") that together summarize the largest amount of the original space, we can use them to plot a two-dimensional graph, showing something of the relationships between our texts.

---

[12] In fact, the version of DocuScope we use here counts 113 LATs, but we use only 72 in the analysis because we discard LATs with very low frequencies.

[13] We give a fuller account of PCA in Basu, Hope, and Witmore (forthcoming). Most standard statistics textbooks cover PCA (and factor analysis, to which it is closely related); we have found Field (2013) useful. Literary scholars will probably get most out of Alt (1990), which is a brief and very clear conceptual account of what the statistical procedures are trying to achieve.

Figure 2 is a plot of the 554 plays which make up early modern drama in what we can call "PCA space." When you look at it, remember that you are looking at a huge simplification: 72 dimensions have been reduced to two. We have thrown away all but 26 per cent of the information in our 72-dimensional space. Let's hope we kept the important bits.

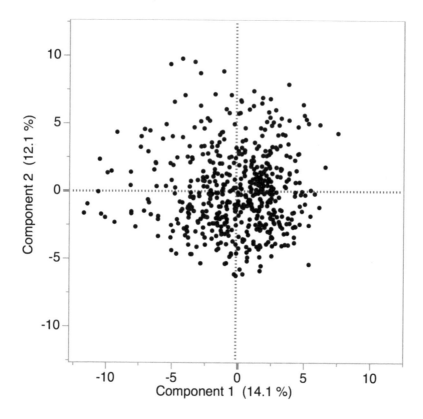

Figure 2. The corpus of 554 early modern plays visualized in PCA space.

So Figure 2 is our visual summary of a set of linguistic relationships in the corpus of early modern drama. Each dot is a play, and we can say, roughly speaking, that plays appearing next to each other use similar types of linguistic features at similar rates (and avoid similar groups of features). Conversely, plays a long way from each other will differ linguistically (we are hedging this because our statistical simplification may warp space, so we need to check things, but let's pretend all this is true for now). We now have an overview of early modern drama in linguistic space, and we can start to "read" it. One thing is clear: the dots are not evenly distributed across the space. There are empty areas, especially in the bottom half, free of dots. So there are combinations of language that just don't get used in drama—why not? One interesting avenue of further research would be to add texts from different genres to this corpus to see if they occupy the spaces drama avoids: perhaps sermons, or a nascent genre like the novel would show up here—or perhaps these combinations are simply not used in English writing.

Another thing: the dots seem to clump in a circle roughly centred on the point of origin, with a sparse cloud of less densely-packed plays to the upper left. It looks as though most plays are broadly similar to each other, using the identified set of linguistic features at more or less similar rates. When playwrights vary from this norm, they do so by moving into only certain areas. Again, why should this be?

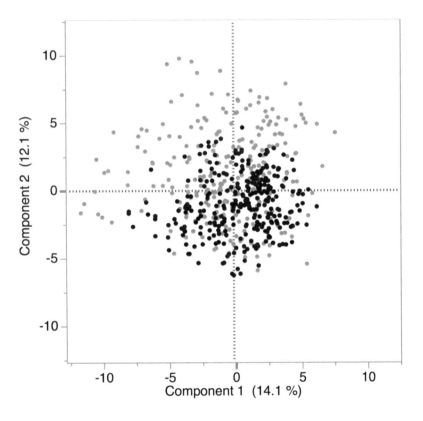

Figure 3: Early modern drama with 299 "career plays" highlighted.

We have been playing around with this data, and one thing we did was to pick out the plays in this sample written by "career" playwrights. These are writers who have written large numbers of plays, five or more—"professional" playwrights in the true sense. The big names of the period—Shakespeare, Fletcher, Dekker, Massinger, Middleton, and so on. When you add up the plays by these men, they total about 299. The rest of the plays in the sample are written by people who have written only one or two—or they are translations not intended for the professional stage. Figure 3 shows those 299 professional plays highlighted in black, with the rest in grey.

We found this result surprising—the clustering effect we noted above becomes even more pronounced. What this result suggests to us is that professional early modern playwrights were exactly that: professional. Professional in the sense that they knew what a professional play sounded like, and could hit the target every time. They are not, on this evidence, a set of daring experimenters, despite including some of the most celebrated names in English literature. The experimenters are the grey dots out in the upper left regions of the graph—these turn out to be translations from the classics, and mavericks either not writing for the stage at all, or writing once, and never being employed again.

Once again, this overview suggests several paths for future research. What is the relationship between date of writing and position on the graph? (We suspect that many outlier plays are either very early or very late.) What about genre? (We begin this discussion in a forthcoming paper.[14]) Can we learn anything about "successful" playwrights by looking at the outlier plays? Do certain types of play group in the outlier areas?

The answer to that final question is, in some cases, yes. For example, 12 of the outlier plays grouped along the horizontal axis to the left turn out to be translations of Seneca's tragedies (see Figure 4). Any history of the early modern stage will tell you that these plays are the foundation of, the key influence on, early modern drama as a whole, and early modern tragedy in particular. What do we make of the linguistic space separating Seneca from the professional dramas he is universally held to have influenced? Does the extreme language of the Senecan plays mean we need to reassess these claims for "influence"? Or can we read the distance from the Senecan region to that of the "core" early modern tragedies as "influence" or adaptation?

---

[14] Witmore, Hope, and Gleicher (forthcoming).

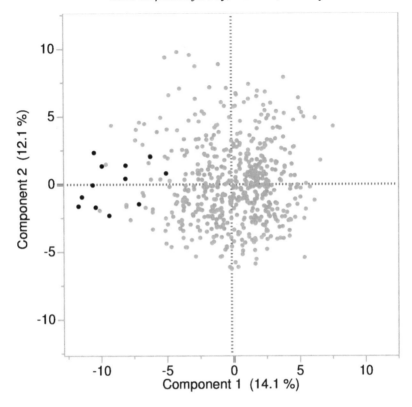

Figure 4. Early modern drama with translations of Seneca highlighted.

And what about Shakespeare? Shakespeare clusters with his professional colleagues in the central mass of dots—resolutely average, doing similar things to them, at similar rates. How do we reconcile this with our notion that Shakespeare is, by about as many orders of magnitude as you care to name, "better" than everyone else? Either our sense of Shakespeare as exceptional is wrong, or, whatever it is that makes Shakespeare great, we ain't counting it yet.

These two findings—that early modern professional dramatists stick together, and that Shakespeare sticks with them—chime with something other digital scholars have found in other periods. Ted Underwood, who works on nineteenth-century literature, has noted (2013) that the narratives of traditional literary history focus on rupture and revolution: break points triggered by the emergence of radically new individual genius. When he looks for shifts in literary history in large digital corpora, however, he does not see sudden shifts. Instead, he sees similarity and continuity.

Viewed through a large-scale lens, the subject itself becomes a different thing—a series of slow developments, incremental changes—and the story of genres slowly emerges from other types of text over decades, rather than springing fully-formed from the brow of one exemplary genius. Where does the novel really come from? What about scientific writing? How do sermons, by far the largest text-type for most of the period, but read by almost no one now, fit in? Up to now, literary history has worked a bit like PCA, by cutting away the unnecessary and hoping to leave behind the important. We had to do that because we didn't have the time, or the brain capacity, to read everything. But if you chuck most of the books away, it is hardly surprising that the ones remaining start to look like exceptional peaks rising above the plain, appearing without preparation. It is as if we had gone round the library pulling most of the books off the shelves, leaving only the odd one to represent whole subject areas or periods.

Now, as the paradigm shifts, we are going round again, putting the books back—thousands of them, hundreds of thousands of them, most unread since they were published, containing who knows what. The only thing we have to do is learn how to find them, and then read them again.

WORKS CITED

Alt, Mick. 1990. *Exploring Hyperspace: A Non-Mathematical Explanation of Multivariate Analysis.* London: McGraw-Hill.

Balsamo, Luigi. 1990. *Bibliography: History of a Tradition.* Translated by William A. Pettas. Berkeley, CA: Bernard M. Rosenthal.

Basu, Anupam, Jonathan Hope, and Michael Witmore. Forthcoming. "The Professional and Linguistic Communities of Early Modern Dramatists." In *Community-Making in Early Stuart Theatres: Stage and Audience*, edited by Roger D. Sell, Anthony W. Johnson, and Helen Wilcox. Farnham, UK: Ashgate.

Blair, Ann. 2003. "Reading Strategies for Coping with Information Overload ca.1550–1700." *Journal of the History of Ideas* 64 (1): 11–28.

———— 2010. *Too Much to Know: Managing Scholarly Information before the Modern Age*. New Haven, CT: Yale University Press.

Campbell, James W.P., and Will Pryce. 2015. *The Library: A World History*. London: Thames and Hudson.

Field, Andy. 2013. *Discovering Statistics Using IBM SPSS Statistics: And Sex and Drugs and Rock 'n' Roll*. 4th ed. London: Sage.

Gameson, Richard. 2006. "The Medieval Library (to c. 1450)." In *The Cambridge History of Libraries in Britain and Ireland*, edited by Elisabeth Leedham-Green and Teresa Webber, 13–50. Cambridge: Cambridge University Press.

Hill, Geoffrey. 2013. "'Legal Fiction' and Legal Fiction." Oxford Professor of Poetry Lecture, March 5. http://media.podcasts.ox.ac.uk/engfac/poetry/2013-03-21-engfac-poetry-hill-2.mp3 (audio only).

Hope, Jonathan, and Michael Witmore. 2014. "Quantification and the Language of Later Shakespeare." *Actes du congrès de la Société française Shakespeare* 31: 123–49. https://shakespeare.revues.org/2830.

Kaufer, David, Suguru Ishizaki, Brian Butler, and Jeff Collins. 2004. *The Power of Words: Unveiling the Speaker and Writer's Hidden Craft*. London: Routledge.

Krajewski, Markus. 2011. *Paper Machines: About Cards and Catalogues, 1548-1929*. Cambridge, MA: MIT Press.

Leedham-Green, Elisabeth, and Teresa Webber, eds. 2006. *The Cambridge History of Libraries in Britain and Ireland*. Vol. 1, *To 1640*. Cambridge: Cambridge University Press.

Magnusson, Lynne. 2009. "A Play of Modals: Grammar and Potential Action in early Shakespeare." *Shakespeare Survey* 62: 69–80.

McKitterick, David. 2006. "Libraries and the Organisation of Knowledge." In *The Cambridge History of Libraries in Britain and Ireland*, edited by Elisabeth Leedham-Green and Teresa Webber, 592–615. Cambridge: Cambridge University Press.

Nelles, Paul. 2009. "Reading and Memory in the Universal Library: Conrad Gessner and the Renaissance Book." In *Ars reminiscendi: Mind and Memory in Renaissance Culture*, edited by Donald Beecher and Grant

Williams, 147–69. Toronto: Centre for Reformation and Renaissance Studies.

Norris, Dorothy May. 1939. *A History of Cataloguing and Cataloguing Methods 1100-1850: With an Introductory Survey of Ancient Times.* London: Grafton and Co.

Petroski, Henry. 1999. *The Book on the Bookshelf.* New York: Alfred A. Knopf.

Rissoan, Bastien. 2014. "La gestion de l'information au XVIe siècle (2/2): La bibliographie universelle de Conrad Gesner." *Interfaces/Livres anciens de l'université de Lyon* blog post, July 24. https://bibulyon.hypotheses.org/4825.

Rosenberg, Daniel. 2003. "Early Modern Information Overload." *Journal of the History of Ideas* 64 (1): 1–9.

———. 2013. "Data before the Fact." In *"Raw Data" is an Oxymoron*, edited by Lisa Gitelman, 15–40. Cambridge, MA: MIT Press.

Sargent, Clare. 2006. "The Early Modern Library (to c. 1640)." In *The Cambridge History of Libraries in Britain and Ireland*, edited by Elisabeth Leedham-Green and Teresa Webber, 51–65. Cambridge: Cambridge University Press.

Underwood, Ted. 2013. *Why Literary Periods Mattered: Historical Contrast and the Prestige of English Studies.* Stanford, CA: Stanford University Press.

Webster, Keith. 2015. "Redefining the Academic Library—Revisiting a Landmark Report." *Library of the Future* blog post, April 13. http://www.libraryofthefuture.org/blog/2015/4/13/redefining-the-academic-library-revisiting-a-landmark-report.

Witmore, Michael, Jonathan Hope, and Michael Gleicher. (Forthcoming) "Digital Approaches to the Language of Shakespearean Tragedy." In *The Oxford Handbook of Shakespearean Tragedy*, edited by Michael Neill and David Schalkwyk. Oxford: Oxford University Press.

Plotting the "Female Wits" Controversy: Gender, Genre, and Printed Plays, 1670–1699[1]

Mattie Burkert
Utah State University
madelaine.burkert@gmail.com

At the end of the 1694–95 theatrical season, a group of actors defected from the United Company, housed at the Theatre Royal in Drury Lane, to begin their own cooperative. The split of London's theatrical monopoly into two rival playhouses—Christopher Rich's "Patent Company" at Drury Lane and Thomas Betterton's "Actors' Company" at Lincoln's Inn Fields—generated demand for new plays that might help either house gain the advantage. This situation created opportunities for novice writers, including an unprecedented number of women. Paula Backscheider has calculated that more than one-third of the new plays in the 1695–96 season were written by, or adapted from work by, women, including Delarivier Manley, Catherine Trotter, and Mary Pix (1993, 71). This group quickly became the target of a satirical backstage drama, *The Female Wits: or, the Triumvirate of Poets at Rehearsal*, modeled after George Villiers's 1671 sendup of John Dryden, *The Rehearsal*. The satire, which was likely performed at Drury Lane in fall 1696, was partially the Patent Company's revenge on Manley for withdrawing her play *The Royal Mischief* during rehearsals the previous spring and taking it to Betterton. However, *The Female Wits* was also a broader attack on the pretensions of women writers, whom it portrayed as frivolous, self-important upstarts reviving the overblown heroic tragedy of the 1660s and 1670s with an additional layer of feminine sentimentality.[2]

---

[1] This research was made possible by a grant from the Andrew W. Mellon Foundation to Visualizing English Print, a project led by Michael Gleicher, Robin Valenza, Michael Witmore, and Jonathan Hope. Special acknowledgment is due to Richard Jason Whitt for his initial processing of the corpus from which my sample is drawn; Eric Alexander and Michael Correll for their guidance in using the prototype tools; Catherine DeRose for her invaluable comments on an early version of this paper; and Bret Hanlon for his help in understanding statistical concepts.

[2] Robert Hume argues that the heroic plays of this period, which often defy the tragedy/comedy binary, are primarily defined by their thematics of love, valour, war,

Critics have debated the extent to which the revival of heroic drama in the 1690s was or was not a gendered phenomenon, as well as the extent to which these female dramatists identified as a cohort or community. Some theatre scholars view *The Female Wits* as an attack on a burgeoning feminine literary style and tradition, emphasizing the sense of female solidarity that linked women writers. For instance, Pilar Cuder-Domínguez, following Backscheider, claims that the women writers of the 1690s located themselves not only as a cohort in the present, but as the inheritors of the mantle worn by Aphra Behn and Katherine Philips in the previous generation (2010, 267). Scholars in this school of thought point to the commendatory verses and prefatory materials the women wrote for one another's plays—which contain frequent allusions to "Orinda" (Philips) and "Astrea" (Behn)—as well as to similarities between their works. According to Backscheider, "the plays by women produced in the 1695–96 season illustrate efforts to protect and legitimate many of the revisionary elements in Behn's plays. Most striking in these plays are their depictions of women's utopian dreams of married love and of their fears of loneliness, insult, and rape" (1993, 86).[3] Backscheider thus implicitly links the women's gender and their sense of female literary tradition to particular formal and thematic features of their writing.

---

and politics (1976, 272–73, 286). According to Laura Brown, early Restoration heroic drama is characterized by its expression of aristocratic ideals, its distant settings, and its elevated style. For Brown, the "affective tragedy" of the later Restoration is "directly and closely derived" from earlier heroic forms, but it shifts away from exotic locales and upper-class heroes to become more focused on morality than on social status; consequently, its protagonists are more likely to be women or middling-class men than aristocrats (1981, 3–4, 69–70). Hume similarly connects the resurgent heroic of the mid-1690s to the earlier mode but sees the changes as largely superficial: "The standard formula for a serious play in these years is simple: graft pathos and sentiment onto an old play" (424). Although he cites plays by both male and female authors as examples of this "grafting," he notes that "Female writers, especially, seem consistently to have favoured the heroic mode" in the 1690s, pointing to Trotter's *Agnes de Castro* and Pix's *Ibrahim* as examples (423).
[3] Along the same lines, Laurie A. Finke argues that the female playwrights of the 1690s were "conscious of themselves as an intellectual group" (1984, 64), and Laura J. Rosenthal discusses the prefatory materials Trotter, Pix, and Manley wrote for one another's plays as an attempt to construct "a distinct female literary estate" (1996, 178). Tanya Caldwell likewise states that "Behn had paved a path down which her successors were careful to follow, consciously placing themselves in a tradition and creating a female lineage" (2011, 21).

Some recent critics, however, have pushed against this view, claiming instead that the women's plays display a diversity that belies the notion of a specifically feminine dramatic form. Jane Milling, for instance, shows how female writers operated within the same structures and faced many of the same obstacles as their male counterparts; she emphasizes the variety of ways they responded to these market forces through their writings (2010a and 2010b). Marcie Frank similarly resists the narrative of a nascent female literary tradition, pointing to Dryden rather than Behn as the most powerful influence on playwrights like Trotter and Manley (2003). Thus, scholars continue to debate whether these women's plays are qualitatively different from those produced by their male counterparts, suggesting that late-seventeenth-century debates about the changing generic landscape of drama crystallized around gender in ways that still inform criticism today.

We can begin to disentangle gender and genre by viewing the drama of the period computationally. Quantitative techniques allow us to see the statistically perceptible changes that took place in heroic drama over the course of the final decades of the seventeenth century, and to compare the affective and formal dimensions of male and female authors' language. This paper therefore undertakes a quantitative reexamination of generic shift in late Restoration drama. Using the text-tagging program DocuScope and principal component analysis (PCA) of 272 plays printed between 1670 and 1699, I find that the heroic tragedies of the 1690s are, in fact, measurably different from those of the first wave. In place of the reasoned deliberation that characterizes earlier heroic tragedies, those of the 1690s are marked by higher levels of negativity, contingency, and denial—features that mirror the overblown satire of Manley's dramatic language in *The Female Wits*, as well as Backscheider's characterization of women writers' "fears," above.[4] While several plays by women fit this pattern, so do many written by men. My findings therefore challenge the assumption that this new form of heroic tragedy was primarily the purview of female playwrights. I argue instead that critics who lambasted the "female wits" were responding to more widespread changes in dramatic form, and I suggest that women were used as scapegoats for broader shifts in the language and focus of heroic drama. This case study serves as an example of modern critics' tendency to unwittingly reproduce

---

[4] For the purposes of this analysis, I am not concerned with one prominent difference between the two generations of heroic plays: those of the 1670s are written in heroic couplets, while those of the 1690s are written in blank verse. Rhyme (or the lack thereof) is not perceptible by DocuScope, and is therefore outside the scope of this paper.

the assumptions embedded in early modern criticism, as well as the potential of computational methods to help us challenge these inherited narratives.[5]

While advancing my primary argument about generic shifts in late Restoration drama, I also take the opportunity to reflect on the benefits and drawbacks of performing this kind of research computationally. I wish to make explicit the kinds of access statistical methods provide to large corpora of texts, as well as the limits of that access. Just as importantly, I wish to suggest the new perspectives we can gain by using techniques like PCA, while attending to the ways that the corpus curation, experimental design, and interpretation of results necessarily reproduce certain field-specific assumptions about the objects of analysis.[6] To that end, I provide a detailed account of my methods of corpus curation and statistical analysis before delving into the results and their implications.

Corpus Curation

This analysis draws on a corpus of 272 plays printed between 1670 and 1699, a period that spans the heyday of Restoration theatre through the beginnings of Augustan drama and is often seen as a period of transition in public tastes and sensibilities. The texts were taken from the Early English Books Online-Text Creation Partnership (EEBO-TCP), a collection of hand-keyed electronic transcriptions of facsimiles from EEBO.[7] In addition to hand-selecting the

---

[5] As Catherine Ingrassia observes, "The act of scholarship often replicates the cultural relations it seeks to dissect" (1998, 9).

[6] D. Sculley and Bradley M. Pasanek rightly argue that prior assumptions come to bear on all stages of the experimental process, and that this issue may be even more acute in the humanities: "The literary critic ... can interpret anything, but such interpretation may well be another name for overfitting the data" (2008, 421). They propose that humanities scholars who use data mining methods should be more scrupulous about recording their assumptions, a call I attempt to answer here.

[7] Specifically, I used an archive of EEBO-TCP captured in early 2012 by members of the Visualizing English Print research team. The entire archive contains 20,885 texts published between 1470 and 1700, including 6,588 texts dated 1670–1699, from which I hand-selected my corpus. Each text was assessed individually to determine whether it was dramatic. It is worth noting that this step involved selection criteria about what counted as a "Restoration play" that were necessarily driven by my research questions. For example, I chose to include whole operas but not individual songs or collections of songs from operas. I did not include prologues and epilogues printed separately as broadsides or pamphlets, nor did I include translations of classical plays or Lord Mayor's Day pageants, because these texts were unlikely to provide insight

dramatic texts from this larger archive, I also collected metadata about each play, including genre, publication year, and author name. This metadata was taken from the texts themselves wherever possible and supplemented with information from historical sources like Gerard Langbaine's *A New Catalogue of English Plays* (1688), as well as from EEBO-TCP records. My goal was to use the terms and categories available to contemporaries wherever possible in order to avoid imposing modern scholarly assumptions about authorship and genre that might color the results.[8]

My corpus is necessarily an incomplete and limited set of plays from the decades under consideration, for several reasons. First, I began from an archive of texts that had been collected and modernized in early 2012, when EEBO-TCP was roughly half the size it is as of this writing. Second, the TCP collection itself is an incomplete subset of the more than 125,000 texts available in EEBO.[9] Third, even EEBO does not include every text printed—only those extant texts that have been scanned. Finally, because EEBO texts are dated by publication rather than by composition or performance, the corpus includes reprints of older plays and revivals and therefore may not map neatly onto theatrical trends as they appeared on stage. However, publication of plays frequently followed shortly on their performance, usually in a matter of weeks or months, and new plays and revivals were more likely than repertory plays to appear in new editions (Milhous and Hume, 1974). It is therefore fair to assume that the corpus under analysis represents a reasonable approximation of dramatic tastes across the period.

---

into major genres of the public theatre and would introduce unnecessary noise into the analysis of the data. However, I did choose to include collections of several plays bound together, as well as plays that appeared to have been printed but never performed. Given the scarcity and unreliability of performance records from this period, determining which plays were intended for performance or actually performed would be difficult and time-consuming and would likely only result in the elimination of a small handful of items from the corpus.

[8] For example, I did not use generic categories like "sentimental heroic tragedy" that are found in today's criticism. Instead, I labeled plays according to seven basic genres that correspond to those used in the title pages and criticism of the period: tragedy, comedy, tragicomedy, history, opera, masque, and farce. Similarly, I carefully preserved information about the anonymity of plays at their moment of publication, even in cases where the authorship is now agreed upon.

[9] For more on the TCP collection, please see http://www.textcreationpartnership.org/tcp-eebo/.

The texts were tagged according to the DocuScope dictionary built by David Kaufer at MIT, using the Ubiqu+Ity tagging program developed by Visualizing English Print.[10] DocuScope contains entries for millions of words and phrases, each of which has been preassigned to one of more than 100 Language Action Types (LATs).[11] DocuScope's LATs represent a functionalist approach to language, which is to say that the dictionary tags phrases based on what they do—passing judgment, invoking communal authorities, or describing the physical world, for example—rather than by what they are (e.g., parts of speech). This approach to language makes DocuScope exceptionally good at helping us identify the functional linguistic underpinnings of dramatic genres, as Jonathan Hope and Michael Witmore have demonstrated.[12] It is therefore an excellent tool for exploring how the language of heroic drama may have changed over the final decades of the seventeenth century.

Principal Component Analysis

Tagging 272 texts with more than 100 features results in thousands of data points. In order to analyze the interactions of these features, it is necessary to aggregate them into a manageable number of interpretable phenomena. For this study, I used CorpusSeparator, a program designed by Michael Correll at the University of Wisconsin-Madison, to perform PCA on the collections of LAT measurements for all texts in the collection. CorpusSeparator generates

---

[10] Ubiqu+Ity can be accessed at http://vep.cs.wisc.edu/ubiq/. In order to improve tagging accuracy and frequency, the texts underwent several pre-processing steps. First, when they were initially collected in 2012, they were stripped of their SGML/XML encoding and modernized using VARD 2, an open-access text processor designed by Alistair Baron at Lancaster University to reduce spelling variation in early modern English corpora (see http://ucrel.lancs.ac.uk/vard/about/). For many of the plays under consideration, this modernization step had the added benefit of removing speech prefixes, which were not recognizable as words. I selected my dramatic corpus from this already-modernized archive, according to the criteria outlined in note 6. I then used a batch text editor to remove all vertical bars (|), which indicate words split across line or page boundaries in TCP transcriptions.

[11] DocuScope assigns each word to one string and each string to a single LAT; as such, it does not account for the multiple overlapping functions of linguistic units. For more on the rhetorical and phenomenological underpinnings of the DocuScope dictionary, see Kaufer et al., 2004.

[12] See Hope and Witmore (2004, 2010, and 2014), as well as Witmore and Hope (2007). Susan David Bernstein and Catherine DeRose have also shown that PCA of DocuScope-tagged texts reveals signals of seriality in Victorian novels (2012).

principal components (PCs), which are mathematical representations of how different features' presence or absence tends to co-occur in texts. Each PC has a positive or negative weight for each LAT, indicating the feature's relative presence or absence in that PC. The first components account for the majority of variation across the corpus, with PC1 capturing the most variation. In the case of my 272-play corpus, as we will see, PC1 roughly separates the interpersonal energy of comedy from the descriptive language of tragedy and other genres, while PC2 differentiates the abstract verbal texture of heroic drama from the sensory spectacle of masques and burlesques.

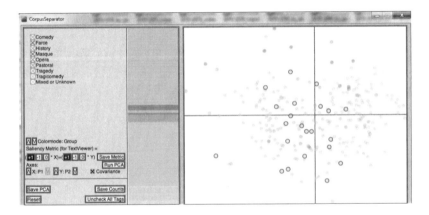

Figure 1. The 272-text corpus color-coded by genre, where the x-axis represents PC1 and the y-axis represents PC2. Comedies, labeled blue, group on the right side of the plot, while tragedies, labeled green, group on the left side. Tragicomedies, circled in red, tend to hover around the y-axis.

Once it has generated PCs that best account for variation between the texts, CorpusSeparator next scores each text against those PCs and projects the results onto a scatterplot.[13] The plot in Figure 1 shows the 272-play corpus color-coded by genre. In this visualization, the x-axis represents PC1 and the y-axis represents PC2. Texts whose LATs best align with the PC's priorities are positioned furthest along the axis in the positive direction, while texts that abound in tags against which the PC selects appear furthest along the axis in the negative direction. The distance between points in this space reflects the texts' difference from one another based on how DocuScope has tagged them. As this scatterplot makes immediately clear, genre is responsible for

---

[13] CorpusSeparator generates this score by using linear combination to multiply the individual text's weights for each LAT by the PC's weight for that LAT.

the most variation in the corpus. The majority of tragedies appear on the left side of the y-axis, while comedies tend toward the right side; tragicomedies fall around this dividing line.

Having established that PCA was, in fact, detecting features that separated the corpus roughly by genre, the next step was to analyze plays from each decade separately.[14] The goal in separating the corpus by decade was to determine how genre did or did not change across the period, in order to better understand why contemporaries experienced the revival of heroic tragedy in the 1690s as being different from the first wave in the 1670s, and also in order to determine whether gender had a role in that change.

PC1: Comedy vs. Tragedy

I began by analyzing PC1, the component that clearly separates comedy from tragedy in the 272-text corpus, in order to see whether the features of

---

[14] For the analyses described in this paper, I used DocuScope version 3.83, which contains 134 LATs. Because the number of features (LATs) exceeds the number of observations (texts) in my two sub-corpora (109 texts from the 1670s and 103 from the 1690s), there may be concerns about the limited generalizability of my findings. I address this concern in several ways. First, the findings in this paper are quite similar to those that were obtained using DocuScope 2.2, an older version of the dictionary that contains only 101 LATs, on the 1690s corpus. The earlier study was presented at the 2013 annual meeting of the Modern Language Association, and the newer study largely mirrors its findings for the 1690s. In addition, I ran a post-hoc feature selection test on the 1670s sub-corpus in which I eliminated the 40 LATs that had the least effect on the first two principal components. Doing so did not change my scatterplot, which suggests that the least prominent LATs were not introducing significant noise into my initial PCA results using the full 134-feature dictionary. This makes sense, since some of the least important LATs were clearly irrelevant to early modern drama. "Reporting_Geography_US_States," for instance, is a poor fit for historical reasons, while variations of "Academic_Citation" track attributional moves of academic prose that are unlikely to occur frequently in drama. Furthermore, in this paper I consider only the highest-weighted LATs for the first two (and therefore most important) principal components, thereby reducing the dependence of my analysis on potentially noisy or irrelevant weightings. Finally, it is important to note that my claims are descriptive rather than predictive: I do not offer a statistical model for classifying all Restoration drama based on this subset. Rather, I take the statistical analysis of this specific corpus as a point of departure for interpretive claims about particular texts grounded in the methods of literary analysis and theatre history.

tragedy as a whole changed across the Restoration period. Figure 2 shows the scatterplot of just the 109 plays published between 1670 and 1679, color-coded by genre. Again, tragedy and comedy are divided along the x-axis, with most comedies scoring positive for PC1 and therefore falling to the right of the y-axis, and most tragedies scoring negative for PC1 and falling to the left. In fact, PC1 appears to differentiate comedies from nearly everything else: most histories, operas, masques, farces, and pastorals are, like tragedy, negative for PC1.

What linguistic features create this separation? The weights that define the "comedy component" suggest that PC1 selects most strongly for the LAT Character_PersonProperty, a tag that is assigned to words like "sir," "servant," "kinswoman," and "wife" that define individuals' social roles. It also selects strongly for the LATs "Interactive_You_Attention" (phrases like "have you," "I hope you," and "do you") and "Interactive_Question" (words and phrases like "who," "why," and "may I," as well as question marks). Comedies of the 1670s, then, appear preoccupied with people and their interactions, a finding that corresponds roughly to Jonathan Hope and Michael Witmore's discovery that Shakespearean comedy prioritizes interpersonal exchange (2010, 373). The agreement between my findings and theirs suggests that the markers of comedy are consistent from the Elizabethan period up to the Restoration.

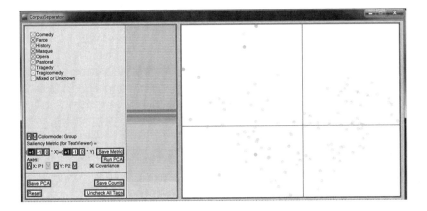

Figure 2. The subset of 109 plays published 1670–1679, color-coded by genre as in Figure 1.

On the other hand, non-comedic plays that score low for PC1 are full of SenseObjects—concrete nouns like "sword," "periwig," "house," "shop," and "books"—as well as SenseProperties, a tag applied to adjectives and adjective phrases that describe physical objects: "swift," "sweet," "dark," "piercing eyes," "tempestuous winds," etc. (The discussion of Thomas Duffet's *The Mock-Tempest*, below, provides a more detailed analysis of a specific passage in which SenseObjects and SenseProperties interact.) What PC1 may tell us, then, is that the dividing line between comedy and other genres in the 1670s is the relative presence or absence of elaborate descriptive language—not, as we might expect, the presence of words suggesting positive or negative emotion.

PC1 again appears to distinguish between comedy and tragedy for the subset of 103 plays from the 1690s, although the points on the scatterplot group more loosely than in the 1670s, perhaps a signal of the experimental nature of much 1690s drama (Figure 3). In the 1690s, the comedy component continues to select for features related to people and conversation and against features related to description. In other words, the linguistic characteristics distinguishing comedy from other genres are similar in both decades (see Table 1).

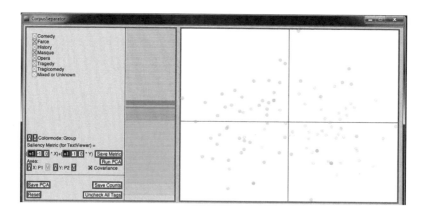

Figure 3. In the 1690s, tragedy (orange) and comedy (light blue) are still separated along PC1, but more loosely.

While PC1 remains similar across the period, however, PC2 shows quite a bit more change from the 1670s to the 1690s. It therefore offers a better explanation of the generic change identified by contemporaries at the end of the seventeenth century—the turn to sentimental heroic tragedy that was associated with the emergence of women dramatists.

Language Action Type (LAT)	Weight (1670s corpus)	Weight (1690s corpus)	Change in Weight	Absolute Value of Change
Descript_SenseObject	-0.883	-0.598	0.285	0.285
Public_Values_Negative	-0.032	-0.152	-0.12	0.12
Interactive_You_Attention	0.435	0.554	0.119	0.119
Descript_SenseProperty	-0.443	-0.341	0.102	0.102
Elaboration_Prep	-0.443	-0.543	-0.1	0.1
Interactive_You_Focal_Subject	0.297	0.389	0.092	0.092
Reporting_Events	-0.19	-0.1	0.09	0.09
Elaboration_That_Wh	-0.056	-0.146	-0.09	0.09
Relations_Resistance	-0.07	-0.151	-0.081	0.081
Relations_Inclusive	-0.188	-0.109	0.079	0.079

Table 1. The LATs that define PC1 (what I am calling the comedy component) change from the 1670s to the 1690s sub-corpus. These ten features show the most change across the period.[15]

PC2: Heroic Plays vs. Sensory Spectacles

As we have seen, the distribution of texts along the PC1 axis is easy to recognize: it is immediately apparent that this component distinguishes comedy from other genres, particularly tragedy. In other words, the phenomenon identified algorithmically by PCA is easily correlated to human-generated categories of genre, in part because the basic dramatic genres (tragedy, comedy, farce, and so on) are a form of metadata offered up by the texts and coded for each item in the corpus. PC2, conversely, does not separate plays into groups that are immediately interpretable in terms of the basic generic metadata provided by title pages and EEBO entries. Rather, it produces groupings that must be understood through an iterative movement between corpus-level patterns and individual texts, informed by domain knowledge of Restoration theatre. As this movement reveals, PC2 can help us discern

---

[15] I first took the top 25 most important LATs for PC1 in each decade (determined by the absolute value of the LAT weights), a total of 29 LATs for the two decades. I then calculated the change for all 29. The average of the absolute value of change across all 29 of these top LATs is 0.069.

the linguistic features of heroic drama, a label not commonly applied on title pages of the period, but nonetheless detectable using PCA.

On the scatterplot of plays printed in the 1670s, the three that appear highest on the y-axis (reflecting the fact that they score highest for PC2) are burlesques by Duffet of serious plays: *The Mock-Tempest*, *Psyche Debauched*, and *The Empress of Morocco: A Farce*. The other high-scoring plays for PC2 are a masque (*Beauties Triumph*, also by Duffet), a tragicomedy (John Crown's *Juliana*), an opera (Monsieur P. P.'s *Ariadne*), and a comedy (Edward Ravenscroft's *The English Lawyer*). Although this is a heterogeneous group, the confluence of farces and farcical comedies, masques, and operas suggests a shared emphasis on physical, visual, and aural entertainment over verbal play. The texts that score the lowest for PC2, on the other hand, are tragedies and tragicomedies that have been labeled "heroic" at various points in history: Edward Cooke's *Love's Triumph*; Frances Boothby's *Marcelia*; Dryden's *Aureng-Zebe* and *The Conquest of Granada*; and John Dancer's translation of Quinault's tragicomedy *Agrippa*.[16] Turning to the weights, we can see that the burlesques that score highest for PC2 are characterized by the frequent occurrence of the LATs Descript_SenseObject, Character_PersonProperty, Directives_Imperative, Character_OralCues and Descript_Motions. The heroic plays against which PC2 selects have a fairly low frequency of those tags, but score high for features such as Public_Values_Positive, Reporting_States, Relations_Resistance, and Reason_Contingency.

To better understand this combination of features, we need to see how they work together in passages that are exemplary of this "heroic component." A single-text viewer tool developed by Michael Correll and Eric Alexander makes it possible to zoom in, not only on the individual tagged texts, but even on passages that push a text toward or away from a particular PC. This step is crucial to humanities inquiry. Although PCA can help researchers identify groups of texts within a corpus that have similar scores relative to a component that accounts for variation within a data set, the algorithm cannot tell us what phenomenon the PC—that set of mutually occurring and/or mutually exclusive features—might represent. Understanding why the algorithm identified those features as significant and why it grouped those particular texts is an act of interpretation. In order to facilitate this interpretation, we can project a "mask" onto individual texts, a kind of mathematical snapshot of the features that define a PC or combination of PCs. Visualizing

---

[16] *Marcelia* was the first play by a woman to be staged professionally in London; see Hughes (2004).

the text's interaction with the mask in this way makes it possible to grasp the meaning of the LATs, not only in terms of their quantitative weights, but in terms of the corresponding qualitative textual features and effects they produce when combined. This movement from corpus-level insights back to the text enables the crucial toggling between "close" and "distant" reading that makes algorithmic criticism meaningful and legible within intellectual traditions like literary studies and theatre history. Importantly, although this step requires field-specific knowledge, it does not require that the researcher should have read all, or even most, of the texts in the collection; rather, it allows us to extrapolate what we know about individual texts and passages to understand patterns within larger corpora.

Figure 4 shows the text viewer tool displaying a passage from *Love's Triumph* that pushes the play low on the PC2 axis for the 1670s corpus. The passage exemplifies the features that, according to my interpretation, characterize the language of heroic plays from the 1670s. The tagged words and phrases are color-coded according to the LAT's importance to PC2. The passage is an exchange between two characters, Perdiccas and Roxana. It reads, in part:

> ROX. Of my deserts in Love, if I might boast,
> I best deserve him, cause I love him most.
> PER. And, Madam, if your Love for him be such,
> Can you for him think any thing too much?
> ROX. On this a dangerous consequence ensues,
> Therefore, my Lord, I justly may refuse.—
> He to destroy my Love, this Boon required.
> PER. Was then this favour by the Prince desired?
> ROX. Yes, but I did deny him that request,
> So much destructive to my Interest.
> PER. At first demand of it my Love did start,
> And all my blood went to support my heart.
> But forcive reason me did plainly show,
> There could no disadvantage from it grow.
> To fair Statira's will I did submit,
> And promised her I would indeavour it. (Cooke 1678, 24)[17]

These lines seem to fit a conventional view that heroic drama centers on a conflict between love and politics or duty. This initial impression is nuanced,

---

[17] The passage and those that follow are presented as modernized by VARD 2, but with speech prefixes from the original edition reintroduced for clarity.

however, by an analysis of the tagged phrases that pull this passage toward other heroic plays of the decade. "Careful," "merit," "justly," "favour," and "reason" are all tagged Public_Values_Positive; "hear," "love," and "think" are tagged Reporting_States; "but," "yet," "refuse," and "deny" are tagged Relations_Resistance; and "it would," "might," "may," and "there could" are tagged Reason_Contingency. Put together, these tags suggest that what makes this passage exemplary is the way it dramatizes individuals weighing multiple possible courses of action, deciding between opposing values, and describing or anticipating conflict. In other words, the heroic component is defined primarily by its emphasis on reasoned deliberation, not necessarily an explicit concern with themes of love and valor.

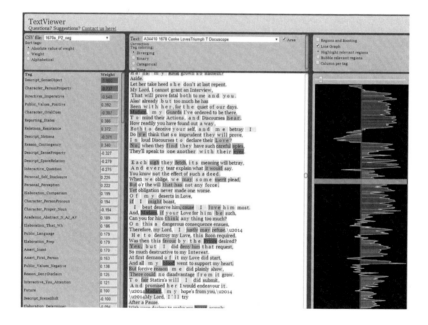

Figure 4. A passage from ***Love's Triumph*** that pushes the play low on the PC2 axis, as seen in the single-text viewer tool. The line graph on the right charts the strength of what we might call an "anti-PC2 signal" over the course of the play. The weights on the left are a mask of "anti-PC2" weights. The passage in the middle pane illustrates the kind of language against which PC2 selects and therefore exemplifies the language of heroic drama that helps pull the play lower on the y-axis in Figure 3. Words highlighted in blue are tagged with LATs that make this passage particularly "heroic," while words highlighted in red are more strongly associated with the sensory language of burlesque, farce, and masque that opposes heroic drama in this corpus.

On the other end of the spectrum, a passage that scores exceptionally high for PC2, and therefore exemplifies the sensory language of burlesque and masque, is a song sung by a devil in Duffet's *The Mock-Tempest*:

> *[A noise of horrid Musick; a Devil arises with a Crown of Fire.]*
> Sings.
> Arise, arise, ye Subterranean Fiends,
> Come claw the backs, of guilty hinds:
> And all ye filthy Drabs, and Harlots rise,
> Which use to infect the Earth with Puddings, and hot Pies;
> Rise ye who can devouring glasses frame,
> By which Wines pass to the hollow Womb, and Brain;
> Engender Head-akes, make bold elbows shake;
> Estates to Pimples, and to deserts turn.
> And you whose greedy flames mans very entrails burn,
> Ye ramping queens, who rattling Coaches take,
> Though you've been fluxed until Head and Body shake.
> Come Clap these Wreches until their parts do swell:
> Let Nature never make them well.
> Cause Legs, and Arms to pine, cause loss of hair,
> Then make them howl with Anguish, and sad groans.
> Rise and obey, rise and obey, Raw head and bloody bones.
> *[Exit Devils]* (Duffet 1675, 20)

This passage abounds in Sense_Objects, including food items ("Puddings," "Pies," "Wines") and body parts ("Brain," "Womb," "entrails"), described by SenseProperties ("hot," "hollow," "Raw"). Directives_Imperative ("Come," "Rise," "make") work together with Character_OralCues (the repeated "ye") to produce the structure of the song, a series of directives to the other devils to wreak havoc on humans. Along with other passages high in PC2, this song highlights how common this combination of imperative syntax and sensory catalog is in songs as well as stage directions—both of which dominate printed texts of masques, farces, and operas that PC2 pushes far away from the heroic plays discussed above.

Examining passages from plays that score exceptionally high and exceptionally low in PC2 shows that this component is roughly tracking the divide between highly sensory plays and highly intellectual or verbal ones. This distinction is, of course, overly schematic, as heroic plays like Dryden's *The Conquest of Granada* were also visual spectacles. Nonetheless, in this collection of plays from the 1670s, language that evokes the senses—whether lines of spoken dialogue that ignite the audience's sensory imaginations or stage directions painting the image of spectacle in readers' minds—tends not to co-occur with the verbal deliberation and evaluation characterizing heroic plays.

Turning to the 1690s sub-corpus, we find that PC2, like PC1, tracks roughly the same phenomena that it did for the collection of texts from the 1670s. Once again, PC2 appears to differentiate between heroic plays on the one hand and farces and musical extravaganzas on the other.[18] However, whereas many of the most extreme examples of the "heroic" component in the 1670s were tragicomedies, here they are all tragedies—suggesting, perhaps, that the heroic mode is more closely associated with tragedy in the later decade. The plays that score highest for PC2 are Trotter's *Fatal Friendship* and *Agnes de Castro*, followed by William Philips's *The Revengeful Queen*, Dryden's *The Fatal Discovery*, and the anonymous *The Unnatural Mother* (by "a young lady"). At the other extreme, we have Elkanah Settle's operas *The World in the Moon* and *The Fairy Queen*, William Mountfort's farcical adaptation of *Doctor Faustus*, and Ravenscroft's comedy *The Canterbury Guests*.

Heroic drama in the 1690s displays several of the same linguistic features as heroic drama in the 1670s, and it similarly tends to avoid the sensory texture of masque, opera, and burlesque. The heroic component here selects against plays like *The World in the Moon*, which abounds in rich descriptions of scenery that include concrete nouns ("Palace," "Marble," "Bases," "Girdles," "Foliage," "Coronets," "Gold," "Roof," "Panels," "Mouldings," and "Flowers"), imperatives ("Enter"), descriptions of spatial relations ("The Scene," "near Thirty Foot high" and "Arch"), and allusions to specific people ("Man," "Sir," "Queen," "Emperor," "Mogul"). This kind of language has no place in heroic tragedies like Trotter's *Fatal Friendship*, the heroic tragedy that scores the highest for PC2 in the 1690s sub-corpus.

However, the LAT weights that define PC2 in the 1690s reveal a much higher degree of change over time than we saw for PC1; whereas the features distinguishing comedy from tragedy did not show much shift from the 1670s to the 1690s, the features distinguishing highly sensory spectacles from highly verbal heroic plays do change from early to late Restoration drama (see Table 2).

---

[18] The direction is reversed, with the "heroic" plays scoring high rather than low for PC2, but this reversal is an arbitrary artifact of PCA.

Language Action Type (LAT)	Weight (1670s corpus)	Weight (1690s corpus)	Change in Weight	Absolute Value of Change
Public_Values_Positive	0.392	0.095	-0.297	0.297
Public_Language	0.179	-0.034	-0.213	0.213
Interactive_Question	-0.275	-0.066	0.209	0.209
Academic_Abstract_N_AJ_AV	0.189	0.045	-0.144	0.144
Descript_Motions	-0.371	-0.228	0.143	0.143
Elaboration_Prep	0.179	0.039	-0.140	0.140
Past	0.089	0.225	0.136	0.136
Reason_DenyDisclaim	0.125	0.259	0.134	0.134
Emotion_Negativity	0.012	0.137	0.125	0.125
Elaboration_Determiner	-0.094	-0.212	-0.118	0.118

Table 2. This table shows how the LATs that define PC2 (heroic drama) change from the 1670s to the 1690s. These ten features show the most change across the period. Compared to PC1 (see Table 1), PC2 demonstrates much more change over time, suggesting that the features distinguishing heroic drama from sensory spectacle are less stable than those differentiating comedy from tragedy.[19]

As Table 2 shows, the LAT Public_Values_Positive has become much less important to heroic drama in the 1690s, its score dropping dramatically from +0.392 to +0.095. Recall that Public_Values_Postive, a key feature of heroic plays from the 1670s, includes words like "merit," "justly," and "favour," terms that suggest positive norms against which individuals' behaviors might be assessed. In the 1690s, these norms have faded away, and PC2 now selects against, rather than for, the related LAT Public_Language, which includes terms like "highness," "power" "royal," and "army" that relate to authority and governance. This shift seems to bear out the notion that heroic drama in the 1690s was less concerned with aristocratic ideals and heroes than its earlier Restoration counterpart.

---

[19] I first took the top 25 most important LATs for PC2 in each decade (determined by the absolute value of the LAT weights), a total of 31 LATs for the two decades. I then calculated the rate of change for all 31. The average of the absolute value of change across all 31 of these top LATs is 0.078. Note that these weights are those that PC2 selects against in the 1670s and for in the 1690s; the directionality of the principal components is arbitrary.

While this kind of public and normative language fades out of the picture, however, other LATs gain importance for heroic drama in the 1690s: Emotion_Negativity's weight rises from 0.012 to 0.137, and Reason_DenyDisclaim's weight rises from 0.125 to 0.259. These weights suggest that in the heroic drama of the 1690s corpus, public and positive values have given way to more negative language. As Table 2 shows, the characters in these plays speak more about the past and are more prone to express denial; simultaneously, their speech has become less abstract and less elaborative. It might appear from this analysis that the heroic tragedies of the later Restoration are more about emotion than intellect, more about conflict than deliberation, more about pathos than the display of nobility. These conclusions, however, must be tested against specific passages that exemplify the characteristics of later heroic drama, as represented by PC2.

The heroic component for the 1690s is strongly defined by what it selects against: the top five most important LATs are all negative weights. As a result, it is challenging to track the features that exemplify it. However, their cumulative, qualitative effects are evident from the language of the climactic confrontation scene between the friends alluded to in the title of Trotter's *Fatal Friendship*, Castalio and Gramond:

> CAST. Take back the shameful Ransom; I'll to Prison,
> And resume my Chains; bestow the Purchase
> Of your Treachery on Knaves, I'll none of it.
> GRAM. Stay, stay, my Lord, there's yet a surer way
> To clear your Fame, the Blood of him that stained it:
> Take, take my Life, it is a just Sacrifice,
> You owe it to your self, to Honour,
> And the Name of Friend so long abused.
> CAST. Is this the Man
> I called my Friend! And was I thus deceived!
> I find indeed Lamira well observed,
> There's the least Truth, where most it does appear.
> Ha! that thought has roused one that alarms my Heart;
> She said it was one esteemed my Friend that wronged her;
> Is it possible that he, the Man whom I
> Preferred to all the World, should be ordained
> The Ruin of the only thing besides
> That could be dear to me!
> GRAM. What said you, do you love her?
> CAST. Whom, what her? It is not Lamira thou'st abused.

> GRAM. Nothing but this could aggravate my Crime,
> Or my Remorse; and was it wanting, Heaven!
> Must every Blow which I, or Fate strikes for me,
> Fall heavier still on him! Why, why is this!
> CAST. That I alone may have the right of Vengeance,
> Which now my Injuries are ripe for: Traitor,
> Defend thy Life. (1698, 51–52)

This passage includes several phrases tagged Reason_Contingency, including "possible that," "That could be," "this could," and "may have." Reporting_States is another frequent tag for words and phrases like "observed," "love," and "wanting." As discussed above, the LATs Reason_DenyDisclaim and Emotion_Negativity become more important to PC2 in the 1690s, and both are present here. "Nor is it," "it can't," "none of," "It is not," and similar phrases are tagged DenyDisclaim, while Emotion_Negativity is assigned to words like "Ransom," "Prison," "Knaves," "abused," "alarms," "The Ruin of," "aggravate," "strikes," and "Burden." This passage, exemplary of heroic drama in the 1690s, feels quite different from the passage from *Love's Triumph* that exemplified heroic drama in the 1670s (Figure 4). The language is still elevated, and the drama continues to occur on a plane apart from the sensory world; this is still the drama of the individual's thoughts and feelings and remains opposed to the sensory language found in printed texts of spectacles. Now, however, the form this highly verbal, highly abstract dramatic language takes is more about negativity and emotion than about deliberation and the weighing of values. The conflict here is darker, angrier, and more emotional. This affective shift aligns with the critical consensus that the resurgence of heroic tragedy in the 1690s took a sentimental or pathetic turn. In addition, two plays by Trotter—one of the "female wits" so maligned by contemporaries—have the highest scores for PC2.

Does the exceptionality of Trotter's plays bear out the association between women writers and the increasing pathos of heroic drama in the 1690s? Not quite. Crucially, the plays that fall closest to Trotter are *The Revengeful Queen* and *The Rival Sisters*, both written by men, and *The Unnatural Mother*, purportedly by "a young lady." What these plays have in common, in addition to their deployment of the language associated with PC2, is not female authors; rather, it is the fact, suggested by the titles, that they feature prominent female characters. Perhaps, then, women came under fire for writing in this style because it was associated in the cultural imagination with female protagonists, the tastes of female audience members, and by extension the productions of female authors.

Just as importantly, some plays by "female wits" that critics typically label sentimental heroic drama score quite low or even negative for PC2. Pix's *Ibrahim*, for instance, is often considered a paragon of sentimental heroic tragedy (note 2), yet it falls *below* the x-axis, unlike most others of the genre. The play has the expected plot structure, but its language is highly concrete and sensory. This exception suggests that PC2 is selecting for a particular way of using language that tended to occur in heroic drama but was not a necessary precondition for writing in that genre. The kind of dramatic language that characterizes PC2 may have been associated with femininity in the theatre, then, but it was not necessarily the exclusive purview of women writers—nor did all female authors of heroic drama choose to write in this way.

Conclusions and Implications

Critics have debated how the character types, common plot structures, and tropes of heroic drama may have changed throughout the late seventeenth century, but my study suggests that at least part of the shift registered by contemporaries—and blamed on women writers—had to do with changes in the linguistic and rhetorical fabric of heroic tragedy. In fact, some contemporary critics did register the importance of language to the changes they were witnessing in dramatic genre. *The Female Wits*, while ostensibly a send-up of a genre and a gender, undertakes a highly specific and pointed satire of the overblown verbal style of Manley's *The Royal Mischief*. As Laura J. Rosenthal puts it, "the play-within-the-play characterizes *The Royal Mischief* as excessive, histrionic, and in violation of even heroic tragedy's generous boundaries of verisimilitude (1996, 173).[20] Lucyle Hook has pushed against

---

[20] For instance, during the rehearsal of Marsilia's play, a young lover laments: "Give me your Heart! your Arms! Oh! give me all! see at your Feet the wretched *Amorous* falls! Be not more cruel than our Foes. Behold me on the Torture! *Fastin* cannot Punish me with half the Racks denying Beauty lays on longing Love" (W. M. 1704, 56). Elsewhere, his forbidden beloved Isabella bemoans her fate, crying "Thou, Mother Earth, bear thy wretched daughter, open thy all-receiving womb, and take thy groaning burden in!" (49). The overwrought pathos of these lines comes from words like "wretched," "cruel," "Torture," "groaning," and "burden," as well as negations like "Be not" and "cannot." According to my interpretation, this is the same kind of language that PC2 selects for in the corpus of 1690s plays: emotional, negative, and contingent.

the satire's view of "female" dramatic style, arguing that the only trait distinguishing *The Royal Mischief* from other heroic tragedies and tragicomedies of the day is the presence of a female protagonist, and that the play's mockery is therefore grounded in antifeminism rather than in any real generic or stylistic shift (1967, viii). Conversely, my statistical analysis suggests that there *were* real changes taking place in the language of heroic tragedy at this moment. This finding, however, does not diminish Hook's larger point that the attacks on women playwrights made them gendered scapegoats. In my corpus, plays by both men and women fit the contingent, emotionally negative profile of 1690s heroic tragedy, while at the same time, there are several plays by "female wits" that do not fit this pattern. It appears, therefore, that women were blamed for a larger theatrical trend followed by men and women alike.

While my findings do not support the existence of a specifically female literary tradition or style at this time, they lend credence to the sense—expressed by women playwrights themselves—that they were coming under specifically gender-based attacks that went beyond the ordinary sparring between writers and critics.[21] In her preface to *The Lost Lover*, for instance, Manley famously blamed her gender for the play's failure: "I am satisfied the bare name of being a woman's play damned it beyond its own want of merit" (quoted in Caldwell 2011, 291). However, the fact that *Fatal Friendship* (1698) scores even higher for the 1690s heroic component than *Agnes de Castro* (published 1696, but likely performed in 1695) suggests the possibility that Trotter responded to *The Female Wits*, not by backing down, but by turning up the volume on what made this new brand of plays unique. My study thus offers a potential corrective to the common assumption that female playwrights were discouraged and set back by the satirical attack.[22]

---

[21] Rosenthal makes this distinction in her discussion of the play: "Writers, of course, pilloried each other all the time during these years, but the actual impersonation and public performance of an author's intimate vulnerabilities, produced by the company for which these women wrote, expresses a particular vehemence" (1996, 173).

[22] Hook argues that *The Female Wits* "had its impact on women playwrights in 1696," pointing out that "Manley did not produce another play until *Almyna*" in 1706 and that Pix and Trotter began writing for Lincoln's Inn Fields rather than Drury Lane (1967, x). Rosenthal, citing Hook, agrees that "*The Female Wits* may have had real effects on the careers of these women," although she emphasizes that the satire is evidence of their importance in the world of professional theater and further points out that Manley went on to become well known for her political and prose

Ultimately, the idea that these women were a "triumvirate," or that they had a specific way of writing, was constructed by contemporaries because female playwrights were associated with the larger cultural threat of changing tastes. The notion was later reclaimed by feminist scholars who sought evidence of female solidarity in this moment of increased output by women writers. The assumptions made about women's plays in the 1690s thus continue to inform criticism today. Computational analysis makes it possible to separate ourselves, albeit incompletely, from some of these assumptions, and to see individual texts of interest within the larger context of literary output at the time. This productive estrangement from familiar materials and narratives is not the same thing as "objectivity," of course. As I have tried to make transparent throughout this discussion, even unsupervised analysis of a human-curated data set is subject to the assumptions embedded in its collection: biases of survival and selection for inclusion in the larger archive, biases researchers bring to the process of picking items to examine and assigning metadata, and biases inherent in the tools themselves. Nonetheless, if approached cautiously and with these limitations in mind, the results of quantitative analysis reveal the inherited categories and narratives that we bring, sometimes unawares, to our materials of study.

WORKS CITED

Backscheider, Paula. 1993. *Spectacular Politics: Theatrical Power and Mass Culture in Early Modern England*. Baltimore: Johns Hopkins University Press.

Bernstein, Susan David, and Catherine DeRose. 2012. "Reading Numbers by Numbers: Digital Studies and the Victorian Serial Novel." *Victorian Review* 38 (2): 43–68.

Brown, Laura. 1981. *English Dramatic Form, 1660–1760: An Essay in Generic History*. New Haven, CT: Yale University Press.

Caldwell, Tanya. 2011. "Introduction." In *Popular Plays by Women in the Restoration and Eighteenth Century*, edited by Tanya Caldwell, 9–34. Peterborough, ON: Broadview Press.

---

writings (1996, 173). Robert Day offers the most extreme assessment of the legacy of *The Female Wits*, suggesting that Manley, Pix, and Trotter, "after shining briefly, were ignominiously extinguished, dying poor and obscure" as a result of their work's hostile reception (1980, 62).

Cooke, Edward. 1678. *Love's Triumph, or, The Royal Union a Tragedy*. London. Early English Books Online.

Cuder-Domínguez, Pilar. 2010. "Gender, Race, and Party Politics in the Tragedies of Behn, Pix, and Manley." In *Teaching British Women Playwrights of the Restoration and Eighteenth Century*, edited by Bonnie Nelson and Catherine Burroughs, 263-74. New York: Modern Language Association of America.

Day, Robert Adams. 1980. "Muses in the Mud: The Female Wits Anthropologically Considered." *Women's Studies: An Interdisciplinary Journal* 7 (3): 61-74.

Duffet, Thomas. 1675. *The Mock-Tempest, or, The Enchanted Castle Acted at the Theatre Royal*. London. Early English Books Online.

Early English Books Online (EEBO). eebo.chadwyck.com.

Eighteenth Century Collections Online (ECCO). www.gale.com/ecco.

Finke, Laurie A. 1984. "The Satire of Women Writers in *The Female Wits*." *Studies in English Literary Culture, 1660-1700* 8 (2): 64-71.

Frank, Marcie. 2003. *Gender, Theatre, and the Origins of Criticism from Dryden to Manley*. Cambridge: Cambridge University Press.

Hook, Lucyle. 1967. "Introduction." *The Female Wits*. Los Angeles: William Andrews Clark Memorial Library, University of California.

Hope, Jonathan, and Michael Witmore. 2004. "The Very Large Textual Object: A Prosthetic Reading of Shakespeare." *Early Modern Literary Studies* 9 (3): 6.1-36. http://purl.oclc.org/emls/09-3/hopewhit.htm.

_____ 2010. "The Hundredth Psalm to the Tune of 'Green Sleeves:' Digital Approaches to Shakespeare's Language of Genre." *Shakespeare Quarterly* 61(3): 357-90. doi: 10.1353/shq.2010.0002.

_____ 2014. "Quantification and Language of Later Shakespeare." *Actes des Congrès de la Société Française Shakespeare* 31:123-49. http://shakespeare.revues.org/2830.

Hughes, Derek. 2004. "Boothby, Frances (fl. 1669-1670)." *Oxford Dictionary of National Biography*, edited by Lawrence Goldman. Oxford: Oxford University Press. doi:10.1093/ref:odnb/68250.

Hume, Robert. 1976. *The Development of English Drama in the Late Seventeenth Century*. Oxford: Clarendon Press.

Ingrassia, Catherine. 1998. *Authorship, Commerce, and Gender in Early Eighteenth-Century England: A Culture of Paper Credit*. New York: Cambridge University Press.

Kaufer, David, Suguru Ishizaki, Brian Butler, and Jeff Collins. 2004. *The Power of Words: Unveiling the Speaker and Writer's Hidden Craft*. Mahway, NJ: Lawrence Erlbaum.

Langbaine, Gerard. 1688. *A New Catalogue of English Plays Containing All the Comedies, Tragedies, Tragi-Comedies, Opera's, Masques, Pastorals, Interludes, Farces, &c. Both Ancient and Modern, That Have Ever Yet Been Printed, to This Present Year 1688*. London. Early English Books Online.

Milhous, Judith, and Robert D. Hume. 1974. "Dating Play Premieres from Publication Data, 1660-1700." *Harvard Library Bulletin* 22 (4): 374-405.

Milling, Jane. 2010a. "The Female Wits: Women Writers at Work." In *Theatre and Culture in Early Modern England, 1650-1737: From Leviathan to Licensing Act*, edited by Catie Gill, 119-29. Farnham, UK: Ashgate.

⎯⎯⎯⎯. 2010b. "Working in the Theater: Women Playwrights, 1660-1750." In *Teaching British Women Playwrights of the Restoration and Eighteenth Century*, edited by Bonnie Nelson and Catherine Burroughs, 15-28. New York: Modern Language Association of America.

Rosenthal, Laura J. 1996. *Playwrights and Plagiarists in Early Modern England: Gender, Authorship, Literary Property*. Ithaca, NY: Cornell University Press.

Sculley, D., and Bradley M. Pasanek. 2008. "Meaning and Mining: The Impact of Implicit Assumptions in Data Mining for the Humanities." *Literary and Linguistic Computing* 23 (4): 409-24. doi: 10.1093/llc/fqn019.

Trotter, Catharine. 1698. *Fatal Friendship a Tragedy, As It Is Acted at the New-Theatre in Little-Lincolns-Inn-Fields*. London. Early English Books Online.

W. M., Mr. 1704. *The Female Wits: or, the Triumvirate of Poets at Rehearsal. A Comedy*. London. Eighteenth Century Collections Online.

Witmore, Michael, and Jonathan Hope. 2007. "Shakespeare by the Numbers: On the Linguistic Texture of the Late Plays." In *Early Modern Tragicomedy*, edited by Subha Mukherji and Raphael Lyne, 133–53. Woodbridge, UK: D. S. Brewer.

A Bird's-Eye View of Early Modern Latin: Distant Reading, Network Analysis, and Style Variation

Maciej Eder

Institute of Polish Language, Polish Academy of Sciences
Pedagogical University of Kraków
maciej.eder@ijp-pan.krakow.pl

Classical philology—including neo-Latin studies—is usually claimed to be rather conservative in adopting new trends and methodological paradigms to the study of ancient languages and literatures. Quite surprisingly, however, this claim does not really apply to the "digital turn" in the humanities, which from the very beginning was supported by historical corpora of ancient languages: the *Perseus Project* might serve as the most representative and the most comprehensive example (Crane 2014). Probably more surprising is the fact that Latin and Greek were the first languages to be assessed with statistical methods. Moreover, the same applies to *stylometry* (both the term and a non-traditional method of authorship attribution), which was, in fact, introduced by classical scholars.

As early as the fifteenth century, Leon Battista Alberti published a treatise on frequencies of particular vowels in different Latin genres. Having scrutinized occurrences of specific vowels in various Latin texts, he concluded that *a* and *e* are particularly frequent in poetry, while the other vowels are typical for rhetorical orations (Ycart 2014). In a study that became both a famous example of early modern *empirical philology* and a milestone in authorship attribution using quantitative methods, Lorenzo Valla dealt with the *Donation of Constantine*, a putative edict in which Emperor Constantine transferred authority over Rome and the western part of the Roman Empire to the Popes. Valla's *De falso credita et ementita Constantini donatione declamatio* (Valla 1440, translated 2007) was a detailed analysis of syntax, morphology, and lexis, showing without any doubt that the text could not have been written in the fourth century due to numerous anachronisms and grammatical idiosyncrasies typical for a much later period.

The nineteenth-century advent of non-traditional authorship attribution based on statistical analysis of language is also strongly connected to Greek and Latin philology—more than one could expect. Among the foundations of stylometric theory are not only Shakespearean studies published by Augustus de Morgan and Thomas Mandelhall, but also approaches to the Pauline Epistles as conducted by William Benjamin Smith, also known as Conrad Mascol (cf. Holmes 1998, 112; Rudman 1998b, 354). As early as 1867, Lewis Campbell, Professor of Greek at the University of St. Andrews, devised a series of statistical tests for a new chronology of Plato's *Sophist* and *Politicus*; these included word order, rhythm, avoidance of hiatus, and what Wincenty Lutosławski (1897, 88–89) termed "originality of vocabulary" as measured by the frequency of once-occurring words (*hapax legomena*). Using these methods, Campbell ascertained the two dialogues as late works by Plato. Campbell's discovery went unnoticed for the next thirty years; it seems likely, therefore, that Constantin Ritter's study on Platonic chronology of 1888, which presented similar methods, was conceived independently of Campbell's. Lutosławski, who proposed a novel method of inferring the chronology of Plato's dialogues, can be counted as one of the seminal founders of quantitative authorship attribution, and in fact invented the term *stylometry* (Lutosławski 1897, 145–61; cf. Pawłowski and Pacewicz 2004). Even if Lutosławski's works are known today only to the most sophisticated experts in Plato, the impact of the term *stylometry* has never been questioned.

While a full analysis of recent scholarship in Latin and Greek is beyond the scope of this essay, it is important to note that approaches to authorship cover many different genres and literary periods. They include the Greek New Testament (Greenwood 1995); the collection of biographies of the Roman Emperors known as Historia Augusta (Gurney and Gurney 1998, Rudman 1998a); medieval visions ascribed to Hildegard of Bingen (Kestemont et al. 2013); the philosophical work entitled Consolatio, allegedly written by Cicero and actually by Carlo Sigonio (Forsyth et al. 1999); and John Milton's De doctrina Christiana (Tweedie et al. 1998), to name but a few studies.

A vast majority of these approaches aimed to answer precisely defined questions: "Who wrote a given work?" or "What was the exact chronological order in a set of extant texts?" For decades, literary scholars expected statistics to provide an objective verification of non-literary facts (such as detecting plagiarism); the questions specifically related to literature itself were somehow kept away from computational algorithms. However, even if the ability of solving mere yes/no problems does not really contribute to our knowledge of literature, statistical techniques might become an attractive extension to the repertoire of time-proven heuristic routines when they support *interpretation* of observed phenomena. As Hugh Craig asks: "If you can tell authors apart, have you learned anything about them?" (Craig 1999; Craig 2004). Probably the most seminal study that brought stylometry from the field of language technology back to literature was John Burrows's monograph on the novels of Jane Austen (1987), which introduced "computation into criticism"—a concept later extended to that of "algorithmic criticism" (Ramsay 2011). Further expansion of exact methods in the field of literary studies was substantially stimulated by Franco Moretti's "distant reading" (Moretti 2007; Moretti 2013).

The present study is directly inspired by the above theoretical propositions. It aims to examine the ways in which digital techniques can support scholarship on early modern Latin in its relation to earlier literary epochs. Particularly, the questions of imitation, intertextuality, stylistic differentiation affected by genre and/or chronology, the impact of authorial stylistic idiosyncrasies on the general picture of Latin literature, etc., will be addressed. To this end, a new stylometric method will be applied (Eder 2015b). The method, discussed below in detail, allows fitting numerous mutual text relations in a single network-like visualization. Like most of the stylometric procedures that have been introduced in recent decades, the method in question is closely related to authorship attribution, and so only a brief presentation of theoretical foundations in computer-assisted authorship inference is required here. Next, the idea of scaling up stylometric techniques will be presented. Finally, the new method will be tested on a relatively large collection of 150 Latin works covering the span of a dozen centuries.

From Attribution to Computational Stylistics

Function words (articles, particles, prepositions, pronouns) are far more frequent than content words, and thus their frequencies can be reliably assessed using statistical procedures (Zipf 1949; Burrows 1987). Apart from their measurability, function words are used subconsciously (Chung and Pennebaker 2007), which makes them a perfect feature betraying individual writing habits. Since they are beyond authorial control, they cannot be easily plagiarized; it is difficult to manipulate the usage of the words that no one really notices. In a groundbreaking study on the collection of essays—some of them written collaboratively—known as the *Federalist Papers*, Mosteller and Wallace counted the frequencies of selected function words and, using Bayesian inference, determined with a very high probability which of the disputed essays had been written by Alexander Hamilton, and which by James Madison ([1964] 2007). Since then, most frequent words (MFWs) have become a classic way to distinguish authorial "fingerprint." The list of MFWs is fairly similar across different languages. For English, these are: *the, and, to, of, a, I, in,* and so forth. For Latin, the frequency list begins with the following words: *et, in, non, est, ut, ad, cum, quod,* ... (a list of the first 250 MFWs in Latin is provided in Appendix B).

Needless to say, frequencies of the function words are a counter-intuitive indicator of stylistic differentiation, regardless of their actual strong discriminative power. It is counter-intuitive because style, as defined by stylometry, is entirely different from our usual understanding of this term. Being profoundly different, however, both points of view—computational and traditional—in fact represent two sides of the same coin. The style in its entirety is a multifaceted phenomenon. Traditional stylistics tries to capture it holistically, while computational stylistics is focused on few aspects, or just one aspect: in most applications this is the usage of function words. Interestingly, this is the very aspect of style ignored by traditional scholarship. Thus, bearing in mind the obvious fact that, say, the elaborated Ciceronian style cannot be simply turned into countable numbers, the present study will apply statistical measures to find out whether stylometric similarities confirm a few time-proven stylistic hypotheses concerning the relations between Cicero and his early modern followers.

There are several statistical techniques used in authorship attribution. Some of them focus on one phenomenon carefully retrieved from a corpus—as do different indexes of vocabulary richness—while others rely on a large number of features (e.g., word frequencies) computed at once. The latter are called *multidimensional*, and they are claimed to be much more sensitive to nuanced differences between samples. The reason for their attributive value is the fact that they aggregate the impact of many linguistic features of individually weak discriminating strength (Nerbonne 2007, xvii). Minute differences in word usage are combined into an overall difference between two texts, or a *distance*; this measure of similarity is computed for every single pair of samples in the corpus. Now, if an anonymous text is compared against a set of samples written by known authors, the sample showing the closest (abstract) distance is considered to be stylistically the most similar to the anonymous one, and hence probably written by the same author. Such a concept of *nearest neighbors* is fundamentally important in authorship attribution.

Stylometric methodology, developed to solve authorship problems, can easily be extended and generalized to assess different questions in literary history. Explanatory multidimensional methods, relying on distance measures and supported with visualization techniques, are particularly attractive for this purpose. Namely, the underlying idea of tracing stylistic neighbors between texts can be extended to capture textual relations in the entire corpus: since attribution involves comparing an anonymous (or disputed) text against a selection of works written by known authors, it can be also used to compare the level of similarity between, say, poetry and prose. An example of such an approach to early modern Latin is a statistical comparison of the epistolary style of Francesco Filelfo versus the style of Cicero (Deneire 2016).

The recent advances in stylometric methodology show that two general directions of extending the attribution techniques attract a good share of attention. Following the famous metaphor of digital humanities as a telescope for the mind, one might compare these two methods of stylometric investigation to two different optical instruments (McCarty 2012, 113). One direction can be compared to using a microscope, while the other is like using a telescope. The microscopic approaches try to look inside a single literary work, represented as a sequence of text chunks, in order to examine the stylistic consistency of these chunks. This way of reasoning has been applied to reveal the nature of collaboration of Joseph Conrad and Ford Madox Ford (Rybicki et al. 2014), to find authorial takeovers in the medieval Dutch poem entitled *Roman van Walewein* (van Dalen-Oskam and van Zundert 2007), and to inspect stylistic breaks in *Roman de la Rose* (Eder 2015a).

The telescopic way of extending stylometric methods involves scaling up the amount of input data many times in the belief that a large-scale perspective will betray some new literary phenomena unnoticeable in close reading. Apart from Franco Moretti, the pioneers of this approach include Matthew Jockers, with his "macroanalysis," or massive comparison of 3,346 English novels from the nineteenth century (2013). In his own words, "the literary scholar of the twenty-first century can no longer be content with anecdotal evidence, with random 'things' gathered from a few, even 'representative', texts" (2013, 8).

The applicability of this vision to the study of early modern Latin will be examined below. However, since macroanalysis has somewhat underestimated the problem of validation of the obtained results, a novel method will be applied. One of its core features is the ability to evaluate particular links between analyzed texts. The underlying idea of the method is quite simple. If an anonymous text scrutinized using an attribution procedure turns out to be the most similar to one of the texts from the reference corpus, then, by extension, the same rule applies to any collection of texts of known authorship. In short, every single text from the corpus must have its own nearest neighbor.[1] The neighbors of the neighbors are neighboring with further neighbors, and so on, resulting in a dense network of mutual links. These links can then be used to map the relations between groups of texts.

From Computational Stylistics to Network Analysis

Distant-reading approaches to stylometry face at least three methodological issues. The most important one is the question whether a given method, reasonably effective for a collection of, say, 25 texts, can be scaled up to assess dozens or hundreds of texts without any significant side-effects. The performance of multidimensional methods depends on the number of frequent words taken into analysis (Koppel et al. 2009, Jockers and Witten 2010, Rybicki and Eder 2011); the results can be further affected by the choice of linkage algorithm (e.g., in hierarchical cluster analysis), and generally by a variety of other input parameters of the experiment. In attribution studies, the problem of unstable results must be even more inescapable in large data sets.

Secondly, fitting the relations between, say, a hundred texts becomes problematic for typical explanatory methods used in stylometry, such as multidimensional scaling or principal components analysis. When the number of texts exceeds a few hundred, the urgent need for new ways of visualization becomes obvious. Also, a balance between reliability, informativeness, and visual attractiveness should be kept in mind.

The third issue is even more important, at least from the viewpoint of stylometry beyond attribution: namely, when one tries to represent nearest neighbor relations between samples, authorial groupings are most likely to appear, but other stylistic signals will probably be filtered out. It has been shown that the authorial voice usually overwhelms other stylistic layers, related to genre, topic, translation, and so forth (Craig 2004, Rybicki 2012, Schöch 2013). However, while it is interesting to see Bacon's *Historia regni Henrici Septimi* linked to his *Novum Organum*, and Erasmus's *Moriae encomium* recognized to be stylistically similar to *Institutio principis Christiani*, these results are neither novel nor surprising. Instead, one would like to know what the relation (if any) is between Erasmus and Bacon. In other words,

---

[1] Note that "neighbor" means, in this case, the most similar text regardless of the absolute value of that similarity. Unnoticeable in a vast majority of cases, this issue might (sometimes) lead to biased results. To give an example: when one collects a random selection of works written in, say, different languages, these works will have their nearest neighbors anyway. Harvesting a corpus, one should be aware of this commonly-known drawback of nearest neighbor methods.

computational stylistics should reveal not only the obvious connections between the works written by the same authors, but also subtle stylistic links that cannot be discovered with the naked eye.

The new method applied in this study aims to overcome the aforementioned three issues: unstable results, inability to create large-scale visualizations, and authorial voice overriding other stylometric signals. The technique combines the concept of *network* as a way to map large-scale literary similarities (Jockers 2013), the concept of *consensus* to find statistically significant relations (Lancichinetti and Fortunato 2012), and the assumption that textual similarities usually go beyond mere nearest neighborhood.

The new technique relies on the assumption that particular texts can be represented as nodes of a network, and their explicit relations as links between these nodes. The most significant difference, however, between the approaches applied so far and the present study is the way in which the nodes are linked. This new procedure of linking is twofold. One of the involved algorithms computes the distances between analyzed texts and establishes, for every single node, a strong connection to its nearest neighbor (i.e., the most similar text) and two weaker connections to the first and the second runner-up (i.e., two texts that get ranked immediately after the nearest neighbor). The second algorithm replicates the original test many times with slightly altered input parameters of the experiment, in order to generate multiple "snapshots" of the corpus (e.g., the test is replicated using 100 MFWs, then 200, and then, at increments of 100, all the way to 2,000 MFWs). Finally, all the connections produced in particular "snapshots" are added, resulting in a consensus network. The weights of these final connections tend to differ significantly: the strongest ones mean robust nearest neighbors, while weak links represent secondary and/or accidental similarities. The results are self-validated because they rely on the consensus of many single approaches to the same corpus that sanitizes robust textual similarities and filters out spurious clusterings. The above twofold procedure of linking text samples is implemented in the package "stylo," an open-source stylometric library written in the R programming language (Eder et al. 2013).[2]

The next crucial step in network analysis is to arrange the nodes on a plane in such a way that they reveal as much information about linkage as possible. In

---

[2] The package is available at the Computational Stylistics Group website (https://sites.google.com/site/computationalstylistics/) or directly from the CRAN repository (http://cran.r-project.org). A detailed description of the whole technique of establishing and evaluating the links is provided in a separate study (Eder 2015b).

the present study, one of the force-directed layouts was chosen, namely the algorithm ForceAtlas2 embedded in Gephi, an open-source tool for network manipulation and visualization (Bastian et al. 2009). Force-directed layouts perform gravity-like simulation and pull the most-connected nodes (i.e., the ones that have several links, and/or links that are very strong) to the center of the network, while the least connected nodes are pushed outward. The two-dimensional plot representing the nodes—or particular texts—can be interpreted by a closer look at the network; any clear groupings and visual separations between clusters are meaningful.

Data Set

Large-scale stylometric methods seem to be particularly applicable to the study of Latin literature in its long term, for several reasons. First and foremost, Latin is one of very few languages that has been used for many centuries in a relatively fixed form. This gives us a unique opportunity to compare (stylistically) Latin literature from different epochs, and to trace development of style over time. Next, the situation of written Latin was quite different (i) in antiquity, when it was basically a high register of a spoken mother tongue, (ii) in the Middle Ages, when mastering Latin required some effort to escape one's own vernacular language, (iii) in the Renaissance, when the humanists undertook the task to artificially bring Latin back to its ancient flavor, and (iv) in the subsequent two centuries, when Latin became a *lingua franca* in scholarly and scientific contexts. This gradual shift from a native to a foreign language offers the perfect fodder for stylometric investigation.

Another reason Latin is particularly appropriate for large-scale stylometric research is the relatively good availability of Latin texts on the Internet in high-quality critical editions—especially when compared with those in other historical languages. Furthermore, unlike modern literatures, the entire corpus of Latin is fixed: the total number of texts ever written can be reliably estimated, which makes any extrapolations feasible, even if numerous early modern works are not digitized yet. Last but not least, Latin literature has been thoroughly researched by generations of scholars. Such abundant close-reading evidence is a great help for a stylometrist because it can be used to validate the results of a computational approach.

Ideally, a distant-reading study should involve large amounts of textual data. The present paper offers a first step toward such an experiment in Latin, and thus it is based on a subset of 150 texts rather than on an exhaustive corpus. The texts were harvested from three open-access databases: the *Perseus*

Digital Library (http://www.perseus.tufts.edu), *The Latin Library* (http://www.thelatinlibrary.com), and the *Bibliotheca Augustana* (https://www.hs-augsburg.de/~harsch/augustana.html). They include a selection of ancient writers (such as Seneca, Tacitus, and Cicero), early Christian and medieval authors (such as Arnobius, Ambrosius, and Thomas Aquinas), as well as early modern ones (such as Erasmus, More, Petrarch, and Kepler). The complete list of authors and works is provided in Appendix A. It has to be emphasized, however, that the selection of texts is far from complete or fully representative. This is partly because the availability of Latin literature varies widely across the centuries (also for copyright issues), and partly because historical corpora will never be fully balanced.

Even if 150 texts is not a very big corpus, it shares some of the characteristics of big data collections. In particular, one cannot inspect all the texts manually and/or emend the transcription. Also, one cannot reliably exclude all external quotations, passages copied from other sources, and similar intertextual links; the same applies to any instances of (hidden) mixed authorship. Last but not least, one cannot normalize all spelling variants, such as *qua re* vs. *quare*, *quam ob rem* vs. *quamobrem*, *obedientia* vs. *oboedientia*, and similar instances. In such cases big data means big noise. Some attempts to normalize the spelling variation were undertaken, however—the most important one being the automatic replacement of all the letters *v* with *u* in the entire corpus, in order to neutralize the impact of different scribal and editorial traditions.

Analyzing Latin text with stylometric methods, one should also remember that the medieval authors relatively often cite the Bible and related sources, while the humanists' treatises are full of explicit and/or implicit quotations from classical literature. More importantly, the humanists consciously tried to avoid medieval vocabulary in favor of words that were used by Cicero. For that reason, a stylometric comparison of medieval and early modern Latin brings some additional issues, intensive text reuse being one of the most important (Eder 2013).

Results

Using the consensus network procedure discussed above, a 150-node graph was computed (Figures 1–4). A few groupings and separations can be identified at first glance: a distinct cluster for Seneca in the top-left corner, Caesar's oeuvre clearly distinguishable at the right side of the network, and so on. As expected, the strongest connections are within authorial clusters, which confirms a predominance of authorial signal in the data set. Identifying these and similar connections might lead to interesting observations. However, the real strength of network analysis comes to the fore when a graph is contrasted against metadata (prior literary knowledge) and supported by literature-oriented research questions.

St. Jerome, the early Christian writer and translator of the Bible, claims that he had a dream in which God accused him: "You are a Ciceronian, not a Christian!" (*Epist.* 22, 30), because he had paid too much attention to the beauty of the Ciceronian style. St. Jerome's famous dream reflects Christian antiquity's general attitude to classical literature: pagan texts were claimed to be generally dangerous, and thus they were rarely imitated (Bolgar 1954). Centuries later, Renaissance humanists "discovered" the classical authors again, and they intended to purge the Latin language of medieval traces. It can be hypothesized, then, that these changes in the attitude to the classical literature were followed by style breaks measurable with stylometric methods.

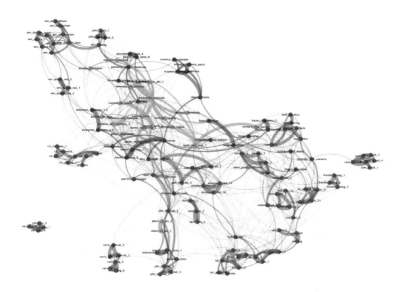

Figure 1. Consensus network of 150 Latin texts, colored according to chronology: ancient texts are colored with blue, medieval texts with green, and early modern ones with red.[3]

---

[3] For a zoomable online version of this figure, see http://ems.itercommunity.org.

On the network represented in Figure 1, the nodes are colored according to chronology (ancient texts marked blue, medieval and early Christian works green, and early modern ones red). The first observation is that the colored samples—or the works written in temporal proximity—tend to form homogeneous regions on the graph. Medieval texts occupy the central area of the network, surrounded by early modern works; these are further surrounded by ancient writers. However, a closer look reveals several exceptions to this apparently simple picture: *Apologia* by Apuleius (ca. 124–ca. 174) has found its place in the center of the medieval cluster, while *Historia Hierosolymitana* by Albert of Aix (floruit ca. 1100) is far away from the Middle Ages, and the same applies to *Passio sancti Edmundi* by Abbo of Fleury (ca. 945–1004). Last but not least, several early modern texts appear scattered around the network. It means that the hypothesis of chronological development of style has not been fully corroborated, even if some temporal patterns can be observed. Alternative stylistic signals should be examined, then.

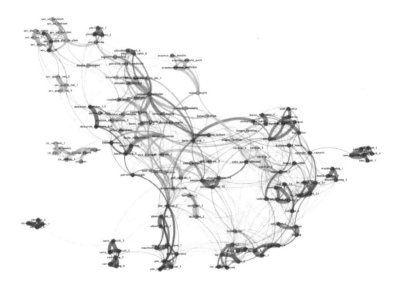

Figure 2. Consensus network of 150 Latin texts, colored according to genre: history writers are colored with red, philosophical works with green, epistles with purple, collections of anecdotes with yellow, other non-fiction works with blue; some works are left unmarked due to their ambiguous genre classification.[4]

---

[4] For a zoomable online version of this figure, see http://ems.itercommunity.org.

In Figure 2, the same network is colored according to genre: history (red), philosophy (green), epistles (purple), "technical" writings such as Vitruvius's *De architectura* (blue), apophthegms and anecdotes (yellow). Some texts were left unmarked due to ambiguous classification or because they represented other genres, such as Apuleius's *Metamorphoses* (literature) or Cicero's *Philippic 1* (rhetorical oration). The pattern seems to be much clearer than in the previous plot. In particular, notice a large cluster of history writers that appears on the right side. This highly homogeneous cluster containing historical works, regardless of which period they represent, suggests that historiography is stylometrically distinct and stable across centuries. Interestingly, the central part of this cluster is occupied by authors of facetiae, apophthegms, and anecdotes: Valerius Maximus, Poggio, Bebel, Tünger. They all seem to be very similar and—what is surprising—they seem to have been successful in imitating the style of the historians.

The other part of the plot is less distinct, even though one can notice that the "technical" writers tend to be linked together, and the same applies to the philosophers. At the same time, however, epistles do not form any cluster: Sallust's letters are grouped with his two historical works, the letters by Pliny the Younger are close to his *Panegyricus*, Seneca's *Epistulae ad Lucilium* are linked to his dialogues, and Ambrose's epistolography is linked to a good share of other authors. The behavior of the epistles clearly indicates the predominance of the authorial signal over genre or topic. Paradoxically, this is also confirmed by the ambiguous results for Ambrose's oeuvre: being a collection of letters both written by him and addressed to him, the text turns out to have a weak authorial signal, which is reflected by multiple yet unclear network connections.

The case of Ambrose prompts another research question: namely, how to interpret centrality and eccentricity of the network. Since the layout algorithm puts the heavily connected nodes toward the center and pulls outward nodes with only a few links, the actual question is how to interpret the density of connections. In Figure 3, colors are applied to the works by the authors traditionally claimed to have mastered Latin style: Cicero, Seneca, Tacitus, Caesar, Petrarch, Pico della Mirandola, Erasmus, and More (early modern authors with red, ancient ones with blue). A clear pattern emerges: these very authors occupy outer regions of the network. Varro is similarly eccentric, and Vitruvius even more so—that is, two authors who have not been particularly appreciated for their style. A provisional interpretation of this phenomenon might be that a small number of connections indicate extraordinary authors—either great stylists, or bizarre writers. Or it could be interpreted the other way around: several connections might mean a less distinct author, with a weak authorial voice. This provisional explanation, however, needs to be confirmed by stronger evidence.

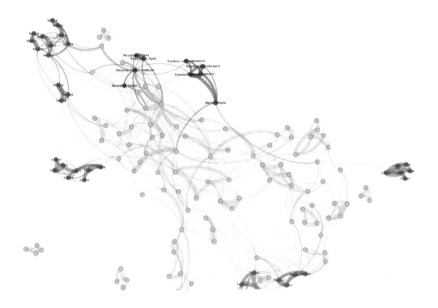

Figure 3. Consensus network of 150 Latin texts, with highlighted works of authors traditionally claimed to be great stylists; early modern authors colored with red, ancient ones with blue.[5]

The density of connections between different authors deserves yet another research hypothesis. Namely, it might reflect some traces of imitation, which is a fundamentally important aspect of early modern stylistic theories. It is a well-known fact that Latin style had a few flavors, the most discussed being the distinction between Ciceronian style and the "Silver Latin" style as represented by Seneca and Tacitus. In the Renaissance, a discussion of which of the two flavors should be imitated turned into the Ciceronian Quarrel, the single most important linguistic debate of that time (DellaNeva 2007). The debate had an immense effect on the development of Latin style. From the perspective of computational stylistics, a particularly interesting task might be an examination of the problem of "Attic" prose and the anti-Ciceronian movement of the late Renaissance and early Baroque, based on the analysis of the style of Justus Lipsius, Erasmus, and other writers such as Puteanus,

---

[5] For a zoomable online version of this figure, see http://ems.itercommunity.org.

Moretus, and Fredro (Croll 1924, Croll 1996; Salmon 1980; Tunberg 1999). The questions to be answered stylometrically might be as follows: are traces of the styles of Seneca and Tacitus indeed noticeable in this modern "Attic" way of writing? Did the "Attic" authors really escape stylistic Ciceronianism? And finally: how influential was Cicero, really?

Figure 4. Consensus network of 150 Latin texts, with highlighted works by Cicero and all the works with which Cicero is directly connected.[6]

While it is still too early to assess the problem of imitation in its entirety using computational techniques, the collected corpus of 150 prose texts allows us to examine the relations between Cicero and his followers. Since consensus networks rely on the concept of nearest neighbor as derived from authorship attribution, any two nodes that are linked are in consequence *very* similar stylistically. Now, if one identifies all the nodes connected directly to Cicero, it should give a relatively clear picture of the circle of his followers. In Figure 4, the nodes linked directly to any of the works by Cicero were colored red. The highlighted texts include: Gellius's *Noctes Atticae*, Apuleius's *Apologia*, Valerius Maximus's *Facta et dicta* Book 4, Arnobius's *Adversus nationes*, and Alcuin's *Disputatio*. This selection of ancient and medieval writers related to Cicero, being noteworthy from a general perspective, is slightly out

---

[6] For a zoomable online version of this figure, see http://ems.itercommunity.org.

of the scope of this paper, however. From the point of view of Renaissance Ciceronianism, early modern authors and texts identified by the procedure are much more interesting: Descartes's *Meditationes*, Pico della Mirandola's *Oratio de dignitate hominis*, Melanchthon's *Vita Lutheri*, and Poggio's *Facetiae*. All these works are surprisingly non-surprising, as is the absence of Erasmus or More among the stylometric followers of Cicero. Pico della Mirandola was a well-known Ciceronianist, and the same could be said about Poggio—even if his *Facetiae*, a collection of obscene and scatological jokes, is hardly comparable to the Ciceronian oeuvre at the first glance. Melanchthon, on the other hand, received a thorough humanistic education, which also included stylistic excellence based on imitating the Ciceronian style. The absence of More, and particularly Erasmus, seems to suggest that their anti-Ciceronianism as a stylistic choice was quite successful. In his well-known dialogue entitled *Ciceronianus*, Erasmus claims that imitating Cicero should be but the first step to achieve actual stylistic excellence; too close an imitation is a form of literary idolatry.

Being somewhat preliminary, the above results definitely deserve to be extended and systematically contrasted with traditional scholarship on Latin writers. Also, future studies should take into consideration a substantially bigger corpus of Latin literature.

Conclusions

Stylometric methods based on the most frequent words turned out to be fairly successful, as evidenced in this reconnaissance study. When combined with attractive ways of visualizing the results—network analysis being one of the possible choices here—they allow the assessment of large amounts of textual data. More importantly, they allow us to ask new research questions. Even if this study was focused on Latin literature, the method presented in this paper can be easily generalized to scrutinize other literary traditions. Also, very similar questions about chronology, genre, topic, sentiment, and other stylistic layers can be formulated in the context of, say, the Shakespeare canon or the medieval French *Chansons de Geste*. The biggest added value of the method applied above, however, is the fact that it can be used to map a given literary tradition in its entirety—no matter how many texts are analyzed, they all can be handled by the method and represented in a form of a network (a map), in which some potentially interesting groups of texts are very likely to appear. Presumably, such a bird's-eye view provides an insight into literary phenomena that is unavailable using close-reading approaches.

Computational stylistics proved a promising addition to the usual repertoire of explanatory methods. At the same time, however, an obvious gap between distant and close reading became embarrassingly evident: from a bird's-eye view, one has no access to the texts themselves. It seems, then, that an optimal approach to literature might be a combination of these two perspectives.

In particular, traditional scholarship might take advantage of the by-products of stylometric experiments to pursue more detailed investigation of stylistic differentiation between authors. For instance, a list of word frequencies, obtained in split-seconds using computational techniques, provides interesting material for further close reading and interpretation. To give an example: if one examines which of the words are shared across the entire corpus of 150 Latin texts, one gets a surprisingly low number of merely 24 shared words (cf. Appendix C). Certainly, the short list contains function words, but a careful comparison reveals that some very common function words are missing (Appendix C vs. Appendix B). In this point close reading comes to center stage, revealing that the conjunction *si*, which is the fourteenth word in the entire corpus, is absent in Matthias of Miechow's *De duabus Sarmatiis tractatus*; the seventeenth word, *aut*, is not used by Apuleius in *Metamorphoses*; Thomas Aquinas (*De ente et essentia*) never uses *ac* and *atque*; and the word *nec* is avoided by Sallust and Caesar. These and similar idiosyncrasies in word usage are usually overlooked by traditional approaches; stylometry helps to bridge the gap.

The relation goes both ways: hypotheses formulated by literary criticism, based on text-centric evidence, are a sine qua non to making stylometric experiments valid. State-of-the-art statistical procedures are very accurate, but they will fail when applied to answer spurious questions. Based on solid scholarship, however, stylometric techniques might be leveraged to answer distant-reading questions, with a belief that at some point, one will be able to produce a map of Latin literature in its entirety. The present study is intended to be a first step toward this vision.

WORKS CITED

Bastian, Mathieu, Sebastien Heymann, and Mathieu Jacomy. 2009. "Gephi: An Open Source Software for Exploring and Manipulating Networks." *Proceedings of the Third International AAAI Conference on Weblogs and Social Media.* http://www.aaai.org/ocs/index.php/ICWSM/09/paper/view/154/1009.

Bolgar, Robert R. 1954. *The Classical Heritage and its Beneficiaries.* Cambridge: Cambridge University Press.

Burrows, John. 1987. *Computation into Criticism: A Study of Jane Austen's Novels and an Experiment in Method.* Oxford: Clarendon Press.

Campbell, Lewis, ed. (1867) 1973. *The Sophistes and Politicus of Plato. With a revised text and English notes.* London: Clarendon Press. Reprint, New York: Arno Press.

Chung, Cindy, and James W. Pennebaker. 2007. "The Psychological Functions of Function Words." In *Social Communication*, edited by Klaus Fiedler, 343-59. New York: Psychology Press.

Craig, Hugh. 1999. "Authorial Attribution and Computational Stylistics: If You Can Tell Authors Apart, Have You Learned Anything About Them?" *Literary and Linguistic Computing* 14 (1): 103-13.

_____ 2004. "Stylistic Analysis and Authorship Studies." In *A Companion to Digital Humanities*, edited by Susan Schreibman, Ray Siemens, and John Unsworth, 273-88. Oxford: Blackwell.

Crane, Gregory R., ed. 2014. *Perseus Digital Library.* Tufts University. http://www.perseus.tufts.edu.

Croll, Morris W. 1924. "Muret and the History of 'Attic' Prose." *PMLA* 39 (2): 254-309.

_____ 1996. "'Attic' Prose in the Seventeenth Century." In *Style, Rhetoric and Rhythm*, edited by J. M. Patrick and R. O. Evans, 51-101. Princeton, NJ: Princeton University Press.

Dalen-Oskam, Karina van, and Joris van Zundert. 2007. "Delta for Middle Dutch: Author and Copyist Distinction in Walewein." *Literary and Linguistic Computing* 22 (3): 345-62.

DellaNeva, JoAnn, ed. 2007. *Ciceronian Controversies.* Translated by Brian Duvick. Cambridge, MA: Harvard University Press.

Deneire, Tom. 2016. "Filelfo, Cicero and Epistolary Style: A Computational Study." In *Filelfo: Man of Letters*, edited by Jeroen De Keyser. Leiden and Boston: Brill (in press).

Eder, Maciej. 2013. "Mind Your Corpus: Systematic Errors in Authorship Attribution." *Literary and Linguistic Computing* 28 (4): 603–14.

―――― 2015a. "Rolling Stylometry." *Digital Scholarship in the Humanities* 30. http://dsh.oxfordjournals.org/content/early/2015/04/06/llc.fqv010.

―――― 2015b. "Visualization in Stylometry: Cluster Analysis Using Networks." *Digital Scholarship in the Humanities* 30. http://dsh.oxfordjournals.org/content/early/2015/12/02/llc.fqv061.

Eder, Maciej, Mike Kestemont, and Jan Rybicki. 2013. "Stylometry with R: A Suite of Tools." In *Digital Humanities 2013 Conference Abstracts*, 487–89. Lincoln: Center for Digital Research in the Humanities, University of Nebraska-Lincoln. http://dh2013.unl.edu/abstracts/ab-136.html.

Forsyth, Richard, David Holmes, and Emily Tse. 1999. "Cicero, Sigonio, and Burrows: Investigating the Authenticity of the Consolatio." *Literary and Linguistic Computing* 14 (3): 375–400.

Greenwood, H. H. 1995. "Common Word Frequencies and Authorship in Luke's Gospel and Acts." *Literary and Linguistic Computing* 10 (3): 183–87.

Gurney, Penelope, and Lyman Gurney. 1998. "Authorship Attribution of the 'Scriptores Historiae Augustae.'" *Literary and Linguistic Computing* 13 (3): 119–31.

Holmes, David. 1998. "The Evolution of Stylometry in Humanities Scholarship." *Literary and Linguistic Computing* 13 (3): 111–17.

Jockers, Matthew. 2013. *Macroanalysis: Digital Methods and Literary History*. Champaign: University of Illinois Press.

Jockers, Matthew, and Daniela Witten. 2010. "A Comparative Study of Machine Learning Methods for Authorship Attribution." *Literary and Linguistic Computing* 25 (2): 215–23.

Kestemont, Mike, Sara Moens, and Jeroen Deploige. 2013. "Collaborative Authorship in the Twelfth Century: A Stylometric Study of Hildegard of Bingen and Guibert of Gembloux." *Digital Scholarship in the Humanities* 30 (2): 199–224. https://biblio.ugent.be/publication/4099638/file/5967103.pdf.

Koppel, Moshe, Jonathan Schler, and Shlomo Argamon. 2009. "Computational Methods in Authorship Attribution." *Journal of the American Society for Information Science and Technology* 60: 9–26.

Lancichinetti, Andrea, and Santo Fortunato. 2012. "Consensus Clustering in Complex Networks." *Scientific Reports* 2: 336: 1–7.

Lutosławski, Wincenty. 1897. *The Origin and Growth of Plato's Logic: With an Account of Plato's Style and of the Chronology of his Writings.* London: Longmans.

McCarty, Willard. 2012. "A Telescope for the Mind?" In *Debates in the Digital Humanities*, edited by Matthew Gold, 113–23. Minneapolis: University of Minnesota Press. http://dhdebates.gc.cuny.edu/debates/text/37.

Moretti, Franco. 2007. *Graphs, Maps, Trees: Abstract Models for Literary History.* London: Verso.

_____ 2013. *Distant Reading.* London: Verso.

Mosteller, Frederick, and David Wallace. (1964) 2007. *Inference and Disputed Authorship: The Federalist.* Reading, MA: Addison-Wesley. Reprint, with a new introduction by John Nerbonne. Stanford, CA: CSLI Publications.

Nerbonne, John. 2007. "The Exact Analysis of Text." Foreword. In Frederick Mosteller and David Wallace, *Inference and Disputed Authorship: The Federalist,* reprint edition, xi–xx. Stanford, CA: CSLI Publications.

Pawłowski, Adam, and Artur Pacewicz. 2004. "Wincenty Lutosławski (1863–1954). Philosophe, helléniste ou fondateur sous-estimé de la stylométrie?" *Historiographia Linguistica* 21: 423–47.

Ramsay, Stephen. 2011. *Reading Machines: Toward an Algorithmic Criticism.* Urbana, Chicago, and Springfield: University of Illinois Press.

Ritter, Constantin. 1888. *Untersuchunger über Plato. Die echtheit und chronologie der platonischen schriften, nebst anhang: Gedankengang und grundanschauungen von Platos Theätet.* Stuttgart: W. Kohlhammer.

Rudman, Joseph. 1998a. "Non-traditional Authorship Attribution Studies in the Historia Augusta: Some Caveats." *Literary and Linguistic Computing* 13 (3): 151–57.

―――――― 1998b. "The State of Authorship Attribution Studies: Some Problems and Solutions." *Computers and the Humanities* 31 (4): 351–65.

Rybicki, Jan. 2012. "The Great Mystery of the (Almost) Invisible Translator." In *Quantitative Methods in Corpus-Based Translation Studies*, edited by Michael Oakes and Meng Ji, 231–48. Amsterdam: John Benjamins.

Rybicki, Jan, and Maciej Eder. 2011. "Deeper Delta across Genres and Languages: Do We Really Need the Most Frequent Words?" *Literary and Linguistic Computing* 26 (3): 315–21.

Rybicki, Jan, David L. Hoover, and Mike Kestemont. 2014. "Collaborative Authorship: Conrad, Ford and Rolling Delta." *Literary and Linguistic Computing* 29 (3): 422–31.

Salmon, John Hearsey McMillan. 1980. "Cicero and Tacitus in Sixteenth-Century France." *The American Historical Review* 85: 307–31.

Schöch, Christof. 2013. "Fine-Tuning our Stylometric Tools: Investigating Authorship, Genre, and Form in French Classical Theater." In *Digital Humanities 2013: Conference Abstracts*, 383–86. Lincoln: University of Nebraska-Lincoln.

Tunberg, Terence O. 1999. "Observations on the Style and Language of Lipsius's Prose: A Look at Some Selected Texts." In *Iustus Lipsius: Europae Lumen et Columen*, edited by Gilbert Tournoy, Jeanine de Landtsheer, and Jan Papy, 169–78. Leuven: Leuven University Press.

Tweedie, Fiona, David Holmes, and Thomas Corns. 1998. "The Provenance of 'De Doctrina Christiana' Attributed to John Milton: A Statistical Investigation." *Literary and Linguistic Computing* 13 (2): 77–87.

Valla, Lorenzo. (1440) 2007. *On the Donation of Constantine*. Translated by G. W. Bowersock. Cambridge, MA: Harvard University Press.

Ycart, Bernard. 2014. "Alberti's Letter Counts." *Literary and Linguistic Computing* 29 (2): 255–65.

Zipf, George K. 1949. *Human Behavior and the Principle of Least Effort*. Cambridge, MA: Addison-Wesley.

APPENDIX A: Analyzed Works

Abbo Floriacensis, *Passio sancti Edmundi regis et martyris*
Abelardus, Petrus, *Ad amicum suum consolatoria*
_____ *Dialogus inter Philosophum, Iudaeum et Christianum*
Aelredus Rievallensis, *De amicitia*
Alanus de Insulis, *Liber de planctu naturae*
Albert of Aix, *Historia Hierosolymitanae expeditionis, Liber I*
_____ *Historia Hierosolymitanae expeditionis, Liber II*
Alcuin, *Disputatio de rhetorica et de virtutibus sapientissimi regis Karli et Albini magistri*
Ambrosius, *Epistulae variae*
Ammianus Marcellinus, *Historiae, Liber XIV*
_____ *Historiae, Liber XV*
_____ *Historiae, Liber XVI*
Anselmus Cantuariensis, *Proslogion*
Apuleius, *Apologia sive pro se de magia liber*
_____ *De mundo*
_____ *Metamorphoseon, Liber I*
_____ *Metamorphoseon, Liber II*
_____ *Metamorphoseon, Liber III*
Arnobius, *Adversus nationes, Liber I*
_____ *Adversus nationes, Liber II*
Augustinus, *De civitate Dei, Liber I*
_____ *Confessionum, Liber I*
Bacon, Francis, *Historia regni Henrici Septimi regis Angliae, Capitulum I*
_____ *Historia regni Henrici Septimi regis Angliae, Capitulum VIII*
_____ *Novum Organum, Liber I*
_____ *Novum Organum, Liber II*
_____ *Sermones fideles sive interiora rerum*
Bebel, Heinrich, *Liber facetiarum*
Beda Venerabilis, *Historiam ecclesiasticam gentis Anglorum, Liber I*
Benedictus Nursinus, *Regula*
Berengarius Scholasticus, *Apologeticus contra sanctum Bernardum*
Bigges, Walter, *Expeditio Francisci Draki in Indias Occidentales*
Buchanan, George, *De Maria Scotorum regina*
Caesar, *Commentariorum de bello civili, Liber I*
_____ *Commentariorum de bello civili, Liber II*
_____ *Commentariorum de bello civili, Liber III*
_____ *Commentariorum de bello Gallico, Liber I*
_____ *Commentariorum de bello Gallico, Liber II*

_____ *Commentariorum de bello Gallico, Liber IV*
Campanella, Tommaso, *Civitas Solis*
Cassiodorus, *De anima*
_____ *Variarum, Liber I*
_____ *Variarum, Liber II*
Celtis, Conradus, *Oratio in gymnasio in Ingelstadio*
Cicero, *Brutus*
_____ *Pro Caecina oratio*
_____ *Pro Cluentio oratio*
_____ *Cato Maior de senectute*
_____ *De natura deorum, Liber I*
_____ *In M. Antonium oratio Philippica prima*
_____ *Pro Sestio oratio*
_____ *Tusculanarum disputationum, Liber I*
Curtius Rufus, *Historiae Alexandri Magni, Liber III*
_____ *Historiae Alexandri Magni, Liber IV*
_____ *Historiae Alexandri Magni, Liber V*
Dante Alighieri, *Monarchia, Liber I*
_____ *Monarchia, Liber II*
Descartes, *Meditationes I-III*
_____ *Meditationes IV-VI*
Einhard, *Vita Karoli Magni*
Erasmus, *Institutio principis Christiani*
_____ *Declamatio de laude matrimonii*
_____ *Moriae encomium*
_____ *Querela pacis*
Florus, *Epitomae, Liber I*
_____ *Epitomae, Liber II*
Frontinus, *Strategemata, Liber I*
_____ *Strategemata, Liber II*
_____ *Strategemata, Liber III*
Galileo Galilei, *Sidereus nuncius*
Gellius, *Noctes Atticae, Liber I*
_____ *Noctes Atticae, Liber II*
_____ *Noctes Atticae, Liber III*
Holberg, Ludvig, *Nicolai Klimii iter subterraneum*
Iustinus, *Historiarum Philippicarum, Liber II*
_____ *Historiarum Philippicarum, Liber XII*
Kepler, Ioannes, *Strena seu de nive sexangula*
Livius, *Ab Urbe condita, Liber I*

########## *Ab Urbe condita*, Liber XXI
########## *Ab Urbe condita*, Liber XXXI
########## *Ab Urbe condita*, Liber XLI
Matthias of Miechow, *De duabus Sarmatiis, Tractatus I*
########## *De duabus Sarmatiis, Tractatus II*
Melanchthon, Philip, *Historia de vita et actis Lutheri*
Mirandola, Giovanni Pico della, *Oratio de hominis dignitate*
More, Thomas, *De nova insula Utopia*
Patricius, Franciscus, *Panaugia*
Petrarch, Franciscus, *Contra medicum quendam*
Piccolomini, Eneas Silvius, *Epistulae* (Part I)
########## *Epistulae* (Part II)
Plinius Minor, *Epistularum, Liber I*
########## *Epistularum, Liber II*
########## *Epistularum, Liber III*
########## *Panegyricus*
Plinius Maior, *Naturalis historiae, Liber II*
########## *Naturalis historiae, Liber III*
########## *Naturalis historiae, Liber IV*
########## *Naturalis historiae, Liber V*
Poggio Bracciolini, Gian Francesco, *Facetiae*
Sallustius, *Bellum Catilinae*
########## *Bellum Iugurthinum*
########## *Epistulae*
Seneca, *Ad Helviam matrem de consolatione*
########## *Ad Marciam de consolatione*
########## *Ad Polybium de consolatione*
########## *Ad Novatum de ira, Liber I*
########## *Ad Novatum de ira, Liber II*
########## *Ad Neronem de clementia*
########## *Ad Serenum de constantia*
########## *Epistularum ad Lucilium, Liber I*
########## *Epistularum ad Lucilium, Liber II*
########## *Epistularum ad Lucilium, Liber XX*
########## *Quaestiones naturales, Liber I*
########## *Quaestiones naturales, Liber II*
########## *Quaestiones naturales, Liber III*
Suetonius, *Divus Augustus*
########## *Divus Claudius*
########## *Divus Iulius*

_____ *Caligula*
_____ *Nero*
_____ *Tiberius*
_____ *Agricola*
Tacitus, *Annalium, Liber I*
_____ *Annalium, Liber II*
_____ *Annalium, Liber III*
_____ *Annalium, Liber IV*
_____ *Germania*
_____ *Historiarum, Liber I*
_____ *Historiarum, Liber II*
_____ *Historiarum, Liber III*
Thomas Aquinas, *De ente et essentia*
Tünger, Augustin, *Facetiae*
Valerius Maximus, *Factorum et dictorum memorabilium, Liber I*
_____ *Factorum et dictorum memorabilium, Liber II*
_____ *Factorum et dictorum memorabilium, Liber III*
_____ *Factorum et dictorum memorabilium, Liber IV*
Varro, *De agri cultura, Liber I*
_____ *De agri cultura, Liber II*
_____ *De agri cultura, Liber III*
_____ *De lingua Latina, Liber V*
_____ *De lingua Latina, Liber VI*
_____ *De lingua Latina, Liber VII*
Velleius Paterculus, *Historiae Romanae, Liber I*
_____ *Historiae Romanae, Liber II*
Vico, Gianbattista, *Oratio VI*
Vitruvius, *De architectura, Liber I*
_____ *De architectura, Liber II*
_____ *De architectura, Liber III*
_____ *De architectura, Liber IV*
Xylander, Guilielmus, *Vita caesaris*

APPENDIX B: The Most Frequent Words

1	et	4	est	7	cum		
2	in	5	ut	8	quod		
3	non	6	ad	9	qui		
10	sed	41	haec	72	ille		
11	quam	42	quibus	73	mihi		
12	a	43	me	74	sine		
13	quae	44	eo	75	magis		
14	si	45	quidem	76	illa		
15	de	46	nihil	77	res		
16	ex	47	uero	78	nunc		
17	aut	48	nam	79	omnes		
18	esse	49	iam	80	post		
19	ac	50	erat	81	quos		
20	se	51	nisi	82	at		
21	ab	52	pro	83	tantum		
22	sunt	53	ea	84	ante		
23	enim	54	sibi	85	itaque		
24	per	55	quoque	86	omnibus		
25	nec	56	quia	87	quis		
26	atque	57	te	88	nos		
27	etiam	58	tam	89	e		
28	hoc	59	quem	90	eorum		
29	quo	60	esset	91	omnium		
30	quid	61	qua	92	igitur		
31	ne	62	ubi	93	rerum		
32	sit	63	omnia	94	ego		
33	uel	64	tum	95	cuius		
34	autem	65	modo	96	tibi		
35	tamen	66	his	97	hic		

36	neque	67	apud	98	sua
37	ita	68	ipse	99	an
38	eius	69	sic	100	causa
39	id	70	eum	101	potest
40	inter	71	fuit	102	quasi
103	inquit	134	ipsum	165	potius
104	minus	135	satis	166	suum
105	dum	136	illis	167	aliis
106	illi	137	ibi	168	quantum
107	suis	138	hanc	169	tunc
108	illud	139	siue	170	nulla
109	ergo	140	tu	171	eadem
110	sint	141	ob	172	animi
111	deinde	142	inde	173	uos
112	eos	143	erant	174	uti
113	nobis	144	quas	175	sicut
114	uerum	145	homines	176	habet
115	ipsa	146	quorum	177	multa
116	rebus	147	simul	178	ipsi
117	contra	148	loco	179	caesar
118	suo	149	hominum	180	illum
119	natura	150	is	181	hunc
120	cui	151	tempore	182	alii
121	primum	152	posse	183	iis
122	maxime	153	propter	184	fuisse
123	sub	154	fieri	185	magna
124	semper	155	parte	186	possit
125	rem	156	die	187	omnis
126	alia	157	aliquid	188	huius
127	re	158	hac	189	una
128	eam	159	aliud	190	solum

129	rei	160	quidam	191	quaedam		
130	unde	161	unum	192	suae		
131	idem	162	suam	193	quoniam		
132	ei	163	bellum	194	usque		
133	sui	164	adeo	195	eodem		
196	omni	215	uitae	234	partem		
197	tempus	216	primo	235	pars		
198	animo	217	dei	236	quin		
199	genus	218	nomen	237	deus		
200	tot	219	super	238	corporis		
201	bene	220	locum	239	aduersus		
202	quippe	221	suos	240	castra		
203	hinc	222	illo	241	populi		
204	mox	223	item	242	facile		
205	fuerit	224	adhuc	243	ideo		
206	saepe	225	erit	244	fere		
207	nomine	226	certe	245	cur		
208	nemo	227	haud	246	terra		
209	ista	228	urbem	247	plus		
210	ipso	229	bello	248	opus		
211	habere	230	naturae	249	statim		
212	rex	231	uelut	250	denique		
213	dies	232	circa				
214	licet	233	exercitus				

APPENDIX C: Words Shared by All the Sample Works

1	et	10	sed	18	per
2	in	11	quam	19	etiam
3	non	12	a	20	quo
5	ut	13	de	21	quibus
6	ad	14	ex	22	quem
7	cum	15	se	23	qua
8	quod	16	ab	24	omnia
9	qui	17	sunt		

Displaying Textual and Translational Variants in a Hypertextual and Multilingual Edition of Shakespeare's Multi-text Plays

Jesús Tronch
University of Valencia
tronch@uv.es

In this essay I explore possibilities for an edition aimed at readers who want to compare differences among texts of the same work, including translations into several languages, beyond the format of the collation notes in single-text critical editions and of the plain parallel-text presentation: specifically, I propose a hypertextual, multilingual edition of Shakespeare's multi-text plays, such as *Hamlet*, capable of displaying different textual versions and of highlighting their variants, both *intra*lingual (that is, among texts in the same language) and *inter*lingual (among translations).[1] This edition is motivated by my collaboration in an online multilingual collection of early modern European theatre, henceforth EMOTHE (Oleza), that offers open-access editions of selected plays from sixteenth- and seventeenth-century English, Spanish, French, and Italian theatre, together with translations into the other languages of the collection, to be read either individually or alongside each other on the screen.

My projected edition, as a digital tool to compare versions, seeks to go beyond the mere presentation of texts in parallel and to implement display methods that draw the reader's attention directly to variants. Displays to appreciate textual differences more economically and efficiently than in parallel texts have been devised in print editions,[2] but this essay is concerned with making

---

[1] Research for this essay benefited from the Artelope project funded by the Spanish government, reference numbers FFI 2012-34347 and CDS2009-00033. I am grateful to Carlos Muñoz, computer engineer on this project, and Cristina Mestre for their patience in answering so many questions.

[2] For print editions, see Gabler, Steppe, and Melchior 1984, Kliman 1996, Tronch-Pérez 2002, and Fotheringham 2006; for electronic editions, see those in the *Internet Shakespeare Editions* series (Best).

them digital while integrating comparison of source texts and translations. To this purpose, I have explored existing digital tools and platforms that display multiple versions of texts, with a view to gleaning what can be useful for the projected hypertextual and multilingual edition. I examine their codification and automation processes to divide texts into segments, align these segments, link variants, and render various visualizations of the data. Then I seek to answer specific questions of how to segment dramatic texts below the level of characters' speech to facilitate alignment, both in translations and in their originals, and how to extend the markup of the critical apparatus proposed by the Guidelines of the Text Encoding Initiative (TEI).

As a working example for this essay, I specifically conceive an edition of *Hamlet* offering readers the following texts for comparison:

- ★ the First Quarto, Second Quarto, and First Folio versions of *Hamlet*;
- ★ Spencer's New Penguin edition; Mowat and Werstine's Folger Digital Texts edition; Bevington's four versions for the *Internet Shakespeare Editions* based on the Second Quarto, on the First Folio, his Editor's Version (conflating these two early texts), and on the First Quarto;
- ★ the historical translations into Spanish by Leandro Fernández de Moratín (1798) and by Guillermo Macpherson (in its four versions, 1873, 1879, 1882, and 1885); into German by Christoph Martin Wieland (1766) and by August Wilhelm Schlegel and Ludwig Tieck (1831); into Italian by Carlo Rusconi (1852) and by Giulio Carcano (1847); and into French by François-Victor Hugo (1865) and by François Guizot (1864).

Taking advantage of the capabilities of the electronic media, users of this edition would be able to read any of these texts individually or in comparison, and to dynamically shift from one text to others, and between one viewing mode to another, not only in parallel texts but also in more variant-enhancing displays, which I explain here.

Looking for Variants Beyond the Parallel-Text Presentation

Useful as parallel-text presentations are, they are not helpful enough for readers looking for variants in Shakespeare's multi-text plays. Since textual versions often differ in inconspicuous single words, speech prefixes, verb inflections, punctuation marks, or transpositions, readers can easily overlook them. Furthermore, most of the textual variation in plays such as *Hamlet, Lear, Othello, Troilus and Cressida*, and *Richard III* is located in single words and short phrases, so that a high percentage of text is identical in the versions being compared, which means that in parallel editions readers uneconomically read most of the text twice.

In print editions, two solutions to this problem can be observed. On the one hand, variants can be highlighted within the parallel texts. For instance, Halio's New Cambridge Shakespeare edition of the Folio-based *King Lear* includes a selection of passages with Quarto and Folio equivalent fragments printed on facing pages, with words, phrases, or lines unique to one text in bold type and with "omissions" indicated by blank spaces (Halio 1992, 82–89), as in the following transcription from 5.3.229–287 (1992, 88–89):

[Quarto: Act 5, Scene 3]   [Folio: Act 4, Scene 3]

LEAR And my poor fool is hanged. No, no, life?   LEAR And my poor fool is hanged. No, **no**, no life?
    Why should a dog, a horse, a rat have life,   Why should a dog, a horse, a rat have life,
    And thou no breath at all? **O thou wilt** come no more,   And thou no breath at all? **Thou'lt** come no more,
    Never, never, never. Pray you, undo   Never, never, never, **never, never.**
    This button. Thank you, sir. **O, o, o, o.**   Pray you, undo this button. Thank you, sir.
EDGAR   He faints. My lord, my lord!   **Do you see this? Look no her! Look, her lips.**
LEAR Break, heart, I prithee break.   **Look there, look there.**   *He dies*
EDGAR   Look up, my lord.   EDGAR   He faints. My lord, my lord!
                                                        KENT Break, heart, I prithee break.
                                                        EDGAR   Look up, my lord.

On the other hand, variants can be highlighted by placing them next to each other in the line itself of a common text, as in the two following samples:

(1)

    313   *Ham.* O that this too too {sallied} <solid> flesh would melt,
    314 Thaw and resolue it selfe into a dewe,

    ...

    319 Fie on't, {ah fie,} <Oh fie, fie,> tis an vnweeded garden
    320 That growes to seede, things rancke and grose in nature,

(2)

    313 *Hamlet.* O that this too too $^{\langle sullied \rangle}_{solid}$ flesh would melt,
    314 Thaw, and resolve itself into a dew,   130

    ...

    319 Fie on't, $^{ah}_{O}$ fie, $_{fie,}$ 'tis an unweeded garden   135
    320 That grows to seed; things rank and gross in nature

Figure 1. Two fragments from *A Synoptic "Hamlet"* (Tronch-Pérez 2002, 102).

The first sample is from Kliman's "Enfolded *Hamlet*" (1996):[3] in this quotation from Hamlet's first soliloquy, a diplomatic transcription of the Second Quarto "enfolds" material variants in the First Folio, prints both variant readings side by side (horizontally) in the line itself, and frames the Second Quarto variant within angular brackets (< >) and the First Folio-only text within curly brackets ({ }). The second sample is from *A Synoptic "Hamlet"* (Tronch-Pérez 2002, 102), a dual-reading text resulting from an imaginary superimposition of two critical and modernized editions, one based on the Second Quarto and the other on the First Folio, with text in common printed once and variants placed next to each other but vertically aligned, the Second Quarto slightly raised above the baseline and the First Folio variant immediately below.

In both samples, variations are shown as "parallel readings" in one text, not as readings in parallel texts. Thus comparison is carried out inline, right before the reader's eyes, within a single text, and not through eye-movement across texts in parallel, a process in which attention to significant variants is less immediately achieved. It should be noted, again, that this "inline

---

[3] The "Enfolded Hamlet" is also available online at the *Hamlet Works* website, http://triggs.djvu.org/global-language.com/ENFOLDED/index.php. Another example of an "enfolded text" edition is *The House that Jack Built* (Fotheringham 2006, 217–314).

variant" system is especially adequate for the kind of variation found in the texts of *Hamlet*, *Troilus*, and *Lear* (to name a few)—that is, a variation characterized by substitutions and omissions of single words and short phrases, omissions/additions of lines and passages, and absence of substitutions or "revisions" of lines and passages.[4]

This "inline variant" method has precedents in the non-lemmatized apparatus of the multi-textual and genetic editorial schools, a well-known example of which is Gabler's "critical and synoptic" edition of James Joyce's *Ulysses*, where variants are printed horizontally in the line and distinguished by means of diacritics. By contrast, the next example prints variants vertically. Here Zeller has nine witnesses of a ballad by C. F. Meyer "registered in synoptic sequence in such a way that the alterations of each as against the preceding one can be read off; only text which varies from that of the directly preceding version is registered in each case. Text deleted in the manuscripts is in [square brackets]" (1975, 250–51):

```
 9  H¹-D⁹     Träg schlich der Tag; dann durch die Nacht
10  H¹-D⁴     Flog Kunde von  geschlag'ner  Schlacht.
    D⁵-D⁹                     verloner
11  H¹         Zuerst begann   der Horgnerthurm;
    H¹         Geüber rief
    H²         [Zuerst begann]
    H²-D⁹      Von drüben rief
12  H¹-D⁹     Bald stöhnten alle Glocken Sturm,
13  H¹-D⁹     Und was geblieben war zu Haus,
    D⁹                                    Haus:
14  H¹         Das stand am See, blickt' angsvoll aus:
    H²D³                       lugt'
    D⁴-D⁹                                      aus.
```

Here the line is divided into "sub-lines," lodging the variants, which are placed vertically below their counterparts in the other witnesses. This multi-reading format is conceived of as a textual apparatus, but it can be a reading text for comparing versions in its own right.

Zeller's tiered format resembles the interlinear glossed texts used in linguistic description and in facilitating the correspondences between words (and even morphemes) of a source text and their translation.[5] An interlinear

---

[4] See a discussion of *Troilus and Cressida* in Tronch-Pérez 2005.

[5] See Bow, Hughes, and Bird 2003.

layout can be helpful in bringing source and translation versions closer to each other than in a parallel-text display, and the digital medium would allow readers to choose and shift from one displaying mode to the other, depending on their needs and/or on the passage being read.

A system with inline variants is used by the online *Internet Shakespeare Editions* to show the collation: as can be seen in Figure 2, variants are placed horizontally one next to the other, distinguished by colors, underlining, and closing superscript sigla. Besides, readers can choose to show or hide the collated witnesses by clicking on the list in the Toolbox.

Figure 2. Screenshot from Bevington's online edition of *As You Like It* for the *Internet Shakespeare Editions*, with inline variants activated (note Toolbox on the left-hand side).

Depending on the users' preference, the projected digital edition could show texts and their variants in several viewing modes, such as parallel texts with or without variants highlighted, interlinear texts, and single texts with inline variants displayed either horizontally (as in the "Enfolded" text and in the apparatus of the *Internet Shakespeare Editions*) or vertically (as in *A Synoptic "Hamlet"*). In order to achieve these variant displays in a hypertext edition (and other visualization formats generated by existing digital tools and editions), the texts must go through several processes, including electronic transcription to produce input data, markup, and division into relevant segments or tokens, alignment of these divisions, and styling of output data.

Exploring Digital Tools and Multiple-Text Editions

Since computers began to be applied to textual endeavors, software has been produced to facilitate more or less automatic text processing, textual collation, and alignment of parallel texts, including bilingual or translation corpora. As regards collation tools, mention should be made of TUSTEP (a toolbox with modules for all stages of scholarly processing of textual data, including text collation), Juxta and its free online workspace Juxta Commons (developed at the University of Virginia), CASE (a nine-program suite developed by Peter Shillingsburg for his Thackeray edition), TEI Comparator, and CollateX;[6] and for tools dealing with alignment of texts in different languages, TCA2 (Hofland and Reigem), Hunalign (Varga et al. 2005), and InterText (Vondřička 2014).[7]

None of these fully accommodate my project's double interest in intralingual versions and translational versions at the same time. The collation tools are obviously ineffectual in dealing with texts in different languages. The aligners can have texts in bilingual or multilingual corpora automatically segmented in sentences, but translations of early modern plays do not always follow the same division of sentences as their source texts. Yet some of these programs contain features that are helpful in the design of a multi-textual and multilingual resource. For instance, Juxta offers various analytic visualizations, among them a "heat map," that could be adapted to the projected edition: this visualization displays the selected text with variants enhanced in blue, and when users click on one of them, a pop-up window in the right-hand margin reveals all the variations in the other collated witnesses. Moreover, colored variants go from lighter to darker shades to signal that there are fewer or more witnesses varying from the base text at this specific point.[8]

---

[6] TUSTEP, Juxta, and CASE are conveniently surveyed by Gabler (2008). Huculak and Richardson (2013) explain their developments, as well as surveying and reviewing TEI Comparator, CollateX, and others (including the Versioning Machine).

[7] These are sentence aligners ultimately based on the *align* program (Gale and Church 1993); InterText is a post-alignment editor and an interface to Hunalign and TCA2. For a full survey of tools to create translation corpora, see Zanettin (2012).

[8] Another interesting visualization is a "histogram," which shows the relative density of textual variation from the base text, which is useful when dealing with long documents.

The digital resources more akin to my projected edition are the Versioning Machine (Schreibman), and the platform *Version Variation Visualization* (Cheesman et al.); the closest digital and multiple-text editions are *Walden: A Fluid Text*, as well as those in the EMOTHE collection and in *Richard Brome Online*. In the following pages I survey these resources.

The Fluid Text edition of Henry David Thoreau's *Walden* at *Digital Thoreau* uses the Versioning Machine to facilitate comparison of nine versions of this novel, showing them in parallel panels or columns as selected by the reader. By means of colors and strikethrough, it also registers variants within the same witness for text interlined in ink, in pencil, interpolations, and cancelled text.

EMOTHE, directed by Joan Oleza at the University of Valencia, is an ongoing collection branching out of the Artelope research project, which has built a database of Lope de Vega's plays.[9] Seeking to facilitate an understanding of Lope de Vega's drama in its European context, and an appreciation of Lope's main contemporary European traditions, EMOTHE collects plays from Italian, French, Spanish, and English drama. With a view to cater to users (both scholars and general readers) who may not be versed in the original language of the selected plays, EMOTHE selects, edits, and shows translations, either individually or simultaneously side by side with the other versions of the same play. At present, EMOTHE's system for simultaneous viewing is inspired by the parallel-text display of the *Richard Brome Online* edition. In this electronic edition, after choosing any of the sixteen Brome plays, users can select the "Both texts" feature from the navigation bar. This brings onto the screen, in parallel columns, a "period text" (a transcription of the seventeenth-century primary text) alongside a critical, modern-spelling, annotated edition. When a reader who has scrolled down any text wishes to see its parallel text, she or he only needs to click on a speech prefix, or on the speech number next to it, and the two versions quickly come into alignment at the top of each column.

Both *Richard Brome Online* and EMOTHE create electronic files of the editions in eXtensible Markup Language (XML) following the Guidelines of the Text Encoding Initiative (TEI). The XML files on the servers are transformed into HTML files for web publication using an extensible stylesheet language (.xsl), a scripting language (.js), and a cascading style sheet (.css). However, with respect to the text's parallel alignment, these editorial projects differ. Not only does EMOTHE show more than two texts, but the texts to be viewed side

---

[9] This database is available (in Spanish) at http://artelope.uv.es/baseArteLope.html.

by side may be translations (and adaptations, to a certain degree) as well as textual versions in the same language (e.g. the A-text and B-text of Marlowe's *Doctor Faustus*). This different situation demands different ways to systematize linkage between corresponding segments in the parallel versions.

Since *Richard Brome Online* offers intralingual versions of the same "text" (the "period text" and the modernized, critical edition), the two texts to be aligned in parallel are bound to contain the speeches in the same sequential order and spoken by the same character. However, total automated matching by means of the sequence of speeches proved to be impossible to achieve, as privately explained by Katherine Rogers, Developer at the Humanities Research Institute of the University of Sheffield, who was responsible for the technical work on the editions. Using the TEI anchor in the modern text to provide the linking points with the early text, "linkage between speeches was done by pattern matching the first line of each speech across the editions, accounting for variant spelling and other anomalies" (Rogers 2014). The computer then created a link whenever there was a match, and when it was unsure the speech was flagged for an editor to fix it manually.

By contrast, EMOTHE editors are likely to work with texts in which sequential correspondence of structural elements (act and scene divisions, stage directions, speeches) frequently varies. The historical Spanish translation of *Hamlet* by Leandro Fernández de Moratín (1798) follows the convention of creating a new scene whenever a character enters or exits (for instance, act three has twenty-eight scenes in Moratín instead of the usual four scenes). At the beginning of the historical Spanish translation of *Macbeth* by García de Villalta (1838), the Second Witch's intervention—which is the second speech in Shakespeare's tragedy: "When the hurly-burly's done, / When the battle's lost and won" (Mowat and Werstine, *Macbeth* FTLN 0003-4)—corresponds to the second and third speech in García de Villalta's translation, here assigned to the Second Witch and the Third Witch respectively (back-translation provided):

BRUJA 2ª Cuando la tierra se zafe   [When the earth frees itself
    del tumulto y rifirrafe.   [of the hurly-burly and the turmoil
BRUJA 3ª Cuando la fiera pelea   [When the fierce fight
    ganada y perdida sea.   [is won and lost

The third speech in sequential order in Shakespeare's *Macbeth*, "That will be ere the set of sun," spoken by the Third Witch, corresponds to the fourth speech assigned to "BRUJA 1ª" in García de Villalta's version:

Antes que se apague el día   [Before the day is gone
cumplirá tu profecía.   [your prophecy will be fulfilled

The series of mismatches goes on three more speeches with the following correspondences.

        Shakespeare   García de Villalta's translation

FIRST WITCH   BRUJA 2.ª
  Where the place?   ¿Y adónde acudiremos esa vez?

SECOND WITCH Upon the heath. BRUJA 3.ª
                               A buscar en los yermos a Macbeth.

THIRD WITCH
  There to meet with Macbeth.

After this line, Villalta adds a stage direction, "Suena una trompeta," absent in Mowat and Werstine's Folger Digital Texts edition, which is the English-text edition used in the EMOTHE collection.

It is therefore clear that neither speech prefixes, stage directions, nor speeches' sequential number can be used as linkage elements to set texts in parallel automatically and that manual alignment at certain points is unavoidable. The same holds true for the *Version Variation Visualization*, a "prototype on-line platform for comparing multiple versions of literary works, in one or more languages" (Cheesman, Laramee, and Hope 2012). For this project, Kevin Flanagan developed two web-based tools: one (called "Ebla") that "stores documents, configuration details, segment and alignment information, calculates variation statistics, and renders documents with segment/variation information"; and another tool (called "Prism") that allows users "to check alignments of pre-created text segments (i.e. speeches), and to create and align additional segments—sentences or phrases within speeches" (Cheesman, Flanagan, and Thiel 2012-2013). Since there is no straightforward, one-to-one correspondence between sequential parts of the play text's structure and their equivalents in a translation, as the *Version Variation Visualization* developers acknowledge, Prism allows users to correct the mismatches produced by the automatic alignment of segments. In Prism's interface (with two parallel WYSIWYG editing displays) these mismatches are clearly shown by colored bars across the parallel columns containing the "original" and the translations.[10] Users can manually remove these links, realign the segments involved, and automatically realign the rest of the text (until a further mismatch is detected by the human eye). Once the aligned blocks are encoded in the text files, the display of versions can highlight the equivalent segments by means of colors and arrows.

The alignment tool currently being developed by EMOTHE offers pre-segmented texts of each version (original and translations), taking advantage of the fact that its base-texts are TEI-conformant XML files, with their corresponding tags for speeches, stage directions, act and scene divisions, prologues, and epilogues. The interface shows these segmented versions in parallel so that encoders click on the appropriate segments in each version to select them and then press on a linking button. The alignment is carried out by hand, but it can be done on more than two texts at once. In time the tool might implement a semi-automatic or machine-assisted process, allowing encoders some freedom to define and align blocks, as observed in the *Version Variation Visualization* project. Yet, as its developer Kevin Flanagan privately communicated, the *Version Variation Visualization* platform does not store data in TEI-complaint XML files because it would prevent dealing with alignment sets in a transactional and synchronous way by multiple users and "generating large amounts of variation dynamically at runtime, so as to reflect changes immediately" (Flanagan 2014).

In EMOTHE's front-end parallel-text display, the linked segments are signaled by means of a "pin" icon at the beginning (Figure 3).

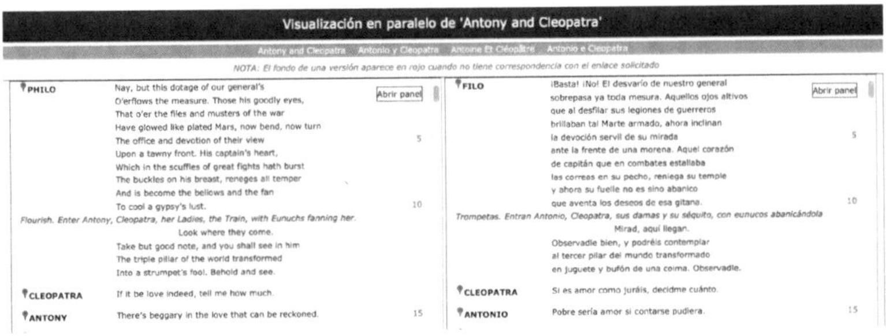

Figure 3. Screenshot from *Antony and Cleopatra* with English text and Spanish translation in parallel, showing the "pin" icon heading each speech.

---

[10] An illustrative video clip is available at https://lecture2go.uni-hamburg.de/konferenzen/-/k/13974.

As in *Richard Brome Online*, users may scroll down any of the parallel texts independently, and may click on the "pin" heading at a given point (a speech, a stage direction, or an act or scene division) whenever they want to see the corresponding sections aligned at the top of their embedded boxes. Within each <div> element and its "class" attribute for speech prefixes, stage directions, and act and scene divisions, a hyperlink <a> tag is inserted with an href attribute to point to a JavaScript function (e.g. <a href="javascript:align(1)"> .... </a>) in an external script file.

In these platforms, a speech is the unit by which the performance text is segmented for synchronous parallel-text presentation. Yet, as stated above, my projected multilingual edition is interested in visualizing variants below the level of a character's speech. In short interventions, such as those by the witches in the first scene of *Macbeth*, it is easy to spot sentence, phrase, or lexical equivalents between original and translations, but in long speeches, such as Hamlet's soliloquies, the task is difficult once the user goes beyond the first line or sentence, even more if the texts are in prose (and even more if lineation in the HTML prose text may change as the screen is resized).

A system for highlighting variants at a lower level can be observed in Susan Schreibman's Versioning Machine—self-defined as a tool (a framework and an interface) for displaying and comparing different versions of literary texts—which in fact visualizes in HTML a previously encoded critical apparatus of the texts in question as defined by the Text Encoding Initiative's Guidelines (TEI, chapter 12)[11] in an XML file. Segmentation and alignment are encoded by means of the <app> and <rdg> elements of TEI's inline critical apparatus. Thus, the source or base text and the other textual versions with which the former is compared are all contained in one XML file. Its display as an HTML page contains a top toolbar section (with drop-down buttons to opt for opening new versions, showing or hiding bibliographic information and an introduction, and showing or hiding pop-up and inline notes) and the main area split into parallel columns for the bibliographic information, the introduction, and the text of the witnesses or versions.

At present, the Versioning Machine display has one scrollbar for the whole HTML page so that texts in the parallel columns are aligned as long as there is

---

[11] The URL for Chapter 12 is http://www.tei-c.org/release/doc/tei-p5-doc/en/html/TC.html.

no omission of one or several lines in one of the versions. A similar problem happens with text transposed to several lines. Omissions and transpositions produce alignment mismatches that destabilize any synchronic viewing of equivalent text in the Versioning Machine. By contrast, Juxta's "side by side" visualization shows the witnesses in synchronously scrolling columns, and uses graphically elastic bars linking variants across the columns to provide an orientation as to their position—certainly helpful functionalities to ideally implement in the Versioning Machine.[12]

Another feature that could be improved is the fact that, in order for the Versioning Machine to highlight variants, users have to click on an encoded segment in any witness for the variants to display in yellow in all texts, but users are not told which segments they have to click on. It would be more convenient to allow users select a display option to have the encoded variants highlighted as a default position when a witness is called. Juxta's "side by side" visualization option "gives a split frame comparison of a base text to a witness text," as explained in the "About" page on its website, and by default shows all variants highlighted in blue. In Figure 3, Juxta Commons conveniently points out variants that can be easily overlooked, such as the plural "soldiers" in Marcellus's speech in the Second Quarto-based edition:

---

[12] This problem can be clearly observed, for instance, in the "Orchard Farming" poem at http://v-machine.org/samples/orchardFarming.html, where the last line in the First witness is not parallel to the equivalent last line in the Second witness; or in the "Economy" chapter in *Walden: A Fluid Text*, where the missing first paragraph in Version A misaligns the rest of the chapter. Although *Richard Brome Online* and the digital *Walden* have independent scrollbars for each column, they are not synchronized with one another.

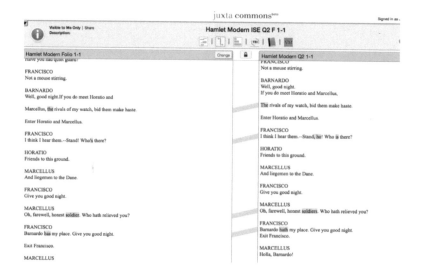

Figure 4. Screenshot from Juxta Commons comparing Folio-based and Second Quarto-based editions of *Hamlet*.

The digital resources for displaying versions that I have explored so far can be grouped in two sets, representing two opposing working models in relation to the way the texts to be compared and their linkage and alignment information are digitally stored: on the one hand, the Versioning Machine stores data on a single, comprehensive electronic file in TEI-conformant format, from which discrete textual versions are reconstructed and eventually visualized in HTML; on the other hand, *Version Variation Visualization* and EMOTHE store separate files of each version in a "back-end" relational database, and the alignment information (relationships among identifiers of segments from all the versions, and the extent of these segments) is registered in a separate third file.[13]

Neither the Versioning Machine nor the TEI's tag set for a critical apparatus are specifically conceived for translations, but they can work with versions in other languages as well, as will be shown below in a test case that uses the first five lines of Hamlet's "To be, or not to be" monologue.

---

[13] This stand-off alignment is also employed by *InterText*, a post-alignment editor and a user interface to the automatic aligners Hunalign and TCA2 (Vondřička 2014).

Encoding Variants

Displaying variants at the level of phrases or short linguistic units involves encoding the intended segments and their interlinkage in a way that predetermines the possibilities of visualization. Again, a good deal of this process is to be carried out manually, with the segments selected at the linguist-encoder's discretion if they are to be meaningful for the purposes of comparison. For intralingual variants, the encoder may decide, for instance, whether the colon after "that is the Question:" in the First Folio text of Hamlet's famous soliloquy is significantly different enough from the Second Quarto's comma ("that is the question,") to be marked up and later visualized. In the case of interlingual variants, for instance, the encoder decides whether to select "To be, … that is the question" as a segment, or to split "To be, or not to be" and "that is the question" as units for comparison with the translations.[14]

In the *Version Variation Visualization* platform, the "Prism" mark-up tool enables "users to freely define and semi-automatically align segments in parallel documents" (Cheesman, Laramee, and Hope 2012). One can create segments within a line or even across verse lines by positioning the cursor on specific points in the text and clicking on the "start marker" button, similarly for the "end marker," and then pressing the "create segment between markers" button; or by selecting a stretch of text and pressing on the "create segment around selected text" button. Of course, a similar procedure is to be applied in the equivalent segment in the translation shown on the right-hand side. At present, Prism's aligner interface accommodates only two texts, but it would be possible to extend it to allow alignment of more than one translation simultaneously on the same screen (as its developer, Kevin Flanagan, privately communicated).

In the Versioning Machine, variants are highlighted when users click on a pre-encoded segment in any witness. These segments correspond to those encoded within TEI's <app> element in the matrix XML file. Readers can try this functionality in another working example that uses the first five lines of Hamlet's "To be, or not to be" soliloquy (TLN 1710-1714), available at <http://www.uv.es/shaxpere/vm4/samples/To_be_multiling2.html>. It is up to the digital editor or encoder to decide on the extent of segments of the intralingual variants, for instance, whether to segment "To be, … that is the question"

---

[14] These segments may coincide with the notion of "translation unit," that is, portions into which the translator segments the source text as a single cognitive unit in order to find an equivalent in the target language (Delisle 1999, 194).

as a unit within the <app> element, or to encode "<app>To be, or not to be</app>" and then "<app> that is the question</app>" in two apparatus entries (as in my working example). Appendix I shows how this is marked up in TEI-XML format. By nesting <app> elements within higher <app> elements, differences involving punctuation among the English-language texts can be encoded with second-level <app> elements, while interlingual variants (between the English-language segment and its translations) can be marked up with higher <app> elements. Thus, these first-level divisions, when clicked on in the Versioning Machine interface, are displayed in yellow.

As the translations sometimes resort to transposition of equivalents, I have used the location-referenced method (TEI P5 Guidelines, 12.2.1) to encode the critical apparatus. For instance, Macpherson placed his Spanish translation of "And, by opposing, end them" (TLN 1714) two lines earlier ("o terminar la lucha,"), next to the equivalent of "outrageous fortune" (TLN 1712). To encode this transposition, the segment "o terminar la lucha" is placed within line 1712 but given the same value for the @loc attribute ("a1714a") as its source segment "And, by opposing, end them." (TLN 1714). Appendix II details how this transposition is marked up. In the Versioning Machine visualization (at the URL given in the previous paragraph), with the witnesses "Mowat and Werstine" and "Macpherson 1873" selected (and "Line numbers" selected), a click on "And, by opposing, end them" (TLN 1714) highlights in yellow its Spanish equivalent even though it appears two lines earlier (at 1712), as shown in the following screenshot.

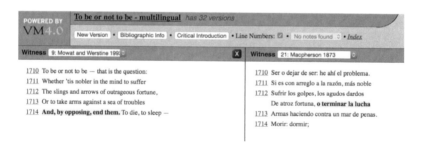

Figure 5. Screenshot from the Versioning Machine display of the working example of the multilingual "To be, or not to be."

As pointed out earlier, the Versioning Machine presents problems of text alignment when an "omission" of one or several lines occurs. One solution to this "omission" problem could be to have the variants marked up as omissions (perhaps using the @type attribute nested in the <rdg> element), and then to encode instructions in the extensible stylesheet language file, and in the JavaScript file, so that whenever a witness containing this type of variant is to be compared with another version without it in the same <app> element, the extent of each omission with respect to the other variant reading is measured in order to provide empty or gap tokens to compensate for the lack of textual space.[15]

I also suggested that the Versioning Machine could have variants displayed as a default position when a witness is called. This would involve the possibility of highlighting <app> elements other than those at the first level, so that individual variants (such as "be," with a comma against "be" without comma) could be easily spotted.

Furthermore, there could be an option to display significant or substantive variants only (as distinct from accidental or orthographic variants) so that users could focus on more meaningful differences and not on textual minutiae.[16] To this end, the @type attribute with the value "substantive" for significant variation within the <rdg> element could be used. For instance, in the first line of Hamlet's soliloquy, one may consider that the only substantive variant among the English texts is the punctuation mark after "question," while variants such as "To be," with comma, against "To be" without comma, are not deemed of substantive importance. This could be codified as reflected in the following example showing the encoding of "that is the question" in the editions by Spencer (abbreviated as "Ham_Spn"), Bevington (in its four editions: Editor's Version, Modern Quarto I ["Ham_BQ1"], Modern Quarto 2 ["Ham_BQ2"], Modern Folio ["Ham_BF"]) and Mowat and Werstine ("Ham_MW"):

<app loc="a1710b">
    <rdg wit="#Ham_Spn #Ham_BQ2 #Ham_BF #Ham_BE #Ham_BQ1 #Ham_MW">
      <app>

---

[15] Similar processes of inserting gap tokens are implemented in the aligners of programs such as Juxta and CollateX ("Documentation"), both using the so-called "Gothenburg model."

[16] Significant or substantive variants are those differing in sense or reference; in morphological, inflectional, or syntactical aspects; in prosody; or affecting staging in a different way. This notion of "substantive" is different from that in Spevack (1980, vii–viii), since he does not include variants (such as "he / a"), contractions ("who's / who is," "never / ne'er"), elisions ("against / 'gainst"), divergences in punctuation, line division, and position of stage directions.

```
    <rdg type="substantive" wit="#Ham_Spn #Ham_BQ2 #Ham_BF #Ham_BE
#Ham_MW"> that is </rdg>
    <rdg type="substantive" wit="#Ham_BQ1"> ay, there's </rdg>
  </app>
  the
  <app>
  <rdg type="substantive" wit="#Ham_Spn">question;</rdg>
  <rdg type="substantive" wit="#Ham_MW">question:</rdg>
  <rdg type="substantive" wit="#Ham_BQ2 #Ham_BF #Ham_BE">question,</
rdg>
  <rdg type="substantive" wit="#Ham_BQ1"> point,</rdg>
  </app>
  </rdg>
</app>
```

If the user selects a comparison of Bevington's Editor's Version and Mowat and Werstine's edition only, the visualization would highlight by default those substantive variants that these editions do not share: in the latter example, Bevington's "question," would be highlighted alongside Mowat and Werstine's "question:" with a colon, but not "that is the question," a reading these editions share against the variant "ay, there's the point," in Bevington's Modern Quarto 1 edition.

Whether the data is encoded and stored in a single matrix XML file (as in the Versioning Machine) or encoded in a user-friendly WYSIWYG interface and registered in distinct files in a relational database (as in the *Version Variation Visualization* platform), the projected online multilingual edition of a play such as *Hamlet* could offer the following displays for text comparison:

- ★ parallel texts, with a further option for highlighting variants within each text;
- ★ interlinear text, also with the option of highlighting variants;
- ★ single text with inline horizontal variants (as in the "Enfolded");
- ★ single text with inline vertical variants; or
- ★ single text with variants in "pop-up" windows.

These displays would be applied to all the texts or versions involved, both English-language editions and the translations. Thus, the user would select from two sets of options: versions and displays. For instance, users willing to compare Bevington's Editor's Version edition with Mowat and Werstine's edition would select these two texts and then select the display mode (for instance, the "enfolded text"); the resulting HTML page would show something like the following (with superscript sigla to indicate sources):

    1710 To be, or not to be, that is the [BE]question,[BE MW]question:[MW]

    ...

    1726 The pangs of [BE]disprized[BE MW]despised[MW] love, the law's delay

(Here Bevington's edition enfolds significant variants in Mowat and Werstine's.)[17]

If users select more than two texts, the projected edition could warn them that visualization problems could occur in the "enfolded" mode, since having three or more variants horizontally side by side would render the line inordinately long and visually ungracious. Apart from the traditional "parallel text" presentation, the variants in the selected versions could be compared in the tiered format with vertical variants, as in the following example with Bevington's Editor's Version (BE), Modern Folio (BF) and Modern Quarto 2 (BQ2) editions, and with Mowat and Werstine's edition (MW) selected:

313	BE BF	O that this too too	solid	flesh would melt
	BQ2		sallied	
	MW		sullied	

Here editions with the First Folio variant "solid" adopted are read against the quarto edition reading "sallied" and Mowat and Werstine's edition adopting the emendation "sullied."[18]

The interlinear layout may be useful to align segments of a translation (or various translations) to their original equivalents in parallel lines (or "sublines") rather than in parallel texts.

1710	MW	To be or not to be — that is the question:
	MORATÍN	Existir o no existir: esta es la cuestión.
	HUGO	Être, ou ne pas être, c'est là la question
	RUSCONI	Essere o non essere, ecco il gran problema....

---

[17] By way of contrast, if Mowat and Werstine's edition enfolded Bevington's, the line would be "The pangs of [MW]despised[MW BE]disprized[BE] love, the law's delay."

[18] The "sullied" emendation is credited to an anonymous conjecture in Clark and Wright (1866, 16) and first adopted by Wilson (1934) and Ridley (1934).

1711 MW	Whether 'tis nobler in the mind to suffer
MORATÍN	¿Cuál es más digna acción del ánimo, sufrir
HUGO	— Y a-t-il plus de noblesse d'âme à subir
RUSCONI	ma è più nobile all'anima il patire

The projected hypertextual and multilingual edition aspires to be a tool to facilitate the investigation of the textual features in variant editions and in translations of plays such as *Hamlet*. By having direct access to textual variants in combination with translational variants, users can study, for instance, the origin of choices in translations in contrast to options adopted by other editions, or the nuances of meaning involved in divergent punctuation (as in versions of the "To be, or not to be" soliloquy). For some translations, such as Moratín's, it is difficult to discern a given English-language edition underlying the translated text, so that a comparative analysis of this target text with its source can be less limited if the source text displays a range of variants available in the selected source-language editions: for instance, Moratín's Spanish translation follows the First Folio variant with Horatio and Marcellus in the speech prefix at 1.5.113. and adopts the Second Quarto variant "dream" at 2.2.10, and either option appears in an "enfolded" text to which Moratín's translation can be compared.

A Brief Concluding Assessment

Using TEI's critical apparatus to bring together various texts into a single file has the advantage of its centrality and its extensibility: on the one hand, changes and corrections need only be made in this matrix file, as in a database. A disadvantage is its requirement of deep, complex encoding (as shown in the appendices below), with which not all editors are familiarized (if the projected edition is to be applied to a series of plays). By contrast, the interfaces of the *Version Variation Visualization* and EMOTHE platforms are more user-friendly, with the former using a WYSIWYG display resembling familiar word processors for editing, segmenting, and aligning texts. In both platforms, the use of a separate or stand-off alignment file preserves its stored essential data from "corruptions" by non-technical editors. The *Version Variation Visualization* platform does not work with TEI-conformant XML files, as do the Versioning Machine and EMOTHE, which may compromise future interoperability.

APPENDIX I

The following sample shows a codification of the first two lines of Hamlet's soliloquy, but, for the sake of economy, it includes only the editions by Bevington (Editor's Version, abbreviated as "Ham_BE") and by Mowat and Werstine ("Ham_MW") alongside Macpherson's 1873 Spanish translation, Hugo's French, Carcano's Italian, and Schlegel and Tieck's German.

```
<l n="1710">
 <app loc="a1710">
  <rdg wit="#Ham_BE #Ham_MW">To
   <app>
    <rdg wit="#Ham_MW">be</rdg>
    <rdg wit=" #Ham_BE">be, </rdg>
   </app>
   or not to
   <app>
    <rdg wit="#Ham_BE">be, </rdg>
    <rdg wit="#Ham_MW">be — </rdg>
   </app>
  </rdg>
  <rdg wit="#Ham_Macpherson_1873">Ser o dejar de ser: </rdg>
  <rdg wit="#Ham_Hugo_2">Être, ou ne pas être, </rdg>
  <rdg wit="#Ham_Carcano">Essere ovver non essere! </rdg>
  <rdg wit="#Ham_Sch_Tck">Sein oder Nichtsein, </rdg>
 </app>
 <app loc="a1710b">
  <rdg wit="#Ham_BE #Ham_MW">
   <app>
    <rdg wit="#Ham_BE #Ham_MW"> that is the </rdg>
     <app>
      <rdg wit="#Ham_MW">question:</rdg>
      <rdg wit="#Ham_BE">question,</rdg>
     </app>
   </app>
  </rdg>
  <rdg wit="#Ham_Macpherson_1873">he ahí el problema.</rdg>
  <rdg wit="#Ham_Hugo_2">c'est là la question.</rdg>
  <rdg wit="#Ham_Sch_Tck">das ist hier die Frage:</rdg>
  <rdg wit="#Ham_Carcano">l'enimma</rdg>
 </app>
```

```
</l>
<l n="1711">
<app loc="a1710b"><rdg wit="#Ham_Carcano"> Qui stà. - </rdg> </app>
<app loc="a1711">
 <rdg wit="#Ham_BE #Ham_MW">Whether 'tis nobler in the mind to suffer</rdg>
    <rdg wit="#Ham_Macpherson_1873"> Si es con arreglo a la razón, más noble </rdg>
  <rdg wit="#Ham_Hugo_2"> — Y a-t-il plus de noblesse d'âme à subir </rdg>
  <rdg wit="#Ham_Carcano"> Se l'alma più sia forte allora </rdg>
  <rdg wit="#Ham_Sch_Tck"> Ob's edler im Gemüt, die Pfeil' und Schleudern </rdg>
 </app>
</l>
```

This sample shows how the different degrees of variation between intralingual and interlingual variants (the phrase "To be or not to be" is more segmented than its translations) can be accommodated in the TEI apparatus, by using <app> elements nested within higher <app> elements. The differences in punctuation between Bevington's "To be, or not to be, that is the question," and Mowat and Werstine's "To be or not to be — that is the question:" are encoded with second-level <app> elements:

★ in the case of the first hemistich,

```
<app>
 <rdg wit="#Ham_MW">be</rdg>
 <rdg wit="#Ham_BE">be, </rdg>
</app>
```

and

```
<app>
 <rdg wit="#Ham_BE">be, </rdg>
 <rdg wit="#Ham_MW">be — </rdg>
</app>
```

★ and in the second hemistich,

```
 <app>
  <rdg wit="#Ham_MW">question:</rdg>
```

```
      <rdg wit="#Ham_BE">question,</rdg>
    </app>
```

The differences between the English versions and the translations are encoded at the first level <app> element: in my example, I have used one <app> for "To be or not to be" and another first level <app> for "that is the question."

APPENDIX II

This example shows how Macpherson's Spanish translation of "And, by opposing, end them" (TLN 1714) appears two lines earlier ("o terminar la lucha," with double underline in the example below). To encode this transposition, the segment "o terminar la lucha" is placed within line 1712 but given the same value for the @loc attribute ("a1714a") as its source segment "And, by opposing, end them." (TLN 1714).

```
<l n="1712">
  <app loc="a1712">
    <rdg wit="#Ham_MW">The slings and arrows of outrageous fortune,</rdg>
    <rdg wit="#Ham_Macpherson_1873">Sufrir los golpes, los agudos dardos <lb/> De atroz fortuna, </rdg>
      <rdg wit="#Ham_Hugo_2">— la fronde et les flèches de la fortune outrageante,</rdg>
    <rdg wit="#Ham_Sch_Tck">Des wütenden Geschicks erdulden, oder</rdg>
      <rdg wit="#Ham_Carcano">Che agli oltraggiosi colpi, alle saette <lb/> Regge della fortuna; </rdg>
  </app>
  <app loc="a1714a"> <rdg wit="#Ham_Macpherson_1873 #Ham_Macpherson_1882"> o terminar la lucha <lb/> </rdg> </app>
    <app loc="a1713"> <rdg wit="#Ham_Carcano"> o quando l'armi </rdg></app>
</l>
<l n="1713">
  <app loc="a1713">
    <rdg wit="#Ham_MW">Or to take arms against a sea of troubles</rdg>
     <rdg wit="#Ham_Macpherson_1873">Armas haciendo contra un mar de penas.</rdg>
```

```
        <rdg wit="#Ham_Hugo_2">— ou bien à s'armer contre une mer de
douleurs</rdg>
        <rdg wit="#Ham_Sch_Tck">Sich waffnend gegen eine See von Plagen,</
rdg>
        <rdg wit="#Ham_Carcano">Impugna contro al mar delle sciagure,</rdg>
    </app>
</l>
<l n="1714">
    <app loc="a1714a">
    <rdg wit="#Ham_MW">And, by opposing, end them. </rdg>
    <rdg wit="#Ham_Hugo_2">— et à l'arrêter par une révolte ? </rdg>
    <rdg wit="#Ham_Sch_Tck">Durch Widerstand sie enden. </rdg>
    <rdg wit="#Ham_Carcano">E affrontandole ardita, a lo dà fine. - </rdg>
    </app>
```

A similar transposition problem appears in Carcano's Italian rendering of "Or to take arms against a sea of troubles," (TLN 1713) with the phrase "O quando l'armi" (in double underline in the above example) encoded within line TLN 1712 but with the value "a1713" for the @loc attribute: " <app loc="a1713">".

WORKS CITED

Best, Michael, coord. ed. *Internet Shakespeare Editions*. University of Victoria. http://internetshakespeare.uvic.ca/.

Bevington, David, ed. *Hamlet*. By William Shakespeare. *Internet Shakespeare Editions*. University of Victoria. http://internetshakespeare.uvic.ca/Library/Texts/Ham/.

_____ ed. *As You Like It*. By William Shakespeare. *Internet Shakespeare Editions*. University of Victoria. http://internetshakespeare.uvic.ca/Library/Texts/AYL/M/default/.

Bow, Cathy, Baden Hughes, and Steven Bird. 2003. "Towards a General Model for Interlinear Text." E-MELD Language Digitization Project Conference, Michigan State University, July 12. http://emeld.org/workshop/2003/bowbadenbird-paper.html.

Carcano, Giulio, trans. 1847. *Amleto. Tragedia di Guglielmo Shakspeare*. Milano: Luigi di Giacoma Pirola.

Cheesman, Tom, et al. *Version Variation Visualization.* http://www.delightedbeauty.org/vvv.

Cheesman, Tom, Kevin Flanagan, and Stephan Thiel. September 2012–January 2013. "Translation Array Prototype 1: Project Overview." *Version Variation Visualization.* http://www.delightedbeauty.org/vvv/Home/Project.

Cheesman, Tom, Robert S. Laramee, and Jonathan Hope. 2012. "Translation Arrays: Version Variation Visualization (Phase 2)." Report to the Arts and Humanities Research Council, October. https://www.scribd.com/doc/122777775/Translation-Arrays-Version-Variation-Visualization-Phase-2-report-to-AHRC-October-2012.

Clark, William George, and William Wright, eds. 1866. *Hamlet.* In vol. 8 of *The Complete Works of William Shakespeare.* 9 vols. London and Cambridge: Macmillan and Co.

CollateX. "Documentation." *CollateX – Software for Collating Textual Sources.* http://collatex.net/doc/.

Delisle, Jean, Hannelore Lee-Jahnke, and Monique C. Cormier, eds. 1999. *Translation Terminology.* Amsterdam and Philadelphia: John Benjamins.

Flanagan, Kevin. "Re: Questions on VVV." E-mail message to author, November 2, 2014.

Fotheringham, Richard, ed. 2006. *The House that Jack Built.* In *Australian Plays for the Colonial Stage 1834-1899*, edited by Richard Fotheringham, 217–314. St. Lucia: University of Queensland Press.

Gabler, Hans Walter. 2008. "Remarks on Collation." https://www.academia.edu/167070/_Remarks_on_Collation_.

Gabler, Hans Walter, with Wolfhard Steppe and Claus Melchior, eds. 1984. *Ulysses: A Critical and Synoptic Edition.* By James Joyce. 3 vols. New York and London: Garland Publishing.

Gale, William A., and Kenneth W. Church. 1993. "A Program for Aligning Sentences in Bilingual Corpora." *Computational Linguistics* 19 (1): 75–102.

García de Villalta, José, trans. 1838. *Macbeth, drama histórico en cinco actos compuesto en inglés por William Shakespeare; y traducido libremente al castellano por Don José García de Villalta.* Madrid: Imprenta Repullés.

Guizot, François, trans. 1864. "Hamlet." In *Oeuvres complètes de Shakspeare, traduction de M. Guizot, nouvelle édition entièrement revue*, 139–282. Paris: Didier.

Halio, J. L., ed. 1992. *The Tragedy of King Lear*. By William Shakespeare. The New Cambridge Shakespeare. Cambridge: Cambridge University Press.

Hofland, Knut. 1996. "A Program for Aligning English and Norwegian Sentences." In *Research in Humanities Computing 5: Selected Papers from the ACH/ALLC Conference*, edited by Giorgio Perissinotto, 165–78. Oxford: Clarendon Press.

Hofland, Knut, and Oystein Reigem. TCA2. Romssa Universitehta [University of Tromsø, Norway]. http://divvun.no/doc/tools/tca2.html.

Hofland, Knut, and Stig Johansson. 1998. "The Translation Corpus Aligner: A Program for Automatic Alignment of Parallel Texts." In *Corpora and Cross-linguistic Research: Theory, Method, and Case Studies*, edited by Stig Johansson and Signe Oksefjell, 87–100. Amsterdam: Rodopi.

Huculak, J. Matthew, and Ashlin Richardson. 2013. "White Paper: A Survey of Current Collation Tools for The Modernist Versions Project." http://web.uvic.ca/~mvp1922/wp-content/uploads/2013/10/WhitepaperFINAL.pdf.

Hugo, François-Victor, trans. 1865. "Le Second Hamlet." In *Les deux Hamlet*. Vol. 1 of *Ouvres complètes de W. Shakespeare*, 201–388. 2nd ed. Paris: Pagnerre.

Juxta: Collation Software for Scholars. University of Virginia. http://www.juxtasoftware.org.

Kliman, Bernice W., ed. 1996. "The Enfolded *Hamlet*." Special issue, *Shakespeare Newsletter*, 2–44.

Macpherson, Guillermo, trans. 1873. *Hámlet, príncipe de Dinamarca*. Cádiz: Imprenta y Litografía de la Revista Médica de Federico Joly Velasco.

———. trans. 1879. *Hámlet, príncipe de Dinamarca*. Madrid: Imprenta de Fortanet.

———. trans. 1882. *Hámlet, príncipe de Dinamarca*. Biblioteca Universal 78. Madrid: Imprenta y Litografía de la "Biblioteca Universal."

_____, trans. 1885. "Hámlet, príncipe de Dinamarca." In vol. 3 of *Obras Dramáticas de Guillermo Shakespeare.* Biblioteca Clásica. 8 vols. Madrid: Perlado, Páez y Cía.

[Moratín (Leandro Fernández de Moratín).] Celenio, Inarco, trans. 1798. Hamlet. *Tragedia de Guillermo Shakespeare. Traducida é ilustrada con la vida del autor y notas críticas por Inarco Celenio.* Madrid: Villalpando.

Mowat, Barbara, and Paul Werstine, eds. 2013. *William Shakespeare: The Tragedy of Macbeth.* Folger Digital Texts. Washington, DC: Folger Shakespeare Library.

Oleza, Joan, dir. EMOTHE: *The Classics of Early Modern European Theatre.* University of Valencia. http://emothe.uv.es/biblioteca/.

Richard Brome Online. Royal Holloway, University of London and University of Sheffield. http://www.hrionline.ac.uk/brome.

Ridley, M. R., ed. 1934. Hamlet. Vol. 6 of *The New Temple Shakespeare.* London: Dent.

Rogers, Katherine M. "Re: On Brome Online edition." E-mail message to author, October 16, 2014.

Rusconi, Carlo, trans. 1852. "Amleto, principe di Danimarca. Tragedia." In vol. 2 of *Teatro completo di William Shakespeare, voltato in prosa italiana da Carlo Rusconi,* 7–96. 3rd ed. Torino: Cugini Pomba e comp. editori.

Schlegel, Wilhelm, trans., and Ludwig Tieck, rev. 1831. "Hamlet, Prinz von Dänemark." In vol. 6 of *Shakspeare's dramatische Werke.* 9 vols. Berlin: Verlag G. Reimer.

Schreibman, Susan. 2010. *Versioning Machine 4.0.* Maryland Institute for Technology in the Humanities. http://v-machine.org/index.php.

Spencer, T. J. B., ed. 1980. Hamlet. By William Shakespeare. The New Penguin Shakespeare. Harmondsworth, UK: Penguin.

Tanselle, George T. 1993. "Textual Criticism." In *The New Princeton Encyclopedia of Poetry and Poetics,* edited by Alex Preminger and T. F. V. Brogan, 1273–76. Princeton, NJ: Princeton University Press.

TEI Consortium. 2014. *P5: Guidelines for Electronic Text Encoding and Interchange. Version 2.7.0.* Text Encoding Initiative Consortium. Last modified

September 14, 2014. http://www.tei-c.org/release/doc/tei-p5-doc/en/html/index.html.

Thoreau, Henry David. *Walden: A Fluid Text Edition.* In *Digital Thoreau.* State University of New York at Geneseo, Thoreau Society, Walden Woods Project. http://www.digitalthoreau.org/fluid-text/toc/.

Tronch-Pérez, Jesús. 2002. *A Synoptic "Hamlet": A Critical and Synoptic Edition of the Second Quarto and First Folio Texts of "Hamlet."* València and Zaragoza: Universitat de València and SEDERI [Spanish and Portuguese Society for English Renaissance Studies].

―――――― 2005. "Dual-Reading Editions for Shakespeare's Two-Text Plays in the Example of *Troilus and Cressida.*" *TEXT: Journal of the Society for Textual Scholarship* 17: 117–144.

Varga, Dániel, Laszlo Nemeth, Peter Halacsy, Andras Kornai, Viktor Tron, and Viktor Nagy. 2005. "Parallel Corpora for Medium Density Languages." In *Proceedings of RANLP 2005* [Recent Advances in Natural Language Processing], 590–96.

Vondřička, Pavel. 2014. "Aligning Parallel Texts with InterText." In *Proceedings of the Ninth International Conference on Language Resources and Evaluation (LREC '14),* edited by N. Calzolari et al., 1875–79. European Language Resources Association. http://www.lrec-conf.org/proceedings/lrec2014/pdf/285_Paper.pdf.

―――――― *InterText: Parallel Text Alignment Editor.* http://wanthalf.saga.cz/intertext.

Wieland, Christoph Martin, trans. 1766. *Shakespear theatralische Werke.* Illustr. Salomon Gessner. Vol. 8. Zürich: Orell Gessner und Comp.

Wilson, John Dover, ed. 1934. *William Shakespeare: The Tragedy of Hamlet, Prince of Denmark.* New Cambridge Shakespeare. London: Cambridge University Press.

Zanettin, Federico. 2012. *Translation-Driven Corpora.* Abingdon, UK: Routledge.

Zeller, Hans. 1975. "A New Approach to the Critical Constitution of Literary Texts." *Studies in Bibliography* 28: 231–63.

Re-Modeling the Edition:
Creating the Corpus of Folger Digital Texts[1]

Rebecca Niles
Folger Shakespeare Library
rniles@folger.edu

Michael Poston
Folger Shakespeare Library
mposton@folger.edu

Jerome McGann's "Rationale of Hypertext" foresaw a world where hyperediting creates complex networks of reference and commentary that are navigated in electronic form. Almost twenty years later, a competing goal is apparent: to forego rich encoding in order to quickly produce large corpora of texts. Where before there was the rush to create digital facsimile collections, now there is a sprint to add searchable transcriptions. We have noticed this trend, both in our work at the Folger Shakespeare Library on transcription projects like Early Modern Manuscripts Online, and anecdotally through conversations with other scholars.[2] Readers, it seems to us, are willing to use

---

[1] This paper evolved from a conference presentation conceived of by Michael Poston and Rebecca Niles, and delivered by Niles on the Modern Language Association Committee on Scholarly Editions panel entitled "The Data is the Scholarship" (at the MLA Convention of 2014 in Chicago). The authors wish to thank Julia Flanders for organizing this panel, and to thank Michael Ullyot, Laura Estill, and Diane Jakacki for their help in preparing this paper for publication in the present volume. Appreciation also goes out to the anonymous peer reviewers who took the time to review our contribution and provide helpful commentary. A special thanks to Dr. Alan Galey, for many insightful discussions. Finally, the authors wish to thank the Folger Shakespeare Library and Eric Johnson (Director of Digital Access) for institutional support.

[2] Julia Flanders identifies a similar phenomenon in the development of digital collections: "The rhetoric of abundance which has characterized descriptions of digital resource development for the past decade or more has suggested several significant shifts of emphasis in how we think about the creation of collections and of canons. It is now easier, in some contexts, to digitize an entire library collection than to pick through and choose what should be included and what should not: in other words, storage is cheaper than decision-making" (2009, par. 5).

whatever digital texts are available.³ Minimal editing of large text corpora is forgiven by some simply because the very fact that the works of non-canonical authors are available at all has revolutionized the field (Hirsch 2011, 574–75). For others, minimal editing may be a feature. Renewed interest in textual witnesses and primary source documents, initiated in part by the social editing/unediting movement, has contributed to the popularity of the digital archive. Others are consoled by the thought that "good enough" data parsed by "good enough" software can still raise meaningful questions.

Regardless of whether coding constitutes a scholarly act,⁴ we firmly believe that *encoding* is necessarily scholarly. Whether by intention or not, textual transmission always involves transformation. Semantic encoding relies on interpretive judgments about the meaning and intent of the text (including not only its literary but also its bibliographic codes). Even in cases that purport to offer mere transcriptions of early witnesses, the fact remains that, as D.F. McKenzie once pointed out, "every society rewrites its past, every reader rewrites its texts, and, if they have any continuing life at all, at some point every printer redesigns them" (1985, 25). It is worthwhile to contemplate, in the creation of digital files, how responsibilities are shared between editors and encoders, and what editorial methods emerge in the process.

When working with a corpus rather than an individual title, the stakes are magnified. Franco Moretti describes the problem of scale in relation to world literature:

> The sheer enormity of the task makes it clear that world literature cannot be literature, bigger; what we are already doing, just more of it. It has to be different. The *categories* have to be different. [...] That's the point: world literature is not an object, it's a *problem*, and a problem that asks for a new critical method: and no one has ever found a method by just reading more texts. (2013, 46)

---

³ Consider all the projects that make use of the EEBO-TCP (Early English Books Online--Text Creation Partnership) corpus, or the propagation of the nineteenth-century Moby edition of Shakespeare through the MIT Shakespeare site, Open Source Shakespeare (OSS), and WordHoard. The former, while admittedly a profoundly influential resource, is composed of texts that receive little individual attention, and the latter is widely accepted as critically lacking in the benefits of modern textual editing.

⁴ For more on this subject, Ramsay and Rockwell (2012) are particularly comprehensive.

It is a problem for editors and encoders as well as readers. Editing for the web is very similar to editing for print; it relies on many of the same techniques, and seeks to achieve many of the same aims. But digital formats facilitate, if not encourage, new ways of reading, which editors must acknowledge and for which they must accommodate. This paper will discuss the ways encoders fulfill the roles of editors when creating individual texts as well as entire corpora, and consider the implications of their decisions.

We use our experiences as the encoders and editors of Folger Digital Texts (FDT) as a guide and illustrate our points, but our analysis is not confined to this project. FDT offers several useful windows into the subject. It deals with Shakespeare's plays, a topic that has been the locus of much of the development of modern editorial methodology. Consisting of thirty-eight plays (as well as 154 sonnets and three longer poems), the sample is small enough that each text can receive significant attention, yet large enough to highlight the difficulties of creating a corpus. Since the source for FDT is the New Folger Library Shakespeare Editions, a modern-edited series, we are keenly aware of the differences in editing methodologies for print and digital audiences.[5] Finally, since the motivation for FDT is to provide access to the text rather than to promote a specific research agenda, its challenges serve to expose the impossibility of a "neutral encoding."

The Digital Edition

Fredson Bowers prophesied that "the final and authoritative form of Shakespeare's text ... awaits an electronic birth" (1966, 178). We've reached the electronic age, and we're still waiting. Editors have largely given up on the goal of the "authoritative form" (in the sense of the form of the text as the author intended, free from the incidents of transmission) as neither achievable nor desirable. But another of Bowers's desires remains relevant: the hope for a complete works of Shakespeare in which an overarching, consistent editorial framework would be applied across the entire corpus. In 1966, he doubted the practicality of such a venture, given the volume of labor required in relation to the economic limits of the publishing model of his day (148–49). This goal has of course been realized without the explicit

---

[5] The series editors, Barbara Mowat and Paul Werstine, generously provided their insights throughout the development of Folger Digital Texts, and were also closely involved, particularly during the early stages of the project. The following analysis of the editorial structure of the Folger Editions and how it was translated into semantic encoding would be impossible without their contributions, for which the authors are continually grateful.

aid of digital technology, as is evident in a number of print edition series and collected volumes. Nonetheless, today's electronic editing and publishing platform provides the opportunity for a hitherto unrealized level of uniform consistency in that it relieves the human editor of a great deal of labor in establishing this consistency, and offers a "late-binding" rather than "early-binding" publication model that affords the flexibility to retroactively impose consistency across a corpus (Sperberg-McQueen 2009, 29–30).[6]

Bowers's desire for consistency is characteristic of a centuries-long trend in editing, particularly of Shakespeare: the systematization and regularization of the text. Alexander Pope attempted it in his 1725 *Works of Shakespear*, and his efforts had a lasting impact on subsequent generations of editors (Murphy 2003, 8). This tendency appears to be symptomatic of the editorial assumption that "behind the obscure and imperfect text is a clear and perfect one and it is the editor's job not to be true to the text's obscurity and imperfection, but instead to produce some notional, platonic ideal" (Orgel 1999, 118). This assumption is embedded in the editorial methodologies of the New Bibliography movement, which was both born of and had a profound effect upon the modern editing of Shakespeare. Under these auspices, editors regularize spelling, speech headings, and sometimes even meter. They systematize layout and reference schemes. They standardize the representation of prose, verse, and poetry within each play. Some of these interventions are explicitly delineated in the form of textual notes, but many decisions are made without comment—a fact that readers who are at home in the architecture of the critical edition often pass without noticing.

We have since witnessed a serious methodological shift as a product of the social editing and unediting movements of the 1980s and 1990s; McKenzie with his "sociology of the text," McGann with his concept of "bibliographic codes," as well as the interventions of scholars such as Leah Marcus and Randall McLeod, and many other investigations into the materiality of the text, have largely demolished the editorial rationale of ideal text in favor of a more inclusive perspective on textual instability. Out of this methodological shift is born the concept of the edition-as-archive, which takes advantage of

---

[6] Adapting terminology from computer science, C.M. Sperberg-McQueen uses the term "early-binding" to describe a publication medium in which variables of content and presentation must be established early in the process, and are therefore relatively unresponsive to the needs of particular audiences, and "late-binding" to describe a medium in which these variables may remain undetermined until potentially as late as the moment the content is retrieved by the user.

the affordances of hyperediting and the technological ability to capture and store large amounts of digital documents. The archive allows the reader to engage with primary materials in ways significantly less mediated by editorial choices. This approach faithfully represents the mutability of literary works by making various early textual forms accessible, thereby avoiding, if not countering, the systematization and regularization of the text. But as Kathryn Sutherland has pointed out, it also trades the old imposition of editorial authority for the "seemingly endless deferral" of interpretation (1996, 13), which transfers a great deal of editorial responsibility onto the reader. Furthermore, on a practical level, even scholars like Marcus (whose work helped popularize the very term "unediting") recognize the edited text's role in encapsulating the research of past textual scholars and providing a stable basis for teaching, reading, and studying literature from previous centuries (1996, 3–4).

Despite the ideological shift away from prescriptive editing toward the theoretical promise of harnessing technology to present the unmediated record of the text, within the reality of digital textual editing remains the fact that every transcription is influenced by editorial biases. Which textual elements are significant, and how structural elements interrelate, are interpretive judgments. Thus, digital editions also function as analytical models for describing the text. Julia Flanders makes the case for this type of textual entity: "Editing, at bottom," she points out, "has always been a way of modeling a text or group of texts so as to make them available for certain kinds of analysis" (2009, 60). The digital text as analytical model is strengthened through abstraction. And, rather than resist or even subvert the rationalization of the text by exposing the mutability of its forms, the edition designed for analytical purposes is most successful when it is made up of "data whose encoding scheme has been carefully constructed to identify textual features unambiguously, to group them into categories which express useful logical distinctions, and to express relationships between features in a way that usefully models the structure of the text" (2009, 57). The basis for the edition that can perform analytical functions is its data model, which governs how an encoding scheme, whether in print or digital, identifies features, makes logical distinctions, and expresses relationships.

Strategies for transcribing, editing, and encoding a text are based on a perceived data model. Flanders defines a data model, as it pertains to the edition, as "a set of functional distinctions, that together express how information is represented, and how it is organized" (2014). Printed texts often reveal an implicit model through textual features, such as layout and formatting.

Sometimes the data model of an edition is formalized, but sometimes it is not. Furthermore, while there is theoretically unlimited flexibility in how a data model is established and applied, there is also a large potential for success or failure in both faithfully representing a text and profitably modeling its features, based on the data model's comprehensiveness and/or appropriateness.

Back in the midst of the print age, Pope himself presents an example of a data model gone wild. In his 1725 edition, he codes Shakespeare's text with symbols distinguishing the "most shining Passages" from the "Low" scenes. He diminishes, literally, the "excessively bad" passages by removing them from the flow of the play-text and printing them in small text at the bottom of the page.[7] Samuel Johnson's criticism of Pope's "asterisks, commas, [and] double-commas," is a rejection of Pope's data model, which represents and organizes the information based on aesthetic judgment rather than any objective rationale.

If it seems as though Pope overreached his editorial mandate, it may be, in part, because his data model is out of time, not out of place. Data models reflect not only the objectives of the editors and the structure and behavior of the text, but also expectations for how audiences will want to read the work. Edmund King suggests that Pope's editorial model may have been informed by classical editing practices (2008, 7); it seems clear from the criticism of Johnson, who complained that Pope's aesthetic interventions "preclude the pleasure of judging for ourselves," that regardless of whether aesthetic taste was a valid editorial rationale in the past, the contemporary audiences of Pope's edition had different expectations.

We can observe in modern editions of Shakespeare how even an act of regularization as seemingly innocuous as modernization of spelling imprints a model of the text that is particular to the editor and her or his time. As Margreta de Grazia and Peter Stallybrass have argued, without the modern sensibility toward rigid standardization of spelling and usage, Renaissance textuality contains semantic indeterminacy, or "slippage," and encourages interpretive free play, promoting not only multiple possible interpretations, but multiple concurrent interpretations. "Modernization," they continue, "requires that this slippage be contained" (1999, 12). In almost every decision pertaining to systematization and regularization, editors apply a slightly new data model, rewrite rather than represent the text, and sacrifice some

---

[7] For an analysis of the historical precedents and contemporary context of Pope's editorial interventions, see King 2008.

features of the early text, all in an attempt to achieve a text that suits the perceived needs of the readers of its day.

⁓

Folger Digital Texts (FDT) makes the plays, poems, and sonnets of the Folger editions (modern critical editions edited by Barbara Mowat and Paul Werstine) freely available online as TEI-compliant eXtensible Markup Language (XML) files.[8] The text, speech headings, and stage directions have been settled, modernized, and regularized, as one might expect. Our task, as we initially conceived of it, was simply to translate the editorial decisions of Mowat and Werstine into semantically descriptive tags. In other words, our initial expectation was that we could simply interpret the data model of the print editions and manifest it seamlessly in a digital format.

A number of outward signs indicate that the Folger editions operate on a data model that privileges and distinguishes certain information, and ontologies of information, over others. The plays are structured hierarchically through acts, scenes, and speeches. Indentation and alignment are used to distinguish between verse and prose. Speech headings are in small capital letters. Stage directions are italicized. Entrances are centered, delivery directions are often in-line, and exits and other stage business are right justified. Songs and poems are centered and indented according to the needs of the stanza. Foreign words and phrases are italicized. Special symbols denote significant editorial interventions. It is productive to semantically recreate the relationships that are expressed typographically in the text, because they provide a solid, and even somewhat nuanced, basis for our data model.

The underlying data model of the Folger editions is surprisingly consistent, given its twenty-year production process. While many other series of editions experience shifts in editorial rationales or style guides as general editors and production staff come and go in the project's life-span, the Folger editions maintain a remarkable level of editorial consistency through having the same two editors—Barbara Mowat and Paul Werstine—throughout the series' twenty-year-long development; the uniform typographical expression of the editions are thanks to the lead production editor, Stephen Llano,

---

[8] As per the Folger Shakespeare Library's agreement with Simon and Schuster, publisher of the Folger Shakespeare Library Editions, Folger Digital Texts includes the play texts, poems, and sonnets, but not the glosses, images, front matter, or back matter of the editions; textual notes from the original print titles are to be incorporated into Folger Digital Texts in the future.

who oversaw the entire series. Thus, deciphering and encoding the underlying data model was, for the most part, a straightforward process. Still, areas of uncertainty have crept in. One issue concerns the representation of the data model in print. Its typographic vocabulary is limited, and overloaded as a result. For example, a single distinguishing method — italic font — is used to indicate no fewer than six categories of content: headings, stage directions, foreign words and phrases, certain songs and poems, names and/or titles, and the content of letters.

Additional complications occur when mistakes are made by the printer, or when formatting is changed to look better, or simply fit, on the printed page. The Folger edition prints prose and short lines in line with the speech prefix, while verse begins a new line. However, there are places where short lines begin on a new line, or verse lines begin in-line. If too many of these disruptions occur, the edition may become an untrustworthy witness of its own data model.

As we began translating the data model of the print edition into a rational encoding scheme for FDT, relationships in the text that seemed clear became less distinct under scrutiny. The print text indicates the character(s) that deliver(s) each speech by including a speech prefix at the top, but precisely where speeches begin and end (and hence who delivers them) can be unclear. Consider the following passage from Much Ado About Nothing:

⌜CLAUDIO, *reading an*⌝ *Epitaph.*

> *Done to death by slanderous tongues*
> *Was the Hero that here lies.*
> *Death, in guerdon of her wrongs,*
> *Gives her fame which never dies.*
> *So the life that died with shame*
> *Lives in death with glorious fame.*

⌜*He hangs up the scroll.*⌝

Hang thou there upon the tomb,
Praising her when I am ⌜dumb.⌝
Now music, sound, and sing your solemn hymn.

Song

> *Pardon, goddess of the night,*
> *Those that slew thy virgin knight,*
> *For the which with songs of woe,*

> *Round about her tomb they go.*
> > *Midnight, assist our moan.*
> > *Help us to sigh and groan*
> > > *Heavily, heavily.*
> > *Graves, yawn and yield your dead,*
> > *Till death be utterèd,*
> > > *Heavily, heavily.*

⌜CLAUDIO⌝

> Now, unto thy bones, goodnight.
> Yearly will I do this rite.

(5.3.3–23)

An editorial intervention assigns the epitaph to Claudio (as is indicated by the superior half-square brackets around "CLAUDIO, *reading an*" at the top of this excerpt). But is the song part of Claudio's speech, or is it a new entity? Our data model, which attributes speeches to speakers, demands that we know. The content suggests that the song is separate: "Now music, sound, and sing your solemn hymn." The editors seem to agree, inserting "Musicians" into the scene's opening stage direction ("*Enter Claudio, Prince, and three or four* ⌜*Lords*⌝ *with tapers,* ⌜*and Musicians.*⌝", SD 5.3.0) and adding a new speech prefix when Claudio resumes his dialogue.

The relationship is less clear in this example from *As You Like It*:

HYMEN

> [...]
> Whiles a wedlock hymn we sing,
> Feed yourselves with questioning,
> That reason wonder may diminish
> How thus we met, and these things finish.

Song.

> > *Wedding is great Juno's crown,*
> > > *O blessèd bond of board and bed.*
> > *'Tis Hymen peoples every town.*
> > > *High wedlock then be honorèd.*
> > *Honor, high honor, and renown*
> > *To Hymen, god of every town.*
> > > (5.4.142–51)

It is unclear, in either the Folger edition or in its base text, the 1623 First Folio, whether or not Hymen sings the song. A conversation with Mowat and Werstine confirmed that they consider the song to be separate from the speech. However, since there are no musicians on stage, FDT cannot identify the "speaker." These examples reveal that the print text does not always supply sufficient detail for the encoding.

As these examples illustrate, ambiguities in the text require encoders to think like editors; not only because they choose how to describe the text, but also because successful encoders first need to recognize contentious areas even before they decide how to interpret them. Even though FDT began with an edited text, the methodology of the encoding raised questions that the print text could not answer. Even when encoders conceive of themselves as mere transcribers, their work will be influenced by assumptions and interpretative judgments. Describing the structure of a document is itself an editorial act.

There has been, in the past, a resistance to thinking of encoding as scholarly work. On the surface, it is difficult to identify the distinction between the work of encoding the edition and, say, typesetting a book. Stanley Wells, writing of the modernization of early modern drama from the spelling and punctuation standards of its day to those of our own, could very easily have been discussing the encoding of a scholarly edition:

> The argument about the scholarly validity of the practice itself has been attended by surprisingly little discussion of how, once modernization has been decided upon as a course, it should be carried out. The assumption appears frequently to have been that it is ... "merely a secretarial task." Anyone who has himself attempted to modernize the spelling of an Elizabethan author in a responsible fashion is likely to know that in fact it calls for many delicate decisions (1979, 3).

Our experience of the FDT project, from its early conception as a straightforward encoding exercise to a deceptively nuanced practice of editorial decision-making, mirrors Wells's observation regarding preparing a new edition of Shakespeare's text: while we ourselves initially conceived of the project as being something akin to "merely a secretarial task," we too soon faced a mounting set of "delicate decisions" that forced us to reconsider the role of the encoder. A closer examination of one unexpectedly problematic area, speech headings, may further illustrate this point.

Case Study: Speech Headings

Regularization of speech prefixes, as well as other textual features like stage directions, is a standard procedure for both print and digital editions. This practice, it seems, is largely based on the belief that these elements are not as sacred as spoken text; they can be rewritten as editors see fit. In defense of standardization over retention of early documentary form, R.B. McKerrow proposes the following approach: "To follow the original texts in this irregularity would [...] be unnecessarily confusing to a reader; and as, after all, the speech-prefixes are merely labels intended to show to whom the various speeches are to be attributed, it seems to me an editor's clear duty to treat them as labels and make the labels uniform" (quoted in de Grazia and Stallybrass 1999, 14). Two factors contribute to McKerrow's assertion: audience needs, and the sub-work status of speech prefixes. If speech prefixes function primarily as paratextual scaffolding to support the literary edifice of the work, the needs of the audience should govern the editor's choices, rather than faithful reproduction of source text(s).

Pope claimed of Shakespeare's characters that, "had all the Speeches been printed without the very names of the Persons, I believe one might have apply'd them with certainty to every speaker" (1725, iii). However, the ways in which speech prefixes were applied in early editions indicate otherwise. Some printings introduce ambiguity caused by indistinct use of speech prefixes; as de Grazia and Stallybrass observe, there are occasions in the *Macbeth* of the First Folio in which both "Seyton" and "Seyward" are referred to as simply "Sey.," making it effectively impossible to distinguish the two (1999, 14). Editors' attempts to disambiguate may only make matters worse; Sidney Thomas cites an example in which an editor of the second quarto of *Romeo and Juliet* appears to have attempted to disambiguate a speech prefix (probably "Man"), and ends up inserting the Nurse's man, Peter, into Act 5, scene 3, when it should be Balthasar attending Romeo to Juliet's grave (1999, 18).

Assuming the editor can positively identify the speaker, there is the matter of consistently conveying that information to the reader. A common problem in early texts is the use of multiple labels to refer to a single character. For instance, Lady Capulet of *Romeo and Juliet* (Q2) is referred to variously as "Wife," "Old La.," "Capu. Wi.," "Ca. Wi.," "La.," "Mother," Mo.," and "M.," (McKerrow 1997, 4). This tendency occurs in other plays: *Midsummer Night's Dream, Love's Labor's Lost, All's Well that Ends Well,* and *The Merchant of Venice* are some of the most notable culprits (McKerrow 1997, 4-5). Standardizing Shakespeare's speech prefixes has thus become a perennial concern. As far back as the First

Folio there are "easily demonstrable efforts to regularize varying labels for the same character" (Long 1997, 21). Even Pope attempted to achieve some consistency in the ways he "apply'd the very names of the Persons."

Regardless of attempts by modern editors, the problem nonetheless persists. There are still many modern instances in which a single token designates more than one character, or multiple tokens designate the same character. These issues are greatly exacerbated if one considers the corpus as a whole, instead of an individual play. For instance, in the Folger edition of *2 Henry IV*, the character known first as "Prince Hal" is later called "King" (and then "King Henry" in *Henry V*). This is doubly confusing, given that there is another character also named "King" in *2 Henry IV*, one who was known first as "Bolingbroke" and then as "King Henry" in *Richard II*. The character called "Clifford" in *3 Henry VI* is not the same character as the senior Clifford, his namesake in *2 Henry VI*; rather, he is called "Young Clifford" there. The elder Clifford is slain by the man known first as "Richard" Plantagenet, and then as "York," in *1 Henry VI*, *2 Henry VI*, and *3 Henry VI*. Interestingly, Richard's son, also named Richard, is always identified as "Richard," even after he becomes Duke of Gloucester, and later King Richard III.

The issue is more complicated for minor, functional characters (those without a given name). McKerrow observed that "the designations of ... persons of secondary importance," as he describes them, tended to vary in the early witnesses (1997, 1). Even in modern editions these minor characters are handled much less precisely than main characters. In the Folger editions, there are several characters referred to simply as "Man," and even one known only as "One" (*Henry VIII*, 5.3). Is the "Messenger" in one scene the same as the "Messenger" in another? Is the "First Citizen" in the first scene of *Coriolanus* the same as the "First Citizen" in Act 2, scene 3? Worse, "Citizens" enter and exit throughout Act 2, scene 3 until "the Plebeians" enter (SD 2.3.170.1). The term "Plebeians" clearly refers to the "Citizens" because the First, Second, and Third Citizens promptly speak (implying that they entered as part of the "Plebeians" group). Are these the same characters as the First, Second, and Third Citizens at the beginning of the scene? Similarly, in *Macbeth*, Act 3, scene 1, "Attendants" enter (SD 3.10.1), but a "Servant" remains (SD 3.1.47.1) and speaks (3.1.50). Thus, it appears that the editors' standardization of speech prefixes does not always extend to character references in stage directions.

In the Folger editions, the methodology for regularization suits the editors' goal: to help the reader imagine the text in performance. As Mowat and

Werstine state in their introduction to each text, "[t]his edition differs from many earlier ones in its efforts to aid the reader in imagining the play as a performance rather than as a series of fictional events."[9] With this in mind, speech headings are comprehensively standardized. Stage directions are added and updated to clarify entrances, exits, and other performed actions. Even so, this did not give us as encoders enough information to fully disambiguate characters. In part, this is because it is impossible to fully disambiguate characters. But more than that, there is a limitation to the methodology. Because the locus of regularization occurs inside the paratextual element itself (the speech heading), it has the effect of privileging only those characters who speak. The "Characters in the Play" at the front of each edition exposes this bias, as it lists speaking characters individually, while others are lumped together in a paragraph at the end, despite natural relationships or affinities. See *Antony and Cleopatra*, where "Soldiers" and "Servants" who have speaking parts are listed separately from "Soldiers" and "Servants" who only have stage business. And, as discussed above, the standardization does not always extend to references to functional characters in stage directions.

FDT preserves the text of the speech prefixes from the print edition, but it also creates a unique, stable XML identifier for every character referenced (in the speech headings as well as in stage directions). The editors' rationalization of character names helped facilitate the creation of universal identifiers, but numerous questions were left unresolved, even after we decided to group effectively identical character types together, such as Plebeians and Citizens, or Servingmen and Servants.[10] The following stage directions highlight ambiguities that, while understood by readers, do not translate well into unique, precise, character references: "Enter the King of England and all his train before the gates" (*Henry V*, SD 3.3.0); "Enter the two French Lords and some two or three Soldiers" (*All's Well that Ends Well*, SD 4.3.0); "Enter King and two or three" (*Hamlet*, SD 4.3.0). FDT struggled to preserve productive ambiguity in a format that favors consistency and precision.

---

[9] This phrase appears in every book in the Folger Editions series; see, for instance, *A Midsummer Night's Dream*, updated edition (New York: Simon and Schuster, 2014), liv.

[10] But are "Plebeians" really equivalent to "Citizens," despite possible linguistic, rhetorical, or even sociopolitical distinctions between the two labels? Are "Servingmen" and "Servants" truly identical, despite the difference in gender distinction? Many researchers would say no, and the very fact that the functional equivalencies that allow us to best represent stage activity programmatically may be unacceptable for other research goals demonstrates one of our central points, that an edition's data model is necessarily shaped by some form of editorial bias.

Imposing consistency and precision is difficult work for one edited play, let alone thirty-eight. Imagine the difficulty starting from hundreds of unedited works. Still, this is necessary work. It is essential for technological reasons; creating reading and searching interfaces is easier for documents with standardized encoding schemes. Moreover, it is essential work for scholarly uses. People need to search not only the text, but elements of the text. The practice of "distant reading," for example, a form of scholarly use that has the systematic analysis of large corpora at its core, depends on the ability to "focus on units that are much smaller or much larger than the text: devices, themes, tropes—or genres and systems" (Moretti 2013, 48–49). The following sections demonstrate how encoding considerations affect the ways digital editions can be used.

Encoding for Distant Reading

In the 1997 article "What's the Bastard's Name?," the author breaks from a detailed analysis of character naming conventions to stage an exchange between a professor and a student. The professor, who is also editor of a new scholarly edition, answers questions about the names of certain Shakespearean characters in the tone of the benevolent dictator: "Our fixing of the speech tags is in all the editions. It's a *system*. It's for your own good." The exchange playfully but pointedly characterizes the attitude of the professor toward his student as a performance of dominance. The author invites us to consider textual editing as an assertion of aggressive control over the text, and the grotesque quality of the satirical exchange sends a clear message that, in the eyes of the dialogue's author at least, the prescriptive manipulation endemic to the regularized, systematized, scholarly edition is hubristic, perverse, even violent. This discussion is doubly relevant as we consider whether to attribute this satire to the name under which it is published—Random Cloud—or to regularize the author's name to Randall McLeod.

Imposing a strict editorial framework can seem aggressive, but it can be performed, we hope, in ways that honor the primary text while meeting the needs of its anticipated audiences. In FDT, our fixing of the speech tags is in all of our play-texts, in the sense that we both correct and stabilize character references in an attempt to make implicit features of the text explicit features of the encoding. It's not just a system, it's our system, designed, in a sense, for your own good: to support potential new uses by readers. That said, FDT is not a digital surrogate for the absent forms of the text that have come before it. Other digital editions can mitigate Random concerns by offering both the original and the edited text, allowing the reader to choose.

Some form of regularization may be needed to extend the range of capabilities afforded by the text. Digital editions may generate reading versions, but they also allow for a range of other functions. Open Source Shakespeare (OSS), for example, introduces a variety of mechanisms for searching and displaying the texts. Beside and beyond the traditional reading interface, the site offers a concordance, character lists, word counts, and numerous other statistical calculations. The success of these functions depends on the considerations that went into the encoding.

The OSS "Character List" function lists every character name, with play title, and the number of speeches attributed to that character. However, twenty-three plays assign speeches to a character named "All." "All" has sixteen speeches in *Coriolanus*, not including speeches assigned to "All Citizens" and "All Conspirators." *Timon of Athens* includes the characters "Some Speak" and "Some Others." More dangerous is when other apparently stable names lose their precision. The OSS character list contains only one "First Lord" in *As You Like It*, though there is one "First Lord" in Duke Senior's camp, and another in Duke Frederick's. In the histories, OSS conflates references to "Lord Clifford" in *2 Henry VI* and *3 Henry VI*, though they refer to separate characters; "Young Clifford" in *2 Henry VI* becomes "Lord Clifford" in *3 Henry VI*. A glaring example concerns Richard Plantagenet, Duke of York, father of Edward, George, and Richard. The list identifies 166 speeches by "Richard Plantagenet (Duke of Gloucester)" in *1 Henry VI*, *2 Henry VI*, *3 Henry VI*, and *Richard III*. Even allowing for the misattribution of "Gloucester" for "York," the elder Richard dies in *3 Henry VI*, making his subsequent appearance in the timeline that continues in *Richard III* impossible. The speeches in Richard III belong to the young prince, his grandson, who also appears in the list as "Duke of York" (with six speeches). "Richard Plantagenet the Younger" has six speeches in *2 Henry 6*; the same character, referred to as "Richard III (Duke of Gloucester)," has 409 speeches in *3 Henry 6* and *Richard III*.

OSS purports to search for all speeches by a "character," but in fact returns speeches assigned to a given token (i.e., the phrase that appears in the location of the speech prefix). "First Lord" does not always refer to the same lord, and "Richard" does not always mean the same Richard. OSS does use an edition where the speech prefixes have been standardized. And the problem is not with the nineteenth-century Moby edition; as shown above, the Folger edition would yield a similar result. The speech headings, perfectly acceptable in print editions, are not sufficient for disambiguation in sites like OSS.

What has changed in the move from the print to the digital environment? It is tempting to say that the computer, as an entity distinct from human readers, constitutes a new audience. But that's not quite right. A traditional edition already caters to several audiences, each of which processes the text differently. Consider again the stage direction, "Enter the King of England and all his train before the gates" (*Henry V*, SD 3.3.0). Who comprises the train? Students may imagine an amorphous group of lords and soldiers attending the king. Scholars may apply their knowledge of the historical context of *Henry V* to include specific characters like the Duke of Bedford. They may examine conventions of printing and performance to consider, for example, if "train" is shorthand for the group members who appeared with the king a few scenes ago. Performers might see this as a direction to fill out the stage with actors who aren't otherwise occupied. Casual readers may not give it much of a thought at all.

How would a computer interpret the direction? Strictly speaking, it wouldn't; computers have neither the intellectual capacity nor experiential basis for such a judgment. People create computer programs to interpret texts in certain, specific ways. These programs, as Susan Schreibman puts it, function as "extra intelligence [that] allows computers to more effectively locate and process semantic textual units" (2013). The distinction between computer-as-tool and computer-as-reader is critical, despite the simplicity and romance of the metonymic conflation of the latter. This holds even as more and more work is pushed into algorithms. Techniques such as unsupervised learning and statistical modeling may appear to allow for computational interpretation, but still the scripts are written, validated, and used by human readers.[11] Therefore, although the creation of the electronic document, its transmission over the internet, its processing by an algorithm, and its graphical rendering to a reader are all performed by computers, we choose to focus on the motives of the human users rather than perceived limitations of electronic environments.

After all, it is possible for computers to make use of ambiguity or multiplicity of information. This is part of the beauty of McGann's original vision for hyperediting, insofar as multiple witnesses can be made "simultaneously present to each other" (1997, 22). Many digital texts (for instance, *Internet Shakespeare Editions*) highlight and unpack textual variants, in much the same way

---

[11] "Unsupervised learning," as it is being used here, refers to a type of machine learning in which an algorithm searches for patterns without being given rules defining the pattern.

as critical editions in print have in the past. Experimental interface forms, such as Alan Galey's "Animated Variants" on his site *Visualizing Variation*, present us with other compelling examples of how thoughtful interface design can be used not only to create human-readable expressions of the power of encoded text to account for textual variation, but also to challenge the very concept of the definitive, or fixed, text (Galey 2014). If digital editions cannot accommodate ambiguity, the fault is not in our stars, but in ourselves.

No, what has changed in the transition to the digital environment is the potential loss of context. Computer processing makes it easier to read or reassemble texts in new ways, facilitating, if not encouraging, various forms of analysis. Texts can be also be taken apart, manipulated, and reconstituted into a variety of formats. As OSS demonstrates, they can appear in a list of search results, or as a concordance. Or they can be displayed as cue scripts or word maps. They can be represented as graphs of dramatic action (such as the FDT "charChart" API function, pictured in Figure 1). They can be scrutinized for metrical, linguistic, or rhetorical features. In short, they can be quantified, qualified, and summarized according to a vast array of criteria. Each of these activities carries the potential for isolating the content from the contexts readers are used to when encountering and interpreting such material.

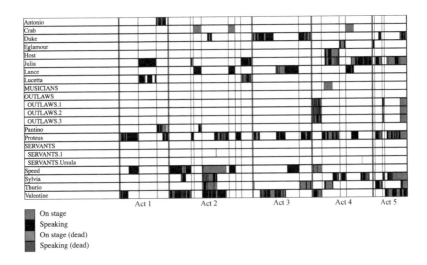

Figure 1. *Two Gentlemen of Verona*: The "charChart" API function gathers and visualizes data from stage directions (gray represents presence on stage, indicated in the text by entrances and exits) and speech attributions (black indicates lines delivered).

The context surrounding a speech provides necessary information to interpret a speech heading. Consider the speech prefix and subsequent line from *Hamlet*: "ALL Treason, treason!" (5.2.354). How should "ALL" be understood outside of the context of the play? FDT asserts that the following characters are on stage: Gertrude, Laertes, Hamlet, Claudius, Horatio, and Attendants. Do they all speak the line, as the speech prefix might suggest? Well, at 5.2.354, Gertrude is dead, Laertes is mortally wounded, Hamlet is committing the so-called treason, Claudius is the victim, and Horatio is unlikely to view it as treason. Therefore, the speech prefix "ALL" is encoded such that the only speakers responsible for the line "Treason, treason!" are the Attendants. It is up to the encoder, as digital editor, to include enough detail to compensate for the loss of contextual information that may occur when the text is abstracted into different forms. The more precise the encoding, the more discerning the encoder, the easier it is for those visualizations to avoid the pitfalls demonstrated by the OSS character list.

For this reason, while a character list generated from the Folger editions (i.e., based solely on the speech prefixes as printed) would look very similar to the OSS character list, the list for FDT is very different. The FDT character referencing system employs a more complex, but robust and flexible, attempt to represent ambiguities in the text while still tagging speeches and stage directions as comprehensively as possible. Named individuals retain their names, and all character identifiers have play-specific suffixes, to distinguish among similarly named characters within the corpus: King Henry V is HenryV_H5. Other characters may be treated as members of groups, with a decimal system allowing for the inclusion of only as much information as is present. FDT's *Henry V* uses SOLDIERS.ENGLISH_H5 and SOLDIERS.FRENCH_H5 to refer to groups, as well as SOLDIERS.ENGLISH.Bates_H5 and SOLDIERS.FRENCH.LeFer_H5 to refer to specific characters who may operate as part of the group or individually. Special designations .X and .0 are used as the lines between individual and indistinguishable members of a group begin to blur.[12]

The challenge is not merely to express ambiguity in an XML framework that prefers (but does not require) certainty. It is necessary to come to terms with the levels of abstraction needed to prepare a usable digital text. For FDT, our source edition is already an abstraction. It combines text from various witnesses, then modifies elements such as speech prefixes and stage directions

---

[12] Please visit www.folgerdigitaltexts.org/fdt_documentation.pdf for documentation of FDT's character referencing system, as well as the rest of its encoding scheme.

to help explicate the dramatic as well as literary functions of the play. Our encoding goes one step further. Converting the Folger editions to a digital form, FDT standardizes the standardizations, creating an abstraction of the abstractions. While the print edition steers the reader toward an imagined performance, the digital edition attempts to represent the range of possible actions that may occur within that performance. It is not so much that digital editions like FDT contain new information, but rather that they take on some of the decision-making that is conventionally left to human processing. Features made apparent in print through formatting (distinguishing between verse and prose, for instance) are explicitly denoted through markup. Questions left unanswered must be contained, if not resolved.

This building of abstraction on abstraction has something in common with distant reading. Moretti sees "distance ... as a condition of knowledge," with distance defined as analysis that is "'second hand': a patchwork of other people's research, without a single direct textual reading" (2013, 48). Our work with FDT was based on Mowat and Werstine's research. We have collaborated with Martin Mueller to create a WordHoard version of the Folger Editions texts mediated through FDT.[13] Future users may annotate Mueller's adornment of our encoding of the Folger editions of Shakespeare's quartos and folios: knowledge increasing with distance.

Pushing the point further, Moretti claims, "Reading 'more' is always a good thing, but not the solution" (2013, 46). We have discussed difficulties encoding FDT, a small corpus comprised of previously edited texts. How can we responsibly transcribe larger collections of primary documents into useful corpora? Is there a corollary here, that editing "more" is always a good thing, but not the solution? Is there a way to reliably standardize and encode collections of texts, without relying on an editor's close reading and analysis of every document?

Distant Editing

Consider "distant editing" an analog to distant reading, where encoders draw on existing texts, devise a data model to describe them, and restructure the texts accordingly. The process need not be automated, but much of it likely

---

[13] WordHoard is a project out of Northwestern University that uses scripts (such as MorphAdorner) to add metadata (such as lemmatization and part of speech tagging) to texts. In the past, as with many other digital Shakespeare projects, WordHoard used the Moby edition as its source text. Visit http://wordhoard.northwestern.edu for more information.

should be, if for no other reason than expedience. Subsequent scholars, editors, encoders, and readers can build on this work, annotating them, linking between them, moving toward a more realized edition. We concede that the initial transformation may not, strictly speaking, result in an "edition," with the full weight that term carries. Perhaps "distant encoding" would be a better term, even if we do believe that encoding is editorial work. Nonetheless, in the same vein that "distant reading" is reading, we will use the term "distant editing."

As we have shown, the practice of encoding is similar in many ways to editing. Both depend on developing abstract understandings of the text, and then grooming the text accordingly. Both devise representations to best express the significant features of the text. Both processes are negotiations among the text, its structure, and the needs of an audience. Or, put another way, both accommodate the various layers of the text: its contents, data models, and style sheets.

The features of the text, balanced by the constraints and affordances of the editorial medium, shape the data model. The editor must be able to abstract the key features of a text into a definable, describable system. This can be difficult when the source material lacks a consistent structure, or if it has competing structures. The blank page has relatively few constraints; marks can be placed anywhere without regard for order or consistency. The lack of structure imposed by the medium may convolute the manifestation of the data model. Expressing the data model in a digital form, on the other hand, offers more transparency, but it must fit within rigid constraints. These constraints may come from the editor, or they may be imposed externally (e.g,. by ISO standards). Nevertheless, with databases and XML files, editors define a schema that at first describes, but then ultimately prescribes, the structure of the model. In Willard McCarty's terminology, once the abstract "model of" a text becomes codified in the schema's "model for" the text, it can be difficult to handle exceptions or oddities as they arise.[14]

Once defined, the data model directs the alterations to the text. The model defines whether elements are substantive or accidental, textual or paratextual. The model grounds the methodology for standardizing the structure or regularizing the text. As an edition transitions from print to digital, the data model may need to be refined. As individual editions coalesce into corpora, the data model may need to be expanded in order to account for a wider

---

[14] That is to say, as McCarty so skillfully adapted it for the digital humanities from Clifford Geertz; see McCarty 2004.

range of textual features in a consistent manner. New "models for" may need to be developed.

The primary contribution of editors is the presentation of the content, filtered through the data model. A style sheet renders the editors' work in a form accessible to audiences. The print edition is the application of one particular style sheet to the edited text. An XML file is both a repository of content and a record of the data model.

Any number of style sheets can be applied to the XML file (or database, or other digital structure), to visualize the text in any number of ways. When the data model is apparent and the content is available, interpretations can be made and recorded on a global scale. Scripted analysis and alternate visualizations can point to areas needing further attention. They can highlight errors in an automated scripting of the text. Or they can facilitate other editorial functions, such as standardization and regularization of the content. We will discuss some specific examples from FDT shortly.

The Folger editions were assembled word by word, with careful deliberation. FDT, at least initially, was mostly a mechanical process. To a large extent, the data model of the print edition was adapted to the digital format in a relatively easy manner. We looked at the collection of plays, cataloged the editorial judgments, and found a counterpart representation in TEI. The bulk of the conversion process was scripted. We copied the text from PDF files, and a computer program used formatting clues to identify act and scene headers, speeches and speech prefixes, stage directions, songs and poems, and verse/prose lines. To generate first drafts, a minimum of human intervention was needed, to identify foreign words, to describe stanzaic structures, and to link shared lines. The character reference information was added later.

The process for other projects appears to be similar. For OSS, Eric Johnson started with digital versions of the Moby Shakespeare texts, analyzed the structure of the plays, created a database schema to contain it, and performed a bulk upload.[15] His "first challenge was to get the texts into a uniform order," to facilitate the function of his parser. The challenge was not to settle textual issues, catalogue variants, or research the editorial tradition. Like FDT, OSS employed a distant editing methodology to interpret and annotate an existing edition. A parallel can be found with WordHoard. Matthew

---

[15] As described in "The Editing and Structure of Open Source Shakespeare" (http://www.opensourceshakespeare.com/info/technicaldetails.php), part of Johnson's M.A. thesis, *Open Source Shakespeare: An Experiment in Literary Technology* (2003–5).

Jockers notes the irony that WordHoard bills itself as "an application for the close reading and scholarly analysis of deeply tagged texts," while enabling the quantification more often associated with distant reading practices.[16]

After the initial scripting, we used a number of techniques to validate FDT files. We created a style sheet to reproduce the appearance of the print edition (by using the same formatting rules to express the data model as we understood it), so that an optical collation exercise could highlight areas where the texts didn't align. It pointed to simple errors (mismatched alignment indicating that verse/prose was incorrectly tagged), as well as to larger deficiencies in our understanding of the Folger data model (for example, can prologues and epilogues be classified as acts or scenes, or must they be handled separately?). Ultimately, by seeing how closely our visual model of the print edition matched the edition itself, we were able to verify not only our transcription and encoding, but our data model as well. In this way we could validate our own work from a distance, proofread without reading.[17]

While this practice merely tested conformance to another edition, other exercises impacted the edition (print and digital) more substantively. As early drafts of the XML files became available, we created alternate visualizations of the text. Unexpectedly, when we used concordances to help identify OCR errors in the transmission process, we discovered inconsistencies in how the source's editors regularized words. The contraction for "thou art" appears as either one or two words ("thou'rt" or "thou 'rt") throughout the print editions. On the advice of the editors, it is regularized to the two-word form in the digital files, and this change will be reflected in the print editions going forward. Similarly, we created graphs of when characters appear on stage, initially to help us understand, and later to help us audit, stage directions like "They exit." In some cases, we even found mistakes in the print edition where characters did not appear to exit (the Sailor in *Othello*, 1.3.20), or when characters entered a second time without having first exited (the Abbess in

---

[16] Jockers writes: "Although it is true that WordHoard provides access to, or tools for, harvesting richly encoded texts, the results being gleaned from the texts are not so much the results of a close reading-like process that aggregates a number of relatively small details into a more global perspective. As such, the process seems to have less in common with close-reading practices and more with Moretti's notion of distant reading. The devil is in the details and in how the details are investigated and aggregated in order to enable a larger perspective" (2013, 22).

[17] This methodology, of course, was used *in addition* to standard practices for preparing scholarly editions, including rigorous word-by-word proofreading.

The Comedy of Errors, 5.1.116). New stage directions were inserted into the digital text, and they will appear in future updates to print editions as well.

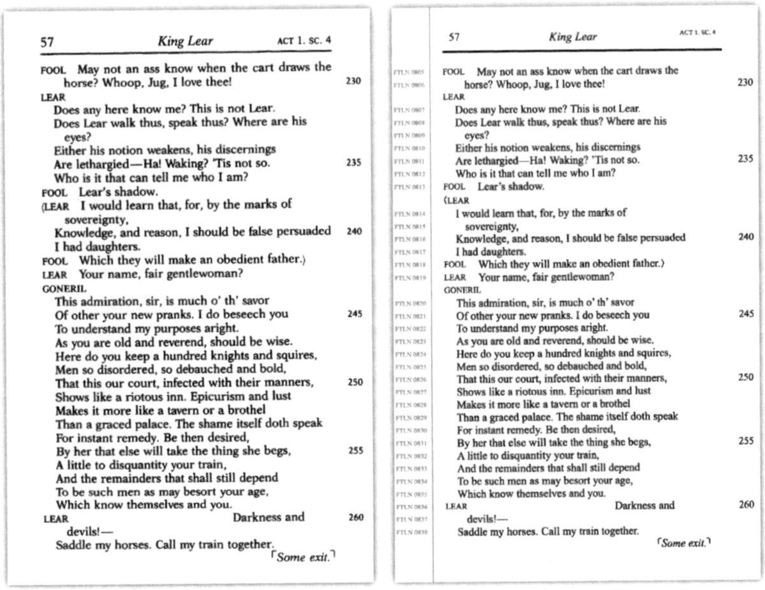
57 *King Lear* ACT 1. SC. 4

FOOL May not an ass know when the cart draws the horse? Whoop, Jug, I love thee!
LEAR
Does any here know me? This is not Lear.
Does Lear walk thus, speak thus? Where are his eyes?
Either his notion weakens, his discernings
Are lethargied—Ha! Waking? 'Tis not so.
Who is it that can tell me who I am?
FOOL Lear's shadow.
(LEAR I would learn that, for, by the marks of sovereignty,
Knowledge, and reason, I should be false persuaded
I had daughters.
FOOL Which they will make an obedient father.)
LEAR Your name, fair gentlewoman?
GONERIL
This admiration, sir, is much o' th' savor
Of other your new pranks. I do beseech you
To understand my purposes aright.
As you are old and reverend, should be wise.
Here do you keep a hundred knights and squires,
Men so disordered, so debauched and bold,
That this our court, infected with their manners,
Shows like a riotous inn. Epicurism and lust
Makes it more like a tavern or a brothel
Than a graced palace. The shame itself doth speak
For instant remedy. Be then desired,
By her that else will take the thing she begs,
A little to disquantity your train,
And the remainders that shall still depend
To be such men as may besort your age,
Which know themselves and you.
LEAR Darkness and devils!—
Saddle my horses. Call my train together.
⌜*Some exit.*⌝

57 *King Lear* ACT 1. SC. 4

FOOL May not an ass know when the cart draws the horse? Whoop, Jug, I love thee!
LEAR
Does any here know me? This is not Lear.
Does Lear walk thus, speak thus? Where are his eyes?
Either his notion weakens, his discernings
Are lethargied—Ha! Waking? 'Tis not so.
Who is it that can tell me who I am?
FOOL Lear's shadow.
(LEAR
I would learn that, for, by the marks of sovereignty,
Knowledge, and reason, I should be false persuaded
I had daughters.
FOOL Which they will make an obedient father.)
LEAR Your name, fair gentlewoman?
GONERIL
This admiration, sir, is much o' th' savor
Of other your new pranks. I do beseech you
To understand my purposes aright.
As you are old and reverend, should be wise.
Here do you keep a hundred knights and squires,
Men so disordered, so debauched and bold,
That this our court, infected with their manners,
Shows like a riotous inn. Epicurism and lust
Makes it more like a tavern or a brothel
Than a graced palace. The shame itself doth speak
For instant remedy. Be then desired,
By her that else will take the thing she begs,
A little to disquantity your train,
And the remainders that shall still depend
To be such men as may besort your age,
Which know themselves and you.
LEAR Darkness and devils!—
Saddle my horses. Call my train together.
⌜*Some exit.*⌝

Figure 2. A page from *King Lear*, in which FDT (on the right) conforms almost perfectly with the original Folger edition page (on the left) save in one respect: line 238. FDT correctly identifies the line as a metrically complete one, and the style sheet sets it below the speech prefix. In this case, the visual difference points out an error in the source text.

These discoveries were unintended. But we quickly realized the value in this process for standardizing and regularizing the text. The creation of character lists (like the one in OSS) pointed to areas where the edition needed to better disambiguate between characters (the First Lords in *As You Like It*, the Cliffords and Richards in the histories). Nesting character identities within larger group identities (as with Bates and the Soldiers group in *Henry V*) prompted interesting editorial discussions: Are "captains" (e.g., *Antony and Cleopatra*, 2.7.16 SD) a type of soldier? Is there any semantic difference between "Mariners" (e.g., in *The Tempest*) and "Sailors" (e.g., in *Hamlet* or *Othello*)? Are "plebeians" and "commoners" also "citizens," as they seem to be in *Coriolanus*?

The ability to shift perspectives, to view the corpus at a glance and then zoom back into a closer reading, is immensely useful. If, as Flanders points out (2009, 57), "the most important area of research in text encoding [...] is the attempt to discover ways to represent all aspects of the text [...] in such a way as to give us a model of the thing that functions, analytically, better than the thing itself," shouldn't we build that analytical functionality back into the very methodology of the digital editing process? Scholars are already taking apart editions and reconstituting them in interesting ways. Why shouldn't editors? Programmatically analyzing their own work may help editors to expose flaws in methodology or implementation, refine judgments, expose outliers, and move toward consensus and consistency. Every visualization enables a new kind of review, potentially revealing errors, weaknesses, or limitations of the text.

∽

Editors have always constructed a data model in order to express to audiences key structural and semantic features in the text. Digital editions help to reveal explicitly, through semantic description, what may have been implicit in print, through typographical cues. And, to varying extents, editors have always adapted the material, regularizing or standardizing the text or structure, to fulfill the needs and goals of the data model. This is a process employed by print and digital editors alike. Some elements resist regularization. There may be a multiplicity of meanings, perhaps but not necessarily caused by ambiguity inherent to the text. Editions, digital and print, can enumerate all variants, all possible options, if that is what is desired.

The preference for an authoritative edition, which privileges one reading above others, is an editorial, not a technological choice. Even such an edition can convey productive ambiguity. It may compromise the data model, blur boundaries, and preserve the interpretive space that such phrases as "Some speak" open up. In these cases, the editor expects that the reader, given the appropriate context, will be able to tolerate, perhaps resolve, or even relish the ambiguity.

Johanna Drucker tells of the moment in the late 1990s when scholars debated the limitations of XML, shrugged, and went back to encoding (2012, 88). It is still an important debate, although perhaps a resolution may not be necessary. True, it may not be possible to create an edition that rises above the limitations of its encoding to be definitive and all-encompassing. But perhaps XML is useful *because* of its limitations.[18] Moretti and McCarty both acknowledge the value of a model by *virtue* of, not *despite*, its reductive

---

[18] We hope it is clear, however, that we unequivocally support Drucker's call for theory to take precedence in guiding the production of texts.

function (Moretti 2013, 49; McCarty 2004). Reductive models exist not as a side effect of ill-conceived technology, but as a realizable aid to scholarly understanding. Editors have always made choices, sacrifices; the job of the editor is to make hard decisions about the text, its content, and its structure. Why do we fault digital editions that offer a reductive model, that promote certain decisions over other ones? It seems we are back to Bowers, except this time, it is the electronic age that is awaiting "the final and authoritative form of Shakespeare's text."

As techniques evolve for reading out of context, or, more properly, for creating new contexts, editor-encoders test the usefulness of the edition, push the limitations of the data model, and expose the irregularities of their source texts. Acknowledging this is crucial if editors are to continue to serve their audiences. Encoding may need to supply context or offer interpretations that were previously deferred onto the reader. A formal distant editing methodology might better prepare texts for addressing these concerns, and would allow corpora to be quickly reconstituted as errors are revealed, as new avenues for standardization become available, as new data models become desirable.

WORKS CITED

Bowers, Fredson. 1966. *On Editing Shakespeare*. Charlottesville: University Press of Virginia.

De Grazia, Margreta, and Peter Stallybrass. 1999. "The Materiality of the Shakespearean Text." In *Shakespeare and the Editorial Tradition*, edited and introduced by Stephen Orgel and Sean Keilen, 1–29. Shakespeare: The Critical Complex. New York: Garland Publishing.

Drucker, Johanna. 2012. "Humanistic Theory and Digital Scholarship." In *Debates in the Digital Humanities*, edited by Matthew K. Gold, 85–95. Minneapolis: University of Minnesota Press.

Flanders, Julia. 2009. "Data and Wisdom: Electronic Editing and the Quantification of Knowledge." *Literary and Linguistic Computing* 24 (1): 53–62. doi: 10.1093/llc/fqn036.

———. 2014. "The Data is the Edition." Presentation at Modern Language Association convention, Chicago, January 10: "Innovative

Interventions in Scholarly Editing," special session arranged by the MLA Committee on Scholarly Editions.

Folger Shakespeare Library. 2014. *Folger Digital Texts*. Last modified June 1, 2016. http://www.folgerdigitaltexts.org.

Galey, Alan. 2014. *Visualizing Variation*. http://individual.utoronto.ca/alangaley/visualizingvariation/.

Gants, David. 2006. "Drama Case Study: The Cambridge Edition of the Works of Ben Jonson." In *Electronic Textual Editing*, edited by Lou Burnard, Katherine O'Brien O'Keeffe, and John Unsworth, 122–37. New York: Modern Language Association of America.

Hirsch, Brett. 2011. "The Kingdom has been Digitized: Electronic Editions of Renaissance Drama and the Long Shadows of Shakespeare and Print." *Literature Compass* 8/9: 568–91. doi: 10.1111/j.1741-4113.2011.00830.x.

Internet Shakespeare Editions. University of Victoria. http://internetshakespeare.uvic.ca/.

Jockers, Matthew. 2013. *Macroanalysis: Digital Methods and Literary History*. Champaign: University of Illinois Press.

Johnson, Eric. 2003–5. *Open Source Shakespeare: An Experiment in Literary Technology*. M.A. thesis, George Mason University. http://www.opensourceshakespeare.org/info/paper_toc.php.

King, Edmund G. C. 2008. "Pope's 1723–25 Shakespear, Classical Editing, and Humanistic Reading Practices." *Eighteenth-Century Life* 32 (2): 3–13. doi: 10.1215/00982601-2007-002.

Long, William B. 1997. "Perspective on Provenance: The Context of Varying Speech-heads." In *Shakespeare's Speech-Headings: Speaking the Speech in Shakespeare's Plays*, edited by George Walton Williams, 21–44. Newark: University of Delaware Press.

Marcus, Leah S. 1996. *Unediting the Renaissance: Shakespeare, Marlowe, Milton*. London and New York: Routledge.

McCarty, Willard. 2004. "Modeling: A Study in Words and Meanings." In *A Companion to Digital Humanities*, edited by Susan Schreibman, Ray Siemens, and John Unsworth, 254–70. Oxford: Blackwell.

McGann, Jerome. 1997. "The Rationale of Hypertext." In *Electronic Text*, edited by Kathryn Sutherland, 19–46. Oxford: Clarendon Press.

McKenzie, D.F. 1985. *Bibliography and the Sociology of Texts*. The Panizzi Lectures. London: British Library.

McKerrow, R.B. 1997. "A Suggestion Regarding Shakespeare's Manuscripts." In *Shakespeare's Speech-Headings: Speaking the Speech in Shakespeare's Plays*, edited by George Walton Williams, 1–4. Newark: University of Delaware Press.

Moretti, Franco. 2013. *Distant Reading*. London: Verso.

Murphy, Andrew. 2003. *Shakespeare in Print: A History and Chronology of Shakespeare Publishing*. Cambridge: Cambridge University Press.

Northwestern University. 2004–2013. *WordHoard*. http://wordhoard.northwestern.edu/userman/index.html.

Orgel, Stephen. 1999. "What is an Editor?" In *Shakespeare and the Editorial Tradition*, edited and introduced by Stephen Orgel and Sean Keilen, 117–123. Shakespeare: The Critical Complex. New York: Garland Publishing.

Pope, Alexander. 1725. "The Preface of the Editor." In vol. 1 of *Works of Shakespear*, iii–xxiv. London: J. Tonson.

Ramsay, Stephen, and Geoffrey Rockwell. 2012. "Developing Things: Notes toward an Epistemology of Building in the Digital Humanities." In *Debates in the Digital Humanities*, edited by Matthew K. Gold, 75–84. Minneapolis: University of Minnesota Press.

Random Cloud [McLeod, Randall]. 1997. "What's the Bastard's Name?" In *Shakespeare's Speech-Headings: Speaking the Speech in Shakespeare's Plays*, edited by George Walton Williams, 133–209. Newark: University of Delaware Press.

Schreibman, Susan. 2013. "Digital Scholarly Editing." In *Literary Studies in the Digital Age: An Evolving Anthology*, edited by Kenneth M. Price and Ray Siemens. New York: Modern Language Association of America. doi: 10.1632/lsda.2013.4.

Sperberg-McQueen, C.M. 2009. "How to Teach Your Edition How to Swim." *Literary and Linguistic Computing* 24 (1): 27–39. doi: 10.1093/llc/fqn034.

Sutherland, Kathryn. 1996. "Looking and Knowing: Textual Encounters of a Postponed Kind." In *Beyond the Book: Theory, Culture, and the Politics of Cyberspace*, edited by Warren Chernaik, Marilyn Deegan, and Andrew Gibson, 11–22. Oxford: Office for Humanities Communication.

Thomas, Sidney. 1997. "McKerrow's Thesis Re-Examined." In *Shakespeare's Speech-Headings: Speaking the Speech in Shakespeare's Plays*, edited by George Walton Williams, 17–20. Newark: University of Delaware Press.

Wells, Stanley. 1979. *Modernizing Shakespeare's Spelling*. Oxford Shakespeare Studies. Oxford: Oxford University Press.

Collaborative Curation and Exploration of the EEBO-TCP Corpus

Martin Mueller
Northwestern University
martinmueller@northwestern.edu

Philip R. Burns
Northwestern University
pib@northwestern.edu

Craig A. Berry
Chicago, Illinois
craigberry@mac.com

EEBO-TCP and the Fate of Early Modern Studies in a Digital World

The topic of this essay is the collaborative curation and exploration of the digital versions of an extraordinary cultural heritage resource: English books printed between 1473—the date of the first English printed book—and 1700. Over the course of the past fifteen years, the Text Creation Partnership (TCP) has created 66,000 digital transcriptions of these books, adding up to approximately two billion words.[1] In January 2015, 25,000 of them moved into the public domain, to be joined in 2020 by the remaining 41,000. Five years from now, anybody with an Internet connection will have access anywhere and anytime to a "deduplicated" library that includes 92 per cent or more of

---

[1] That is a lot of words, but hardly amounts to "big data," as the term is now used. Think of it as 3,000 bibles. After the Civil War, the annual output of newspapers in the United States was about two billion words. Based on a subset of 44,000 texts (1.3 billion words) available to us in 2014, the distribution of the texts by length (in words) is as follows:

min	2.5%	10%	25%	50%	75%	90%	97.5%	max
7	300	600	1,300	6,300	25,000	68,000	197,000	3,100,000

Half the texts are the length of pamphlets. Three quarters are shorter than *Othello*, and only 10 per cent are "book length."

© 2016 Iter Inc. and the Arizona Board of Regents for Arizona State University.
All rights reserved
ISBN 978-0-86698-725-7 (online) ISBN 978-0-86698-557-4 (print)

distinct English books printed before 1700.[2] In fact, you will be able to store all these books on a smartphone or iPad and read them far away from the nearest Wi-Fi hot spot, as long as your batteries last. Hamlet's "table of ... memory" in a digital world is a very different thing from Montaigne in his tower with all his ~2,000 books in full view and at hand.

Digital full-text transcriptions of just about every distinct English book before 1700 will, over time, affect the documentary infrastructure of early modern studies in deep ways that we do not yet fully understand. If you live near a rare book library like the Bodleian, the Newberry, or the Huntington, or if your university subscribes to Early English Books Online (EEBO), you may not feel a need to change your habits of interacting with the "real" books or their facsimile images. But if you live in some small town in the English-speaking world, large chunks of past worlds are suddenly available.

Like other surrogates, digital transcriptions fall short of their originals in some ways but exceed them in others. A cheap reproduction of the Mona Lisa may be a pale shadow of the real thing, but you can look at it any time you want. And the powers of surrogates go beyond convenience: they may offer affordances that add value even under conditions of unlimited access to the original. Which brings us to the "query potential of the digital surrogate," a mouthful of a phrase that will guide the rest of this essay.

What can you do *with* digital surrogates of all English books before 1700 that you cannot do with their originals, and what do you (or somebody else) need to do *to* them so that the distinct potential of the surrogate is fully realized? An IBM executive once observed that people are smart but slow, while computers are dumb but fast. When Lear says that Cordelia's voice was "ever soft, gentle and low," human readers will recognize the sequence of three adjectives as a common literary device, but it would take them forever to

---

[2] "English" and "duplicate" are slippery terms. The English Short Title Catalogue includes many books not written in English or published in English-speaking countries. With early books it is not easy to decide when a book is a duplicate of another, and for some inquiries, every instance of a book is a unique thing. Paul Schaffner, in an email of October 1, 2014 to Martin Mueller, offered as a "good guess" that the project will fall short of "completion" by about 5,000 titles, but warned that much depends on how or what you count. By that calculation, EEBO-TCP (Early English Books Online–Text Creation Partnership) will be 93 per cent complete. Given the redundancy built into any human communication system, the loss of new information from that 7–8 per cent of missing titles is likely to be very small. But if scholars discover shocking gaps, there will always be opportunities for adding texts later.

track all occurrences across some 500 surviving plays by Shakespeare's contemporaries. Given a linguistically annotated corpus and a query language that lets you say, "Look for the sequence 'adjective, adjective, conjunction, adjective,'" a rather ordinary machine will take less than two seconds to range across twelve million words and retrieve 718 matches, including "dull, frosty, and wayward" as a description of the Scottish king in *Perkin Warbeck*.[3]

Readers can look at only one page at a time, although in their minds they may perform complex and almost simultaneous operations that associate the words on the page with many things seen, heard, or read. Machines can race through all pages of a book or many books virtually at once and retrieve patterns that meet precisely specified search criteria.[4] Exploiting the query potential of the digital surrogate is a matter of putting the dumb but fast machine in the service of the slow but smart reader. To maximize the query potential of the 66,000 digital transcriptions of English printed books before 1700, you need to transform them into a human readable and machine actionable corpus.[5] That transformation is a task for collaborative curation. It is likely to take quite a few years and will need to be the work of many hands.

A little of that work has been done by some hands at Northwestern University over the past few years. The three of us have worked with two generations of undergraduates on developing tools for annotation and collaborative curation and on getting ~550 TCP-encoded plays from the mid-sixteenth to the mid-seventeenth century into a shape that unlocks much of their query potential. We call this project Shakespeare His Contemporaries (SHC). The twelve million words of these plays add up to less than one percent of the EEBO-TCP corpus, but their mixture of verse and prose, learned, colloquial,

---

[3] This search was carried out with the BlackLab search engine of the Institute of Dutch Lexicography: https://github.com/INL/BlackLab.

[4] "Precisely specified" covers many things that readers would find difficult or impossible to find with any reasonable investment of time, such as words where the letter "a" is preceded and followed by an equal number of letters or words that occur 51 times, or—more sensibly—words that are relatively more common in this text than some other text(s).

[5] Linguists use "corpus" as a technical term with quite precise requirements for the proportions of texts and their annotation. We use the term more loosely: a corpus is any body of texts that have been organized in some fashion to make it easier to relate one book to another. Books scattered across a floor are not a corpus. Catalogued books are.

and vulgar discourse make them a good microcosm of the TCP collection.[6] Our experience with the SHC corpus has deeply shaped our view of what can or should be done with all the EEBO-TCP texts. Leaving aside outliers of various kinds, we believe that you can with considerable confidence generalize from the SHC corpus about the TCP corpus as a whole.

The Three Cs of Collaborative Curation

Curation of retrodigitized texts can be divided into the three imperatives of Connect, Complete, Correct. The second and third are, in principle, finite tasks. The first is endless: there will always be new ways of connecting or enriching data. While all three forms of curation can proceed simultaneously, it makes sense to focus first on completion and correction: scholarly readers have a very low tolerance for misspelled words or other manifest errors, even (or especially) where those errors do not interfere with their understanding of the text. And there are a lot of manifest errors in the EEBO-TCP corpus. But it would be a mistake to think of data curation primarily in terms of the categories of "fixing" and "cleanup." On the other hand, it is equally wrong to think that these mundane activities do not matter. They are forms of "textkeeping," and there is as much virtue in good textkeeping as in good housekeeping.[7]

"Connecting curation" is the key factor in using the digital medium to enhance as well as extend access. Extended access uses digital technology in an emulatory mode and thinks of it as bringing more books to more readers. Enhanced access thinks of digital technology as providing new tools for a

---

[6] While 1 per cent may not sound like much, the 172,500 distinct spellings found in the SHC corpus account for 97 per cent of word occurrences in our current corpus of 44,000 TCP texts. The SHC corpus is not, however, a balanced sample of the TCP corpus. Of the twenty most common spellings not in the SHC corpus, eight are abbreviations for books of the Bible, and the others are, in descending order, "righteousness," "baptism," "public," "quakers," "sanctification," "moche," "presbyters," "ordination," "sentiments," and "godliness." From which one might infer that the analysis of a corpus of 500 sermons would go quite a ways toward correcting a lack of balance.

[7] The term "dirty OCR" is illuminating in this context. It refers to text that has been transcribed algorithmically from page images. Such transcriptions have many errors. They are typically left uncorrected because, from the perspective of machine-based queries, the improvements in search results are not worth the cost of correction. But the uncorrected OCR is hidden from the reader because it is "dirty." In classical Latin, *laute*, or "washed," is a highly approving word for "elegant" or "sophisticated." Where texts of any value are seen by people, they should be "clean."

more sophisticated analysis of available materials. With enhanced access, the distinction between catalogue information "about" the book and information "in" the books becomes increasingly blurred.

A library catalogue is a paradigm of connecting curation. Six million books scattered in a warehouse are one thing, six million books properly catalogued and shelved quite another. Librarians refer to catalogue data as "metadata." Such metadata are typically about the object as a whole, and they are gathered in the special genre of the catalogue record, which is distinct from the object it describes. There is, however, no reason why such cataloguing should stop at the water's edge, so to speak, and not extend into the object itself.

Structural and linguistic annotation are de facto forms of cataloguing, although they are not usually called by that name. When the chapters of a novel in a TEI-XML version are wrapped in separate <div> elements or lines of verse are wrapped in <l> elements, the parts of a text are catalogued and each part can be marked with a "call number" or unique ID that adds speed and precision to subsequent and machine-based activities, such as retrieving words that only occur in verse. Similarly, when a text is linguistically annotated, every word is identified as a separate token and can be associated with various properties, such as its dictionary head-word (lemma) or its part of speech. Readers will think of a text as words in different locations. For the machine a text is a sequence of locations, each of which can be associated with an arbitrary number of positional attributes, including but not limited to the spelling of the word at that position. A word in a text or a book on a shelf each takes up a defined space that you can associate with arbitrary amounts of properties or metadata.

Structural and linguistic annotation of this kind is produced through some combination of algorithmic and manual work. The added information is typically added *in situ* or inline in a manner that makes it easy for the machine to distinguish between the source text and the annotation. Annotated texts of this kind are very hard for humans to read, but they are not intended for human readers. They are intended to be processed by machines. Think of a catalogued and annotated digital corpus as a triple-decker structure with metadata at the top or bibliographical level, at the mid-level of discursive articulation, and at the bottom level of individual word occurrences. The catalogue of 66,000 titles is also a catalogue of two billion word tokens, each of which has a virtual record that inherits its bibliographical and discursive metadata. For these metadata to be useful it must be possible to use them for crosscutting analyses. The texts may be apples and oranges, but the metadata must be interoperable.

A triple-decker structure of this kind is a bloated monster from any readerly perspective. A bibliographical catalogue record will be on average 0.1 per cent of the size of a text, while the aggregate of structural and linguistic annotation may add up to ten times the size of the text. But monstrous as it looks to a reader, from a technical perspective there is no longer anything especially forbidding about a deeply curated corpus of 66,000 primary texts with its two billion or so distinct bibliographical, discursive, and linguistic catalogue records. Deep curation of this type greatly leverages the query potential of primary source materials. Texts annotated at these three levels support robust searching that let you:

1. search by bibliographical criteria;
2. limit a search to particular XML elements, such as text in <epigraph> elements;
3. do simple and regular-expression searches for the string values of words or phrases;
4. retrieve sentences;
5. search for part-of-speech (POS) tags or other positional attributes of a word;
6. search for phrases that combine string values with POS tags or other positional attributes, such as an adjective followed by "liberty";
7. define a search in terms of the frequency properties of the search term(s), such as words that are unique to a document or occur more than ten times;
8. look for collocates of a word ("Christian," for example, is the most common collocate of "liberty" in early modern prose);
9. identify unknown phrases shared by two or more works (sequence alignment);
10. compare frequencies of words (or other positional attributes) in two arbitrary sub-corpora by means of some statistical routine;
11. perform supervised and unsupervised forms of text classification on arbitrary subsets of a corpus;
12. support flexible forms of text display ranging from single-line concordance output to sentences, lines of verse, paragraph-length context, and full text;
13. support grouping and sorting of search results as well as their export to other software programs.

Some these queries can be answered only by search engines that do not yet exist, or require modest command line skills that many humanities scholars lack. But once there are deeply and consistently curated corpora, more sophisticated and user-friendly search engines will follow.

The TCP texts have been encoded in TEI-XML, and are therefore catalogued both at the bibliographical and the discursive level. You can think of XML as the "containerization" of texts. Conventionally defined parts of text are put into labeled boxes known as "elements." Boxes with the same label contain the same kind of stuff. A TEI-encoded file includes a header element that is like a card in a library catalogue and contains metadata both about the source text and the encoded digital file.[8] The discursive or structural encoding distinguishes the main body of the text from its front and back matter. There are elements for chapters and other large-scale divisions, paragraphs, lines of verse, lists, tables, notes, speeches, speech prefixes, stage directions, and more. These are not neatly separable categories, and the task of making a model of the muddle in the middle of a text will always be an imperfect enterprise. Nevertheless, structural encoding adds much value to the query potential of a digital surrogate, and supports more precisely delimited searches both within a single text or across many texts.[9]

Linguistic annotation is not only, or even mainly, for linguists. Wherever human language is the object of computationally assisted analysis, the machine needs to be told a few things about how language works so that it can reliably establish the boundaries of words and sentence, tell nouns from verbs, identify name phrases, etc. It is a way of explicitly introducing these rudiments of readerly and largely tacit knowledge into the data so that the dumb but fast machine can do some of the things readers do without even knowing that they are doing it. In many scholarly and professional fields, from biomedical

---

[8] The header information is derived from EEBO data, and in most cases maps to entries in the English Short Title Catalog (ESTC). Josuah Sylvester's *Monodia* of 1594, for example, has the file number A13276, the ESTC number S111303, and the STC number 23579. For bibliographically oriented readers, the last is probably the most familiar number. One might dream of a time when the ESTC becomes a control tool for early modern texts, and a click on its catalogue brings up the full text.

[9] The TEI markup of the TCP makes heavy use of type attributes, but the attribute values typically mirror chapter headings and do little to promote corpus-wide interoperability. It would be a useful project to establish a controlled vocabulary of text types that could be applied with some consistency across the corpus and focus on raising the "highest common factor" of shared encoding.

research to political campaigns, Natural Language Processing (NLP) routines undergird forms of analysis by users who have not the slightest interest in linguistics per se.[10]

In the case of the TCP corpus, lightweight linguistic annotation is also by far the most reliable strategy for dealing with orthographic variance. In the summer of 2001 a group of librarians and faculty from five EEBO-TCP subscribing institutions met at Northwestern to discuss questions of interface design for the corpus. The report about the meeting states: "Without question, the task-force members agreed that early modern spelling remains the biggest and most persistent obstacle to easy access. ... Many advocated the addition of a front-end normalizer, which would automatically look for spelling variants of submitted search terms" (Nunn 2001). The variant detector (VARD) program at Lancaster University's UCREL Centre, and the Virtual Modernization Project at Northwestern, have implemented two such front-end normalizers.[11] The latter has been part of the EEBO search environment since 2008. While context-independent mapping has its virtues ("vniuersitie" > "university"), you get richer searching results from a context-sensitive approach: you first identify a word token in its particular context as a lexical unit in a grammatical state, which is what readers tacitly do, then describe that identification in a manner that a machine can process, and finally map that identification to a standardized spelling that may vary with different purposes.

Linguistic annotation of EEBO-TCP texts has been done at Lancaster University and at Northwestern using the CLAWS and MorphAdorner tool kits, respectively. Over the past two years, the Northwestern project has focused on refining the tokenization and annotation of early modern texts in its work on the SHC corpus.

MorphAdorner is a general purpose NLP toolkit designed by Philip Burns.[12] A recent grant from the Andrew W. Mellon Foundation supported its improvement in three areas of special relevance to the TCP texts and their release into the public domain:

---

[10] An excellent discussion of this point is found in Hahn and Wermter 2006.

[11] For VARD, see http://ucrel.lancs.ac.uk/vard/about/. The Virtual Modernization Project is described at http://www.proquest.com/about/news/2008/ProQuest-Partnership-Makes-Searching-of-Old-Texts-Easier-More-Powerful.html.

[12] MorphAdorner is fully documented at http://morphadorner.northwestern.edu.

1. tokenization and the establishment of a framework for word and sentence boundaries with stable IDs to support collaborative curation and improvement of texts over time;
2. perfecting the algorithms and training data for lemmatization and POS tagging of early modern texts;
3. making MorphAdorner functionalities available as web services that will allow an individual to do this or that with a particular text but will also support automatic workflows.

The separation of initial tokenization from subsequent procedures is a key feature of MorphAdorner.[13] The program creates for each text a hierarchy of addresses sufficiently robust to serve as the basis for collaborative work but sufficiently flexible to allow for minor corrections. In addition to supporting the simple correction of manifest errors, this stable but flexible tokenization routine turns every token or token range into an explicitly addressable object suitable for more sophisticated forms of exegetical annotation.

With tokenization separated from annotation, MorphAdorner can be re-run over a part of a document without overwriting corrections that may have been made to another part.[14] In addition, different lists of abbreviations can be used with different sets of TEI elements. This is particularly helpful for paratext like notes, lists, tables, or title-pages.

MorphAdorner maps each word token to a lemma, a part-of-speech tag, and a standardized spelling. It assumes that early modern spelling can be reliably mapped to a lemma as it is found in a dictionary like the OED. There are edge cases: "gentle," "gentile," "genteel," "human," "humane" are distinct lemmata with clearly demarcated semantic boundaries in modern English, but the boundaries of those words in 1600 are more fluid.

---

[13] Word and sentence boundaries are very tricky. Consider "howe so euer," "hovvsoeuer," "How so'ere," "How-so-euer," and similar variations on a word or phrase. Readers hardly ever have trouble deciding whether the humble "." ends a sentence or not. The machine finds it very difficult, especially since early modern practices vary widely. When is "vs." short for "versus," and when is it "us" followed by a sentence-ending period?

[14] This is a critical point. Even the best NLP tools have high error rates, and text annotated with such tools benefit much from some manual review. If you separate tokenization from annotation and allow for subsequent partial runs, you can iteratively improve a given text through some combination of manual and algorithmic work. If a second run over a text changes the tokenization and wipes out what was there before, iterative improvement goes out the window.

Differences in spelling between modern and early modern English are partly rule-bound: the change from "vniuersitie" to "university" can be turned into an algorithm that works with few exceptions. But the bulk of changes do not yield to rules. Syntax and lexical context will often provide enough clues for algorithmic mappings of spellings that are ambiguous in early modern English, such as "loose," "lose," "deer," "boar," "boor," "bore," "heart," "hart," but such mappings need a pair of human eyes to confirm them.

MorphAdorner is completely at home with XML and the latest version of TEI (P5). It can generate results in a variety of different formats. A tabular output can be easily moved into a relational database environment and is particularly helpful with data review by programs like AnnoLex, discussed later in this essay.

Complete and Correct

The TCP texts were transcribed by keyboarders in various Asian countries working from digital scans of microfilm images of printed books. Many things could and did go wrong in that chain. The printed page might have had too much or not enough ink. There were missing or damaged pages. The microfilm image, typically a double page image, was not de-skewed, and the initial or terminal letters at the inner margins were often distorted beyond recognition. Notes placed in the outer page margin and printed in small italics were often too faint to make out with any confidence.

Given these circumstances, the transcribers did remarkably well. The TCP project aims at a standard of accuracy that permits one error per 20,000 keystrokes, which translates roughly into 2.5 errors per 10,000 words, or half a dozen errors in a Shakespeare play of average length. The TCP review staff sampled each text and accepted it if the samples did not exceed the permissible level of error.

One can live with this error rate for most purposes, but it excludes several things. If TCP transcribers could not read a letter, word, or passage, they were instructed to mark it as a gap and define what they could not read. So the TCP texts are full of gap markers like <GAP DESC="illegible" REASON="obscured" EXTENT="1 word">. Extrapolating from our data set of about 44,000 EEBO-TCP texts, we estimate that the final collection of two billion words includes around seven million words that are missing at least one letter or could not be deciphered at all. Roughly speaking, one out of every 200 words is transcribed incompletely or not at all. A span of 10,000 words will on average have thirty-four missing or incompletely transcribed words—about seventy-five errors in a Shakespeare play.

The errors are not randomly distributed, but cluster heavily by books and particular pages in those books—the significant variable being the quality of the image from which the transcriber worked. A quarter of the texts (11,341) have no gaps of this kind at all. The remaining gaps are distributed as follows:

Distribution of errors per 10,000 words

min	0.5%	2.5%	10	25%	median	75%	90%	97.5%	99.5%	max
0.06	0.3	0.8	2	5	14	42	107	277	610	2839

The texts at the ninetieth percentile and above account for a third of all defects. This heavy clustering is both a good and a bad thing. It is a bad thing because people tend to judge a barrel by its worst apples. Many early modern scholars believe that the texts are in much worse shape than in fact they are. They might be surprised to learn that 11,000 texts have no gaps and another 20,000 have on average one such error per page.

The clustering is a good thing because it is easier to identify and fix errors. Scholars or librarians concerned about the quality of EEBO-TCP texts should make them better through collaborative curation or crowdsourcing. The large majority of textual problems in the EEBO-TCP are not of the philologically exquisite kind (was it an "Indian" or "Iudean" who threw away that pearl?). They belong to the world of manifest error, where the correct answer is almost never in doubt and typically not difficult. Undergraduates who like old books can do good work in that field, and have done so. But so can high school students in Advanced Placement classes, and retired lawyers or schoolteachers. And the results of human curatorial labor can become inputs for machine learning techniques that are surprisingly good, although the results need to be reviewed by humans.

In their excellent study *Sustaining the EEBO-TCP Corpus in Transition*, Judith Siefring and Eric Meyer say more than once that in user surveys "transcription accuracy" always ranks high on the list of concerns (Siefring and Meyer 2013, 22–23). They also say that when users were asked whether they reported errors, 20 per cent said "yes," and 55 per cent said "no." But 73 per cent said that they would report errors if there were an appropriate mechanism, and only 6 per cent said they would not (Siefring and Meyer 2013, 23). There is a big difference between what people say and what they do. But given a user-friendly environment for collaboration, it may be that a third or more of EEBO-TCP users could be recruited into a five-year campaign for a rough cleanup of the corpus so that most of its texts would be good enough for most scholarly purposes. It is a social rather than technical challenge to get to a point where early modernists think of the TCP as something that they own and need to take care of themselves. Greg Crane has argued that "digital editing lowers barriers to entry and requires a more democratized and participatory intellectual culture" (2010). In the context of the much more specialized community of Greek papyrologists, Joshua Sosin has successfully called for "increased vesting of data-control in the user community."[15] If the cleanup of texts is not done by scholarly data communities themselves, it will not be done at all. And it is likely to be most successful if it is done along the "adopt a highway" model where scholarly neighborhoods agree to get rid of litter in texts of special interest to them. The engineer John Kittle helped improve the Google map for his hometown of Decatur, Georgia, and was reported in the *New York Times* as saying, "Seeing an error on a map is the kind of thing that gnaws at me. By being able to fix it, I feel like the world is a better place in a very small but measurable way" (Helft 2009).

Compare this with the printer's plea in the errata section of Harding's *Sicily and Naples*, a mid-seventeenth century play:

> Reader. Before thou proceed'st farther, mend with thy pen these few escapes of the presse: The delight & pleasure I dare promise thee to finde in the whole, will largely make amends for thy paines in correcting some two or three syllables.

In 2013, five Northwestern undergraduates (on summer internships, with some help and light supervision from Mueller) reduced the number of incompletely transcribed words in the SHC corpus by about two-thirds: from ~52,000 to 18,000. In the summer of 2015, three students from Northwestern and Washington University visited the Bodleian, Folger, Houghton, and Newberry Libraries, as well as the special collections at Northwestern and the University of Chicago. Consulting the printed originals, they reduced the errors to about 9,000, of which 6,000 are indeterminate punctuation marks. It may take a while to correct the remaining errors, but in the meantime there

---

[15] Quoted in Mueller 2011; see also http://papyri.info/docs/ddbdp.

is a considerable benefit in texts whose median error rate per 10,000 words has dropped from 14 to 2.5.[16]

The students on the 2014 team worked in the same room for eight weeks using AnnoLex, a web-based curation tool, and in the course of their work made valuable suggestions for improving the tool and making it suitable for use in a mode of scattered curation by many hands, anywhere, anytime. Moreover, their ~34,000 mappings, from incomplete to complete spellings, are very valuable inputs for machine learning algorithms. From the machine's perspective, an incompletely transcribed word in a TCP text forms a pattern consisting of known characters interspersed with unknown characters whose number and position are known. You can therefore develop a Language Model (LM) that looks for replacements that match the pattern and its surrounding lexical context. A recent experiment at Northwestern showed that in almost a quarter of all cases the LM offers just one replacement, which is correct 93 per cent of the time. In at least three-quarters of all cases a human reader need not consult the original, but can pick the right replacement from the first three choices offered by the LM. The percentage is even higher if you settle for the standard spelling of a lexical unit in a grammatical state, which is better than nothing and lets you defer the judgment of the precise spelling to a second stage. Machine-assisted curation holds the promise of offering provisionally good enough replacements for 5.5 million of the seven million incompletely transcribed words in the

---

[16] The U.S. Department of Agriculture grades meat on a six-point scale from Prime (the top 2 per cent) through Choice, Select, and Standard to Commercial and Utility. The number of known errors in a TCP text is a good proxy of the overall quality of the transcription. Few errors suggest that the transcribers worked from a legible page image and that their work can be trusted. Something like the USDA scale would work well as quality assurance markers for EEBO-TCP texts, and each text could receive an initial quality assessment, with subsequent upgrades based on measurable forms of curation. Reserve "Prime" for texts that have been edited by individuals with appropriate scholarly credentials. "Choice" and "Select" would be the categories with error rates unlikely to distract readers. "Standard" and "Commercial" are texts whose errors would be unlikely to skew algorithmic routines. "Utility" would be the grade for texts that are deficient both from a human-readable and machine-actionable perspective. Texts with error rates of 3 per cent or more might fall in this category. If the grade of each text and its percentile ranking were prominently displayed to the user, it would build confidence in the corpus because users would see that quite good texts are much more common than very bad ones.

TCP corpora without the enormous time cost of checking millions of words against a page or its facsimile image.[17]

AnnoLex was designed by Craig Berry (Mueller [n.d.]).[18] It builds on data provided by MorphAdorner and models every curatorial act as an annotation that is kept in a separate file, but is linked to its target through the explicit and unique IDs that are established by MorphAdorner and support the logging of the "who," "what," "when," and "where" of every annotation.

AnnoLex derives its name from "lexical annotation," but it may be easiest to think of it as a spell checker turned inside out. Instead of the machine correcting your spelling, you tell the machine to correct a spelling.[19] A corrected spelling in the context of the EEBO-TCP transcriptions means an improved transcription of a word from an early modern text produced by a user who reconsiders and revises the existing transcription. A user with editorial privileges (which could be the same or a different user) can then approve or reject each correction, and finally, approved corrections are applied to the source text.

AnnoLex presents the reader with a verticalized text in which every word appears in the middle of an eighty-character line, which offers enough context to solve most problems. You could proofread an entire text that way one word at a time, but most of the time you will use the default setting of a preselected filter that directs your attention to a list of word tokens known to be incomplete or likely to contain errors. For instance, a list of words unique to a text is likely to contain a majority of incorrectly transcribed words. And pages that contain several incompletely transcribed words are likely to contain incorrectly transcribed words as well.

Clicking on the "Edit" button next to a search result populates an edit form in the lower left of the screen, where a revised transcription of the word instance may be suggested and saved. While it is often easy to infer the correct spelling from the immediate context, the edit form includes a button that, for members of subscribing institutions, will bring up the relevant double

---

[17] For more information see Mueller 2015a.

[18] For documentation about AnnoLex, see http://annolex.at.northwestern.edu/about/.

[19] The editing panel of AnnoLex lets users correct the lemma and part of speech as well as the spelling. In practice, most users will focus on the transcription of spellings, and in this discussion we ignore the place of AnnoLex in the correction of linguistic annotation.

page image from EEBO. The image may be panned and zoomed such that up to two full pages of context are visible or a single character fills the entire screen (or anything in between), thus allowing a full reconsideration of the TCP transcription using all of the information available to the original transcriber. Even with quite mediocre page images the proper solution often becomes visible by zooming in on a word or line at just the right degree of magnification.[20]

We count roughly half a million untranscribed passages—mostly Greek or Hebrew—in about a quarter of our 44,000-text corpus. Most of them are very short. Mathematical or musical notation appears in a much smaller subset. Transcriptions of Greek or Hebrew require domain-specific knowledge and some familiarity with early modern print practice. A classics major might learn quite a bit by transcribing the several dozen Greek passages in Ben Jonson's plays and reflecting on their purpose. Would anybody ever have the patience to transcribe 23,858 Greek and Hebrew passages in the 1.5 million words of Henry Hammond's *Paraphrase and annotations upon all the books of the New Testament, briefly explaining all the difficult places [...]*? From a technical perspective, most of it could easily be done with AnnoLex. Musical and mathematical notation would pose technical problems, but it occurs in relatively few texts.

The prefaces of early modern printers bear eloquent testimony to their belief in getting things right and to their sense of failure to do so. In the aggregate they bear interesting witness to how Shakespeare's contemporaries thought about truth, error, and textual stability. The TCP transcribers were instructed to type the letters they saw, whether or not the spelling was a manifest error. Tacit correction of such errors has been the practice of the TCP project in many, but not all, cases. In the SHC project we corrected about 10,000 printers' errors, many of them falling into these categories (in no particular order):

1. words wrongly joined: "beidle" to "be idle"
2. words wrongly split: "ho pe" to "hope"
3. missing letter: "Colla ine" to "Collatine"
4. "s" for "f": "assliction" to "affliction"

---

[20] While the TCP transcriptions are moving into the public domain, the EEBO page images remain the property of ProQuest, and access to them is restricted to users from subscribing institutions. The full curation potential of AnnoLex is therefore available only to users from institutions that subscribe to EEBO. There are a lot of people with access, then, but even more who are excluded.

5. "f" for "s": "wickedneffe" to "wickednesse"
6. "n" for "u": "hnsband" to "husband"

Unlike incompletely transcribed words, such errors are hidden, although there are some quite effective error-tracking routines. In the twelve million words of the SHC corpus, errors of this type occur once every 1,000 words or so. There are probably two million in the TCP corpus. Many of them can be corrected without reference to context when they are flushed out by lexical analysis: spellings like "hnsband," "hnsband's," "hnsbandes," "hnsbandman," "hnsbandrie," "hnsbands," "hnsbonde," and "hnsbōde" leave no doubt about the writer's or printer's intention. Which is not the case with the "Indian" or "Iudean" in Othello's final speech.

The TEI encoding of the TCP has been largely ignored by end users, for the simple reason that none of the search engines offers full and user-friendly XML-aware searching. As the texts move into the public domain, that is likely to change. People will use the texts for this or that purpose, will find the encoding useful as a point of departure, and will add to it, whether for presentation or search purposes. TCP TEI encoding is mercifully sparse and, given the heterogeneity of materials, reasonably consistent. AnnoLex is of no help in correcting encoding errors, and the collaborative curation of such errors is a tricky thing. Sometimes an error can be corrected merely by changing the name of an element in its opening and closing tag. But if the encoder missed a speaker label and ran two speeches together in a single element, the correction is a multi-step that presupposes on the part of the curator a good understanding of the hierarchical structure of XML and the particular structure of a TEI tree.

Beyond Fixing Things

There is more to curation than fixing simple mistakes that irritate the reader but even in the aggregate rarely cripple understanding. If you do "text mining" in a corpus of two billion words, fixing ten million words (0.05 per cent) is hardly worth the trouble. Why bother? That is a good question from a text mining perspective. But the TCP archive is not merely a quarry to be mined, it is a cultural heritage resource of the first order and deserves to be "protected, cherished, and preserved," words that appear on a plaque near the gate to the Wolverton Common, a meadow near Oxford that has not been ploughed since the Norman Conquest. You want the digital surrogate of the first 233 years of English print to be a machine-actionable corpus. But it is also a window through which human readers look at the past, and you want to keep that window as clean as possible.

That said, curation does not stop with a cleanup, and in this final section we turn to collaborative curation as a form of metadata enrichment at the top level of bibliography, the middle level of document structure, and bottom level of words and sentences. We emphasize again the central role of linguistic annotation. Words and sentences are the fundamental building blocks of written language. If you get a firm handle on them, everything becomes a lot easier. When computerized catalogues entered libraries about 50 years ago, it was a big technical challenge to manage a million MARC records. Today it is entirely possible to envisage two billion word records of an early modern corpus as the foundation for rich metadata structures at higher levels.

In contemplating the weird language of the 1581 edition of Seneca's tragedies, it may be useful to consider the fact that some of the translators did their work as teenagers. It should be easy to look for books written by men or women between the ages of 17 and 21, but it is not. Nor is it easy to look for authors who are female or grew up in Somerset. A project underway at the English Short Title Catalog (ESTC) aims at mapping author data to authority files, breaking data into smaller chunks that can be used to construct complex searches, and encouraging scholars to become users of and contributors to the catalogue.

The cataloguing of early modern books by subject matter is a hit or miss thing for a number of reasons. The ontologies of the LC and Dewey catalogues are children of the late nineteenth century and map poorly to the world before 1700. Top-down cataloguing could be fruitfully combined with cataloguing from below by looking at shared lexical and phrasal patterns. Amazon does this, and in a world of books before orthographic standardization it is greatly facilitated by mapping the surface forms of words to abstract lexical and morphological patterns. There is a lot of promise in the interaction of lexical and bibliographical metadata.

Plays are a good example of what can be done with structural metadata. Rigidly controlled metadata are a constitutive feature of drama as a genre. In Early English drama Latin was the language of such metadata, a practice that lingered for centuries. The TCP texts follow the metadata as they appear in the text, but make no effort to map the widely varying speech prefixes to unique identifiers that let you count, extract, or otherwise manipulate the speeches of characters. We have done this for 93 per cent of speeches in the SHC corpus—enough to visualize any play as a pattern of who addresses whom and at what length. Things become much more interesting if you construct machine-actionable cast lists in which a controlled vocabulary is used to classify characters in terms of sex, age, kinship relations, and social status. The aggregate of such interoperable cast lists can become an input for social network analysis; 550 plays from between 1550 and 1650 may not be a bad mirror of the "struggles and wishes" of that age. We envisage the construction of this prosopography of early modern drama as a crowdsourcing project, starting with undergraduates but opening it up to "citizen scholars" anywhere. The project will use a revised version of the AnnoLex curation tool. For each character there will be a template with some combination of structured and free-form data entry. We hope to engage undergraduates both in the annotation and in the construction of a controlled vocabulary, which should emerge from a lightly supervised version of a bottom-up folksonomy. A generation whose knowledge of genre is shaped by movies and TV shows may have fresh ideas about a taxonomy of sixteenth- and seventeenth-century plays.

Because plays are "natively interoperable," this upward spiral from written metadata in a single text to their digital formalization and corpus-wide application is particularly pronounced. But similar procedures are possible for other genres.

The TCP texts aim at being accurate orthographic transcriptions. They do not pay much attention to layout or typographical detail. Page breaks are marked, line breaks are not. Typography is reduced to a binary pattern of unmarked and marked, where deviations from an unmarked default text are wrapped in <hi> tags. Text inside <hi> tags usually represents italic text in the printed source, but you cannot reliably reconstruct the layout of a page from the current encoding.

On the other hand, if you want to create an edition that is more faithful to the original layout and typographical choices, it would take much more time to transcribe the text from scratch than to edit the TCP text into a shape of your choice. This has been done in a variety of projects, and "upcoding" of this kind is likely to be a major source of future activity. It will typically involve a combination of semantic, structural, and rendition tags: <hi>Caesar</hi> turns into <name rend="italic">Caesar</name>, and <hi>veni, vidi, vici</hi> turns into <foreign xml:lang="lat" rend="italic">veni, vidi, vici</foreign>. Changes of this kind do not make the text more readable, since at any given point in the text the encoding only tells readers what they already know. But such changes make texts more computationally agile and enhance the query potential for corpus-wide inquiries.

The Thesaurus Linguae Graecae (TLG) began its life in the early twentieth century as a project to create a comprehensive dictionary of Greek, but over fifty years morphed into a project for a digital corpus. But a digital corpus that "knows" about each of its words has a dictionary built into it, and from that dictionary you can tell many stories about each word or group of words if you have reliable data about:

1. the function of that word in the immediate context of its sentence;
2. the type of text in which it appears;
3. the creation date of the text; and
4. biographical data about the author(s).

Many users have a particular interest in the names of people, places, and institutions. The construction of social networks from the distribution of names in a corpus is a major research methodology in many fields. Within the EEBO-TCP corpus it is of special interest to the Thomason Tracts, the 22,000 contemporary books and pamphlets gathered by the English bookseller George Thomason during the English Civil War. Clearly there is much to be learned from the construction of solid networks of authors, printers, the contemporary agents, and the historical, literary, or legendary characters to which they refer.

Names are not tagged in the TCP corpus, but they are identified with high precision in the linguistic annotation. Mapping name spellings to particular individuals or places, real or imagined, is another matter. It is a massive indexing task requiring a subtle interplay of human editorial labor and algorithmic procedures.

Closing Thoughts

If your goal is to create clean and readable versions of early modern books, the layers of metadata envisaged under the heading of connecting curation may strike you as excessive. Librarians may shudder at the thought of extending the creation and maintenance of metadata below the bibliographical levels. Professors of English may think that such metadata come between them and the book. But what if you think of the EEBO-TCP archive as a single multi-volume book, or the opening chapter in a Book of English? The corpus-wide metadata for which we advocate have for centuries been part of the envelope of metadata that are constitutive features of certain kinds of books, particularly of editions and reference books.

From a quick look at the Wikipedia entry on the Weimar edition of Martin Luther's works, you learn that its 60 volumes of texts are complemented by 13 volumes that index places, people, citations, and topics. No Luther scholar would want to be without them. The research potential of a scholarly field is enabled and constrained by the complexity and consistency of its (meta) data. In the Life Sciences, GenBank, an "annotated collection of all publicly available DNA sequences," provides the basis for much research, and the maintenance of this database is itself a research project (National Center for Biotechnology Information, "GenBank Overview," 2014).[21] You can think of it as the critical edition of a very peculiar form of text written in a four-letter alphabet. Conversely, you can think of the EEBO-TCP archive as a cultural genome whose research potential will benefit from comparable forms of annotation. Metaphors of this kind take you only so far, but they take you quite a ways toward the identification of collaborative data curation as an important and shared task for humanities scholars, librarians, and information technologists.

WORKS CITED

Crane, Gregory. 2010. "Give Us Editors! Re-inventing the Edition and Rethinking the Humanities." In *Online Humanities Scholarship: The Shape of Things to Come. Proceedings of the Mellon Foundation Online Humanities Conference at the University of Virginia, March 26-28, 2010*, edited by Jerome J. McGann, Andrew M. Stauffer, and Dana Wheeles. Houston: Rice University Press. Also accessible at http://cnx.org/content/m34316/1.2/.

Hahn, Udo, and Joachim Wermter. 2006. "Levels of Natural Language Processing for Text Mining." In *Text Mining for Biology and Biomedicine*, edited by Sophia Ananiadou and John McNaught, 13–41. Boston: Artech House.

Helft, Miguel. 2009. "Online Maps: Everyman Offers New Direction." *New York Times*, November 16. New York edition: A1.

---

[21] See http://www.ncbi.nlm.nih.gov/genbank/.

Mueller, Martin. 2011. "Getting Undergraduates and Amateurs into the Business of Re-editing Our Cultural Heritage for a Digital World." *Scalable Reading*, January 7. https://literaryinformatics.wordpress.com/2011/01/07/getting-undergraduates-and-amateurs-into-the-business-of-re-editing-our-cultural-heritage-for-a-digital-world/.

———. 2015a. "Engineering English: Machine-assisted Curation of TCP Texts." *Scalable Reading*, August 2. https://scalablereading.northwestern.edu/2015/08/02/engineering-english-machine-assisted-creation-of-tcp-texts/.

———. 2015b. "From Shakespeare His Contemporaries to the Book of English." *Scalable Reading*, March 5. https://scalablereading.northwestern.edu/2015/03/05/from-shakespeare-his-contemporaries-to-the-book-of-english/.

———. [N.d.]. "Shakespeare His Contemporaries: Collaborative Curation of EEBO-TCP Texts with AnnoLex." *AnnoLex*. Northwestern University website. http://annolex.at.northwestern.edu/about/.

National Center for Biotechnology Information (NCBI). 2010. "GenBank Overview." *GenBank*. http://www.ncbi.nlm.nih.gov/genbank/.

Nunn, Hilary. 2001. "Early English Books Summer Camp." *Biblio-Notes* 38. http://www.ala.org/acrl/aboutacrl/directoryofleadership/sections/les/bibionotesnum31/biblionotesnum38#camp.

Siefring, Judith, and Eric T. Meyer. 2013. *Sustaining the EEBO-TCP Corpus in Transition: Report on the TIDSR Benchmarking Study*. London: JISC. http://ssrn.com/abstract=2236202.

"Ill shapen sounds, and false orthography":
A Computational Approach to Early English Orthographic Variation

Anupam Basu
Washington University in St. Louis
abasu22@wustl.edu

It was not until the latter half of the seventeenth century that orthographic conventions for English print settled into a recognizably modern form with relatively little variation. This marked the culmination of a long and often contentious process that can be traced back to well before the beginnings of print. The introduction of print, however, has been described as the key "external event" that accelerated the process of regularization and catalyzed debate around linguistic reform (Smith 2012, 174). In fact, the development of print coincided with the overall standardization of English morphology, phonology, and syntax, and the emergence of London English as the standard dialect. The London standard evolved from various Midlands dialects and became associated strongly with the triangle between London, Cambridge, and Oxford (Wright 2000, 1). Print, itself centered in London, consolidated the dominance of the London dialect. George Puttenham could say, writing in 1589, that the correct English to which a poet should aspire is "that of London and the shires lying about London within lx miles," and that this language, "ruled by th' English Dictionaries and other bookes written by learned men," had become the standard of written English throughout the country for "gentlemen, and also their learned clarkes," although it was not yet the language of the "common people" (1589, R3r). It is not surprising, therefore, that orthographic standardization has often been seen as part of a unified narrative of linguistic development—the emergence and dominance of the London standard. Consequently, it has often been conflated with print culture on the one hand and with grammatical and phonological standardization on the other.

Recent scholarship, however, has challenged this model of development and insisted not only on acknowledging multiple competing sociolinguistic standards but on understanding linguistic evolution as non-linear and non-prescriptive. Within such a framework, as Jonathan Hope has argued, the process by which a linguistic standard emerges must be rethought in terms of individual linguistic features selected from parallel and competing dialects. The survival of each feature is determined by "'natural' linguistic processes" of selection, repetition, and circulation rather than the prescriptive imposition of a single hegemonic dialect (Hope 2000, 51–52). Instead of conceptualizing the standardization of English as a monolithic process in which one central dialect gradually imposes itself over all aspects of language, we need to think of multiple axes of linguistic evolution, each relatively autonomous from the other. Each axis of evolution is, in its own turn, the product of multiple competing iterative processes. While standardization might still be thought of as the suppression and eventual elimination of variant possibilities, such a model of linguistic change does not need to reject all internal variation as erroneous, but can accommodate them as "structured variation"—markers of the process of linguistic selection between competing standards (Milroy 2000, 20–22).

Scholars have often noted that spelling, while a part of the overall standardization of English in the early modern period, is "wholly a matter of written language; in addition, it is independent of meaning" and hence develops with relative autonomy as a function of print (Rissanen 2000, 118). However, while the important role of print in spreading and consolidating linguistic standards is widely acknowledged, the role played by the everyday practices of the early printing press in the emergence and evolution of orthographic regularization has received relatively little attention. In fact, it is the shortcomings of the press as a site for regularization that have often been central to discussions of orthographic variation, with a range of factors being cited—from the low literacy of typesetters, and the commercial impetus of churning out volumes quickly, to purely mechanical constraints such as the justification of text on the printed page, which often led to the choice of variant forms. Scholars of early English print have consistently emphasized the importance of theoretical and prescriptive models of orthography over the accumulated effects of the repetitive practices of typesetting and printing. In particular, a set of late sixteenth- and early seventeenth-century texts discussing spelling reform—part of the debates over linguistic change and standardization often referred to as the "inkhorn controversy"—has received widespread attention. These theoretical writings are certainly of immense significance as the earliest rigorous and methodical explorations of the English language and for their attempts at reform. However, while they seem to have had very little demonstrable effect on general orthographic practice, most scholars have credited them with a large influence on standardization, even if not quite of the kind that they actually advocate. Thus, even though linguists can identify distinct phases in which certain major spelling shifts occur, our general model for early modern orthographic standardization has tended to be linear, envisaging early printed output as highly irregular, and steadily progressing toward less internal variation until orthographic standardization finally comes about around the time of the civil war (1642–51). In other words, standardization tends to be identified with modernization on the one hand—moving gradually to the stabilized, mostly modern forms of the late seventeenth century—and regularization on the other.

I

While the availability of large-scale corpora for early modern English print such as the EEBO-TCP (Early English Books Online—Text Creation Partnership) has brought renewed attention to orthographic variation in the period, it has also tended to reinforce the linear model that tends to view early texts displaying non-modern spellings as manifestations of non-standardized orthographic practice. As a relatively well-curated corpus that already incorporates over 50,000 texts consisting of more than 1.6 billion tokens, and eventually aims to include at least one edition of every English text printed before 1700, the EEBO-TCP promises to transform the way in which scholars approach the early modern period. However, orthographic uniformity is a prerequisite for most computational text analysis that might be described under the rubric of "distant reading," since most text-mining and natural language processing algorithms assume a stable one-to-one mapping of orthographic forms to types. While this is a reasonable assumption to make for modern language corpora, in which, for the most part, there is only one "correct" spelling for a given word, it breaks down spectacularly in the case of early English print, in which spelling is at best a set of fluid conventions. It is not surprising, therefore, that instead of leveraging the scale of EEBO-TCP to trace the contours of the long and contested process of orthographic evolution, most computational approaches to texts from the EEBO-TCP corpus have considered orthographic variation primarily as an irritating hindrance to overcome, at least partially, using tools like MorphAdorner or VARD (Burns 2013), before other computational analysis can happen.[1]

---

[1] Even though they have not challenged the linear model, algorithmic attempts at spelling regularization have been responsible for some of the most interesting recent research on orthographic variation in early modern English. This is especially true of the way in which many of these techniques have balanced the unprecedented scale of

Being hand-keyed transcriptions faithful to the orthography of the original texts, and drawn evenly as a sample of the total number of surviving texts noted in the *ESTC*, the EEBO-TCP corpus can be a rich resource for the study of the evolution of print in England and the role it played in the emergence of an orthographic standard. Figure 1 shows the number of texts currently transcribed in EEBO-TCP for each year. The civil war period, posited by the linear model as the central locus for orthographic standardization, witnessed a pronounced spike in the number of volumes printed. While this also corresponds to an increase in the overall bulk of the total print output measured in number of words per year (Figure 2), there is also a marked increase in the number of shorter texts, mostly pamphlets and political tracts, being printed. Even though the likelihood of survival of such tracts from the civil war period is increased due to the unique collection of George Thomason, a similar pattern of a spike in shorter volumes can be seen at various times of political upheaval.[2] Figure 3 plots the length of each text in the EEBO-TCP corpus (in number of words) and fits a median line, producing a pattern of a drop in the median length of texts printed whenever there is an unusual surge in the number of short tracts being printed. While the generally accepted model focuses mostly on the mid-seventeenth century as the time when variant forms are eliminated, presumably under the impetus of the spelling reformers writing mostly between 1570 and 1620, the distinctive patterns of print output should alert us to the possibility that the material processes of the printing press, as much as the prescriptive dictums of the grammarians and orthoepists, might play an important role in the standardization of English orthography. While many important orthographic changes do occur around the mid-seventeenth century, we should ask to what extent they are driven by the higher productivity of the presses and the efficiency incentives of regularization. Moreover, as I will demonstrate, the processes of spelling standardization are complex, and not confined to this period alone. Instead, they are spread out over a wide span of time ranging from the early sixteenth century—well before

---

data with the depth and detail of sophisticated tagging and text-encoding schemes. DICER, for example, is a tool that uses large numbers of word/variant pairs to "learn" the "cost" or "weight" assigned to particular spelling transformations (Baron et al. 2012). The NUPOS tag set and TEI have facilitated the deep-tagging and markup of texts and finely grained part-of-speech tagging (Mueller 2009).

[2] The Thomason Tracts consist of more than 22,000 volumes including some 7,200 pamphlets and newsbooks printed mostly between 1641 and 1660 (British Library).

the first tracts on orthographic reform appear—to the mid- to late seventeenth century.³

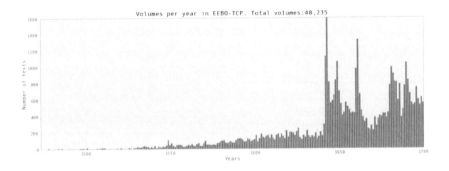

Figure 1. The number of volumes per year in the EEBO-TCP Corpus.

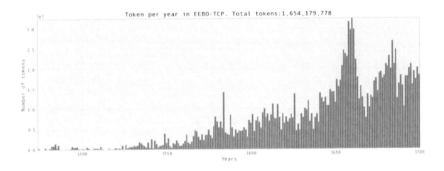

Figure 2. The number of words per year in the EEBO-TCP Corpus. One unit on the Y axis corresponds to 10,000 words.

---

³ All of the data used in this essay is available at http://earlyprint.wustl.edu (Basu 2015).

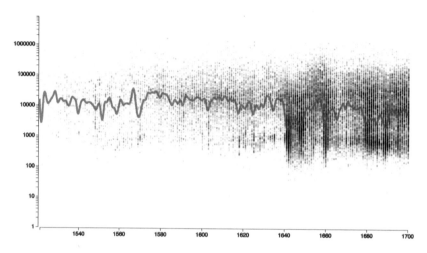

Figure 3. The median length of texts per year, counted in number of words. Each text is represented as a dot, where the X axis denotes the year of publication, and the Y axis denotes the number of words in the text. Note that the Y axis is represented on a log scale. The blue line denotes the median length of texts for each year.

The fact that most computational approaches have encountered orthographic irregularity as a phenomenon to be eliminated through the algorithmic regularization of spelling has tended to privilege "modern" orthographic forms and marked everything differing from that standard as variant. As a consequence, the prominent patterns of print output, as well as the internal consistency of the spelling practices at any given time, have usually been disregarded. For example, when Baron, Rayson, and Archer (2009) attempt to estimate the extent of spelling variation in early modern English, they plot the percentage of variant types in several early modern corpora over time (Figure 4).

"Ill shapen sounds, and false orthography" 173

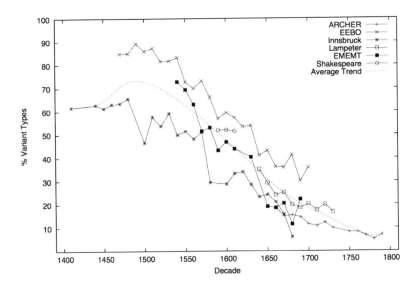

Figure 4. Percentage of variants in various early modern corpora over time (Baron et al. 2009, 11).

However, since their ultimate interest lies in modernizing spelling algorithmically, they define "variant" strictly as not conforming to modernized spelling conventions. Thus, while this indicates that in the case of printed texts (EEBO), the percentage of "variants" hovers at 70 or higher for most of the sixteenth century, this observation does not provide any information about how internally consistent spelling practices were at any given moment. In fact, it might suggest to the casual observer that sixteenth-century spelling practices were in fact extraordinarily irregular and tolerated 70 per cent or more variability. A wide degree of internal variation is obviously prevalent in early print, but once we move away from an understanding of "variant" as simply denoting "not modern" and quantify the internal regularity of spelling conventions over the years, a very different pattern of orthographic standardization emerges. Figure 5 represents type-token ratios for 227,000 samples of 5,000-word chunks drawn randomly from EEBO-TCP—1,000 chunks for each year.[4] It estimates the number of unique words per given number of total words.

---

[4] Type-token ratio is often used as a basic index of the "richness" of language in computational text processing. See Kao and Jurafsky (2012), 9; Jockers (2013), 61.

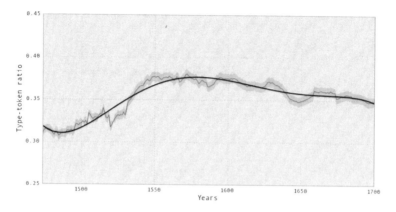

Figure 5. Type-token ratios for 1,000 samples of 5,000 words each plotted over time.

One would expect this ratio to be higher in an orthographic regime that is more tolerant of variation because the same type (word) can map onto multiple orthographic forms, while in a perfectly regularized system of spelling, one type is mapped to one token. The number of texts in the corpus before 1550 is too low to reach our sampling sizes, but once the corpus stabilizes, we get a steady decline for the next century and a half, indicating that there is, in fact, a degree of regularization. What is surprising, however, is that this decline is quite gradual overall. Do we, then, overestimate the degree of internal variation in early texts when we conflate "modernization" with "regularization"?

The evidence from Figure 5 suggests this, but before trying to explain why this might be we need to eliminate another possibility. Might it be the case that the type-token ratios we observe are elevated in later years by the introduction of new words? Of course, the English vocabulary expands greatly in the early modern period, both through borrowing from other languages and through the coining of new word forms. Along with orthographic regularization, the role of borrowed and newly coined words lie at the heart of the "inkhorn controversy." Using an admittedly limited sample of 2 per cent of *OED* entries, Charles Barber estimated at least a quadrupling of the English vocabulary. However, as he and others have noted, this expansion was often due to rarer, more specialized, or idiomatic uses of words, and new words were usually quite short-lived (Barber 1976, 166–67; Görlach 1991, 136). From a quantitative perspective, not only would new words and coinages tend to conform to prevailing orthographic norms when they entered the lexicon,

"Ill shapen sounds, and false orthography" 175

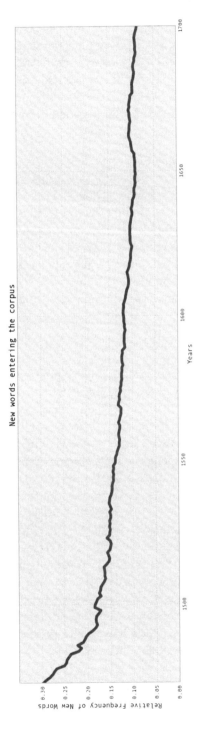

Figure 6. The fraction of words per year in EEBO-TCP that appear for the first time in the corpus.

they would have a limited impact on the type-token ratio due to their rarity. Finally, as Figure 6 shows, as a proportion of the total words per year, the number of new words in the English lexicon remains fairly constant. In other words, even though there is a huge influx of new words and coinages in the language—probably more than Barber's estimate in absolute numbers—that is roughly in keeping with the explosion of print in general, and follows the pattern of growth outlined in Figure 1.

How, then, are we to evaluate the quantitative evidence in Figure 5, which seems to indicate a steady but nevertheless very slow process of internal regularization? Based on the vast sample of early English print that is EEBO-TCP—currently covering at least one edition for about two-thirds of all extant volumes—early modern orthographic practice was a constantly evolving set of conventions. It was, however, more structured and more internally regular than we had previously thought. The emergence of an orthographic standard in England is not simply a movement from chaos to order, from variability to regularization, but the result of accumulated and structured variation.

II

The beginnings of an orthographic standard for English predates the emergence of print by almost half a century. It can be traced to the use of English by the clerks of the royal court at Chancery to send official documents all over the country beginning around 1430. The Chancery standard, as this early orthographic convention has been called, developed into a relatively uniform written convention that superseded local variant forms, much before the consolidation of a spoken standard. It also came to dominate the other incipient written norms of the period, including the North Midland standard followed by much Wycliffite writing, the conventions used in non-official London writing, and the conventions circulated in the literary manuscripts of Chaucer, Langland, Lydgate, Gower, and others (Fisher 1996, 5–6; Burnley 1989, 23–24). The success of this standard—essentially produced by the practices of the London bureaucracy—is owing to its wide geographic distribution, but also because it aligns itself with already emerging communities and practices. For example, the Brewers' Guild began keeping records in English in 1422, and several other guilds followed in the next decades; the standard was also influential in the guilds emerging around that time that focused on legal writing and the book trade (Salmon 2000, 15). Paul Christianson has traced the influence of the Chancery standard on the members of the "Mistery of Stationers," first organized in 1403, who would form the core of the London manuscript and print trade and go on to receive charter as the Stationers' Company in 1557, and the guild of Writers of Court Letters that would receive royal charter in 1616-17 as the Scriveners' Company. These groups existed throughout this period and represented a significant part of the community in London engaged in the "literate crafts" (Christianson 1989, 84). Before the emergence of the Chancery standard, Latin and French provided the only models for orthographic practice. Burnley traces the late medieval commentary on the diversity of the English language and the comparisons to both Latin and French (Burnley 1989, 31-35). However, as both he and Fisher have emphasized, the Chancery standard is primarily the product of bureaucratic practice, rather than the articulation of formal rules or even the conventions of literary manuscript culture. Closely associated with the development of secretary hand in the mid-fifteenth century, this orthographic standard came about through the sheer weight of repetition. The volume of documents being produced in the scriptorium, along with the occasional attentions of a corrector or rubricator, gradually led to the development of a practical standard focused solely on "communicative efficiency" (Burnley 1989, 38): "language is standardized by government and business rather than by literary usage. I do not mean to imply that clerks innovated and expanded the capacities of language but it was their habitual usages that created a 'standard'" (Fisher 1996, 9).

However, while the Chancery standard provided a template for orthographic practice for the early years of print, the earliest printed texts actually seem to have more variation than Chancery documents. Many scholars have followed early modern commentators like Bullokar in ascribing this to the lack of skilled Englishmen in the print trade at the time (Scragg 1974, 64; Görlach 1991, 46; Salmon 2000, 19). Caxton freely admits his difficulties in his introduction to *Eneydos*—"certaynly our langage now vsed varyeth ferre from that whiche was vsed and spoken when I was borne" (Virgil 1490, A1v); his immediate successors were trained and lived large portions of their lives on the continent, and were often dependent on foreign journeymen for setting type. In fact, it was only after the enactment of laws curbing the employment of foreigners in presses in 1515, 1529, and 1535 that a level of regularization comparable to the Chancery standard was achieved (Görlach 1991, 46). Brengelman (1980, 338) compares Caxton's edition of Malory's *Morte d'Arthur* with contemporary manuscripts, and is disappointed to find that "Caxton's spelling is at least as variable" as that of the manuscript tradition (in itself less regularized than Chancery script). Accordingly, for modern scholars, this early lack of standardization and the continued presence of a degree of variation have tended to reinforce the allegations about the laxity of printers that we find in sixteenth- and seventeenth-century spelling reformers like Hart, who lamented "the lak of ordre emongest printers" (1569, 15r). While John Bale's description of the "headie hast, negligence, and couetousnesse" of printers that "commonly corrupteth all bookes" might be characteristically overblown, complaints against printers were not uncommon in early modern books (ca. 1570, 4).

This exasperation with printers' nonchalance, or rather their lack of interest in formal, prescriptive standards for orthography, is echoed in Blake's seminal longitudinal study comparing five versions of *Reynard the Fox* printed between 1481 and 1550. Salmon comments on Blake's analysis that it is "difficult to discern any consistent progress towards a standard orthography" (2000, 25). Nevertheless, even as he observes that certain spelling forms—for example the usage of *y* and *i*—seem "haphazard" and "purely random" (1965, 65), Blake notes a strange regularization with surprise: "One of the most remarkable features of this study of the five versions of *Reynard the Fox* is how the grapheme *ea* appears suddenly and becomes accepted as the standard spelling in some words in such a short time" (1965, 68). This strikes him as an oddity because his conception of orthographic standardization leads him to expect a gradual but simultaneous movement toward modern spelling across all graphemes where variation is slowly but steadily eliminated. I want to argue, however, that the regularity Blake stumbles upon holds the key to the general pattern of orthographic change in the early modern printing press—a pattern whose rough outlines have been discerned by linguists for a while. For example, we know the broad periods for certain shifts, but we can only now begin to fully explore their patterns given the scale of the data at our disposal. Figure 7 shows the sharp regularization that Blake notices in one of his example words—"teach". His other example—"beast"—shows a very similar transition.[5]

---

[5] The time series plots in this essay can be generated using the data available at http://earlyprint.wustl.edu/, which contains relative frequencies by year for 29 million unigram (or single word) records extracted from EEBO-TCP. While the lighter colored curves denote exact frequencies per year, I apply a smoothing function over a 20-year rolling window to emphasize general trends over yearly variations. While I will only consider original spellings in this paper, the database also contains more than a billion algorithmically regularized n-gram records (Basu 2015).

"Ill shapen sounds, and false orthography" 179

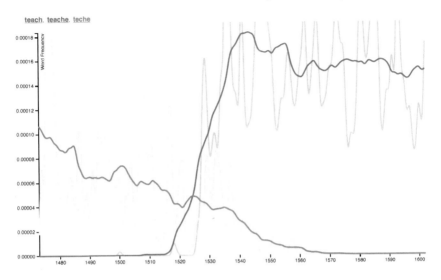

Figure 7. The transition from "e" to "ea" for "teach." The blue line plots the combined frequency for "teach" and "teache." Other words standardizing on the grapheme "ea" show a very similar pattern.

While there are minor variations in the yearly relative-frequency plots, the starkness of the pattern that surprises Blake is undeniable, and the same pattern and time frame for change holds for most words where the digraph *ea* eventually becomes the accepted spelling. Significantly, this pattern of a steady transition within a well-defined period is not limited to the *ea* grapheme that Blake notices. If we take the example of *y/i*, whose usage Blake finds "haphazard," we again find a similarly stark pattern of standardization (Figure 8).

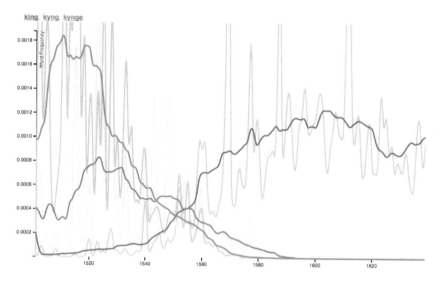

Figure 8. The transition from "y" to "i." "Kyng" and "kynge" are here plotted as distinct lines to show the early popularity of the trailing-e form.

The reason Blake does not notice this distinct pattern at all is that it happens slightly later than the *ea* shift, and after the period covered by his texts. He does not anticipate its possibility because his conception of an overall movement from non-standardized to standardized spelling leads him to expect simultaneous modernization across different graphemes. However, while the patterns of change for individual graphemes do not conform to this model of gradual but overall standardization, these changes are distinctively structured transitions that are far from "haphazard." In fact, each grapheme shift seems to occur autonomously and within its own well-defined time frame.

The process of orthographic change, therefore, was phased and discrete rather than uniform and gradual. The pattern of change that we see above is repeated for different graphemes in different periods. The standardization of *u/v*, for example, happens quite sharply in the years before the civil war (Figure 9), as does the *i/j* switch (Figure 10). However, even though many scholars have tended to speak of spelling change as a move toward "modernization," a description that suggests a linear movement from multiple forms to a state where one form—the modern one—becomes dominant and variants are eliminated, standardization could also lead to forms that were not modern, but nevertheless became the norm at least for a period of time. The trailing *e* in verbs like "doe" or "goe," for example, comes into effect quite distinctively in the mid-fifteenth century, and is an acceptable spelling for a while before it is eliminated equally rapidly a century later (Figure 11).

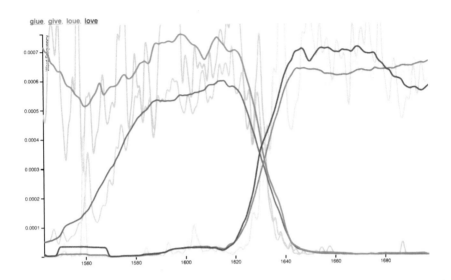

Figure 9. The regularization of "u" to "v."

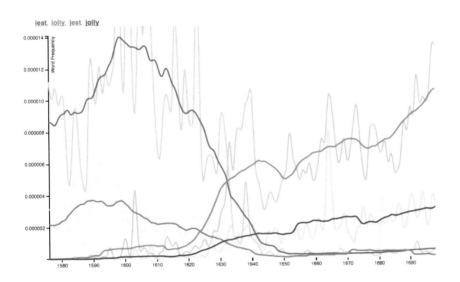

Figure 10. The regularization of "i" to "j."

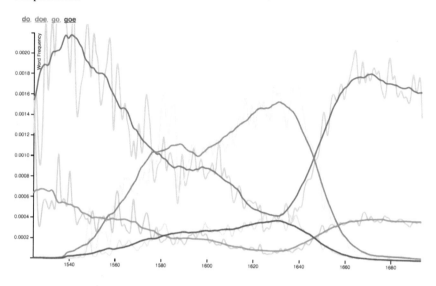

Figure 11. The rise and eventual decline of the trailing "e." Note that in both the instances of "do" and "go," the forms with the trailing "e" rise in popularity for a period although they never completely dominate, and eventually decline in popularity and die out.

The one aspect that is common among all these shifts is the characteristic pattern of rise and fall, usually spanning anywhere between less than two to a little over four decades, where the usage of a grapheme starts to increase or decline. This shift is usually part of a paired switch wherein the increase in one grapheme is matched by a decrease in a corresponding one. Thus, for example, the relative frequency of the word "love" (number of occurrences expressed as yearly percentages of total words in the corpus) doesn't change in the mid-sixteenth century if we take both orthographic forms "loue" and "love," but while the former is the overwhelming standard before the switch, it is almost completely eliminated once the transition is complete.

What drives this pattern of phased orthographic change? Changes for individual graphemes are well defined and seem to occur in independent phases. At least on the evidence of these somewhat arbitrary example words/graphemes, it appears that while individual changes are very distinctive, the overall process of spelling standardization is not concentrated in any one period. Rather, grapheme shifts are distributed throughout the first two centuries of print. In the final section of this essay, I develop an algorithmic approach that will move beyond this small set of selected examples to explore at scale whether there is any pattern to when major orthographic shifts happen. However, we need to look beyond a specific period to understand the processes of standardization. As we have noted, scholars have often tended to point to the decades around the civil war as the moment when the bulk of standardization—conceived as the elimination of variation—happens (Görlach 1991, 55; Howard-Hill 2006, 17). While this period does see a set of well-defined and important orthographic shifts, such a periodization fails to acknowledge the many important changes that occur throughout the sixteenth century. It also tends to privilege the movement toward modern forms as the basis for standardization. It does not recognize the possibility that orthographic forms that were eventually eliminated could nevertheless have been a part of an accepted, if evolving, standard. For example, developments like the prevalence of the trailing *e* or *ie*—which were often used to indicate vowel length and eventually died out from modern English—were temporarily part of the accepted standard (Görlach 1991, 47). As a result, individual words often mutate through multiple shifts in different graphemes to eventually arrive at their final stable form. For example, words like "been" or "teach" (Figures 7 and 12) are subjected to several grapheme shifts and evolve through forms like "bene," "beene," "teche," and "teache," which are eventually rejected.

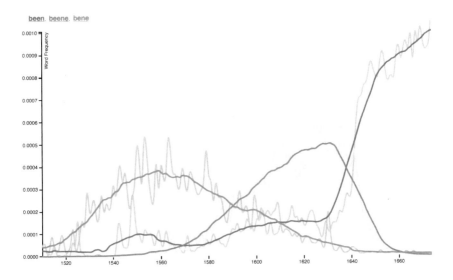

Figure 12. Relative frequencies for "been," "beene," and "bene."

These examples illustrate two central characteristics of the process of early modern orthographic standardization. First, it is not individual words but graphemes that form the basic unit for the process of standardization; second, each grapheme shifts over a relatively short period of time. This might seem to indicate some sort of conscious or concerted effort at standardization, which has usually been understood as stemming out of the intervention of the orthoepists. However, we also observe that these changes are widely spread out over the period, and there is no conspicuous moment when a definitive move to modern spelling happens. This is evident on a preliminary examination of our selected examples, and would align with the pattern of a steady but slow reduction of spelling variation that we observed in Figure 5. In other words, there was no great watershed moment at which early English orthography moved toward modernization, but rather a set of reiterative and independent shifts that gradually resulted in the emergence of modern spelling. Orthographic history, then, is nodal. It is not random, nor is it progressive. It is a history of regularization, to be sure, but it proceeds, not from chaos to order, but from coherence to coherence.

III

In trying to trace the major influences on the regularization of English spelling, recent scholarship has often set the print-houses and their practices in opposition to the works of the spelling reformers. The former are usually seen as a source of disorder and chaos juxtaposed against the prescriptive, ordering impulse of spelling reform. The early modern printing press was an amalgamation of a chaotic set of practices, driven by commercial interest and populated by printers, journeymen, and apprentices who often had little education beyond their practical training in the craft of printing and expressed no interest in any formal scheme of reform for the sake of improving the English language. On the other hand are the writings of linguistic philosophers, grammarians, and educators, including Sir Thomas Smith, John Hart, Richard Mulcaster, William Bullokar, Alexander Gil, and Sir John Cheke among others, produced mostly from the mid- to late sixteenth century to the first decades of the seventeenth that make up the inkhorn controversy. These reformers deserve much credit for giving the first structured account of the development and growth of the English language in the age of print. Besides attempting to construct reformed systems of spelling, they wrote on every aspect of linguistic standardization and change, including the nature of the spoken language and the status and necessity of loaned words entering English. However, their efforts at orthographic reform tended to bemoan the lack of correspondence to the spoken word (and correspondingly, the lack of consistency in pronunciation among regional dialects) and often proposed radical reform. They debated the introduction of new graphemes, accents, or ligatures, among other measures, to achieve phonemic consistency in English spelling. Consequently, it is hardly surprising that their recommendations seem to have had little effect on the commercial press. Many scholars have noted that it is hard to establish any direct influence of the spelling reformers on actual orthographic practice. As Scragg observes, "English spelling seems to have been particularly resistant to the interference of linguistic philosophers" (1974, 81). This resistance is starkly illustrated by the fact that even in the cases of the most important texts of spelling reform, printers and typesetters tended to use the common spelling forms of the day—many of which were disapproved of by the authors of these tracts—rather than the "proper," often phonetic, forms advocated by these texts (Howard-Hill 2006, 16). While this is in keeping with the general disregard of early modern typesetters for copy-spelling, it should make us skeptical about the extent to which the prescriptions of the orthoepists were heeded. However, in spite of the lack of direct evidence, the fact that spelling continued to be modernized has usually been attributed to the influence of spelling reformers.

While a detailed account of either the printing process or the considerable and often conflicting output of the orthoepists is beyond the scope of this essay, I want to ask how far a computational approach can help us understand their relative influences on orthographic change. The opposition between the press and orthoepists in the scholarly literature is significant because it is symptomatic of a tension between praxis and theory, between a conception of change as an apparently chaotic reiterative process and a more intellectual notion of organized, directed change. It is not an overstatement to suggest that the structure of this debate echoes an opposition between commerce and culture and the way the printed book—at once a marketable commodity and a cultural artifact—straddles that divide. Salmon registers this tension even as she underestimates the contribution of the printing press:

> In general, printers demonstrated their views in practice, grammarians in theoretical recommendations, and until recently, when Brengelman offered compelling arguments in support of grammarians and orthoepists, it was a matter of controversy which of the two groups was more influential in determining the eventual form of a standard orthography. (2000, 18)

Brengelman's work focuses attention on the "vast amount of scholarly and pedagogical effort devoted to English spelling during the late sixteenth and early seventeenth centuries" (1980, 333). However, his dismissal of the role of printing stems entirely from a rigidly prescriptive notion of standardization that fails to take accumulative and gradual linguistic change seriously. He notes the lack of interest among printers in developing or disseminating formal and uniform rules of standardization, citing the fact that "not a single printer's style manual survives, if any such manuals existed" (1980, 333). While he alludes to writers like Mulcaster as "resigned to the power of custom," his notion of custom is limited to the conjecture that "accident must have played an important role" in standardization—a course of events he finds highly unlikely, especially when contrasted with the idea that orthographic transformation was "the product of scholarly effort" (1980, 334, 335). Further, he insists that almost all of the standardization of spelling happened in the mid-seventeenth century—"during the middle half of the seventeenth century, English spelling evolved from near anarchy to almost complete predictability"—and uses this observation as evidence of concerted change engineered by "linguistic scholars" (1980, 334). Within this framework, Brengelman argues that "the case for the printers is [...] entirely negative" and feels that all that needs to be done to make the case for scholarly intervention is to demonstrate a "sustained and deliberate effort to rationalize spelling" on part of the orthoepists (1980, 335).

The distinctively structured patterns of phased change in the use of individual graphemes that we have observed (Figures 5 to 10) challenge Brengelman's account of English spelling as being in a state of "near anarchy" and displaying a "high degree of randomness" before a decisive intervention by the orthoepists (1980, 335). Moreover, the evidence of the gradual decrease in type-token ratios over time (Figure 5) suggests that there was in fact no decisive moment of transformation in the mid-seventeenth century. Rather, individual grapheme shifts that make up the process of overall change are distributed widely throughout the period. Thus, the question of whether it was the customs of the printshop or the prescriptive intervention of the linguistic philosophers that formed the epicenter of orthographic evolution is closely aligned with the related issue of when the majority of orthographic change happened. Even scholars who are less convinced than Brengelman about the influence of spelling reformers have often tended to identify the period around the civil war as the critical watershed moment when the bulk of orthographic standardization occurred. For example, Görlach suggests that "after 1630-40, the stabilized conventions of printers succeeded where scholarly effort had failed: they established in practice a set of rules, though not of course the kind of consistent system the reformers had hoped for" (1991, 55). T. H. Howard-Hill, perhaps the scholar who provides the most sophisticated account of printers' conventions as an agent of spelling standardization, conjectures a similarly localized time-format even though he admits that he is working on a relatively small set of texts and that it is not his purpose "to show exactly when standardization occurred and at what rate" (2006, 17). He compares spelling change in the first 242 words across 48 editions of Edmund Coote's *The English School-Master* published between 1596 and 1696, and 200 verse lines drawn from quarto and folio editions of *Titus Andronicus* and *A Midsummer Night's Dream* spanning 1594 to 1685 (the date of the fourth folio). Based on this sample, he suggests that "English spelling was predominantly modern by the end of the seventeenth century, the greatest progress towards standardization in printed works occurring from 1630 onwards" (2006, 16–18). This emphasis on the period of the civil war stems partially from the failure to acknowledge the possibility that non-modern orthographic forms could nevertheless have been the products of highly structured regimes of spelling. Even words that attain their modern forms in the mid-seventeenth century could have existed in relatively regularized earlier forms.

However, if we are to move away from this notion of prescriptive intervention, we also need to understand orthographic change as the accrued effect of the slow changes in individual graphemes that the orthoepists often referred to as "custom." While they saw themselves as striving against the "abuses, and vnperfectnesse, of the olde ortography for Inglish speech, at this day in vse" (Bullokar 1580, 3), the linguistic reformers also acknowledged the power and tenacity of custom. Hart describes custom as "any peoples maner of doings," and says that he is writing for the "good perswasion of a common commoditie" to those not "obstinate in their custome" (1569, 3–4). Bullokar, perhaps the most critical of contemporary usage among the linguistic philosophers, recognizes that linguistic conventions have continuously evolved: "for lack of true ortography our writing in Inglish hath altered in euery age, yea since printing began" (1580, 2). It is Richard Mulcaster, however, who is most receptive to the power and importance of custom, with all its imperfections, as a driver of orthographic regularization: "The vse & custom of our cuntrie, hath alredie chosen chosen a kinde of penning, wherein she hath set down hir relligion, hir lawes, hir priuat and publik dealings" (1582, N1v). Most Elizabethan and Jacobean writers on orthography tend to depict custom as obstinate and recalcitrant, yet hope to supersede it through their recommendations. Mulcaster, however, tends to juxtapose "custom" and "reason" to argue that any prescriptive or reasoned philosophical argument on spelling must necessarily take prevalent scribal and printing practices into account, although he does emphasize writing as the true measure of "custom" compared to the often error-strewn process of print. However, the fact that he acknowledges the strong regularizing impulse of customary usage as early as 1582 indicates a surprising insight into the accumulative nature of linguistic change, and might suggest that, in spite of the occasional rant about the carelessness and ignorance of printers, there was already a notion of an emerging set of standard practices.

Howard-Hill, in an important essay (2006) on the evolution of orthographic practice, has attempted to outline both the social organization of the early print trade and the everyday material practices of setting, printing, proofing, and redistributing type as the central locus from which this notion of "custom" or "common order" emerges. He demonstrates that the impulse toward optimization, efficiency, and ease of use is inherent in the repetitive processes of the early printing press, even if they are never consciously formulated as in the tracts on spelling reform. One of the major benefits touted for orthographic reform is the optimization and the resultant ease for learners—either children or foreigners trying to learn English—and for printers and typesetters. Hart argues that his system of spelling "should not neede to use aboue the two thirdes or three quarters at most, of the letters which we are nowe constreyned to vse, and to saue the one third, or at least the one quarter, of the paper, ynke, and time which we now spend superfluously in writing and printing" (1569, 6). Similar arguments can be found in Bullokar, Gil, and other reformers on the ease of printing and learning that standardization brings. However, the same pressures and benefits are felt in the press from the very advent of printing. Cusack, in a detailed study of the evolution of Caxton's typecase as it is gradually adapted for the demands of English, notes that "the reduction in [his] stock of ligature-characters as his work progressed is one of the clearest trends in the evolution of his founts. For where his first type has about seventy-five different combinations of letters in ligature, his last cuts this figure down by two-thirds" (1971, 43). Such optimizations continue as the printing process matures in England.

Even though it does not result in overt standards or a manual for printing, as Brengelman points out, the demands of assimilating and training apprentices and the need to speed up the setting and redistribution of type has a strong accumulative effect on standardization. Apprentices who were bound to the stationers' company, usually between the age of 13 and 24, came from different regions of England, probably with little more than a mostly Latin-based grammar school education, and brought with them considerable linguistic diversity. Before young apprentices could compose, however, they would learn the layout of the typecases by sorting or distributing type.

They would start by distributing pie—mixed, unordered type—and move on to distributing type set by other, more experienced compositors. This long process involved familiarizing themselves with the complex layout of typecases, which often had well over 200 compartments. However, Howard-Hill argues, just as important for the efficiency of this process was imbibing the orthographic habits of compositors to allow for more efficient redistribution without having to repeatedly look at the set type. In the process, apprentices "trained their eidetic memories to remember the individual spellings of the lines they were assigned to distribute, and visualized the combinations of sorts (ligatures etc.) that made up the line of type" (Howard-Hill, 2006 23). This relatively simple cycle of typesetting and redistribution, repeated countless times in the course of training, formed the crux of a certain kind of "mechanical economy of the printing press" that not only ensured the efficient training of young apprentices to eventually become compositors, but also passed down certain orthographic conventions and inculcated a subtle impulse toward standardization and the elimination of variance (2006, 21). The crux of evolving "custom," within this framework, lies in the mechanical processes of repetition and reinforcement that make up the practices of the early press rather than in the theoretical articulation of a prescriptive standard.

IV

To what extent can we extract and visualize the accumulated effects of such "custom" on early modern orthographic standardization? If, as we have seen in our example words (Figures 7 to 12), spelling regularization happens as a set of phased transformations at the level of individual graphemes, how far can a large-scale computational approach help us in visualizing the broad contours of such change occurring across thousands of words? In this section, I want to outline an approach that will allow us to move from the observation of individual words to an analysis of the cumulative patterns of change over time. Algorithmically identifying words that undergo spelling change, and isolating the approximate time period over which each grapheme-shift occurs, will allow us to explore when orthographic standardization happens in English print. Are the individual grapheme shifts relatively evenly distributed across the sixteenth and seventeenth centuries, or are there watershed moments or phases when a large number of words undergo transformation? Being able to extract the characteristic patterns of spelling change from a vast corpus of texts might allow us to gauge whether the prevalent model of a wave of regularization in the mid-seventeenth century, presumably as an indirect result of the writings of the orthoepists, needs to be reevaluated.

More fundamentally perhaps, the scale of analysis that such computational techniques make possible might allow us to rethink the ways in which the cumulative effects of repetition—actions that are in themselves minuscule, mechanical, and untheorized—aggregate over time to create a well-defined set of standards.

Using a database of n-gram frequencies for all texts in the EEBO-TCP distribution, we can calculate yearly relative frequencies for each orthographic form from 1473 to 1700.[6] Every word in the corpus is thus represented as a time series consisting of values across 227 years, each calculated as the number of instances of the given word divided by the total number of words for that year. As we have seen (Figures 7 to 12), these values can readily be represented as line plots that show the usage rates of an orthographic form. However, while the general trend and any conspicuous shift in a word's usage becomes apparent on visual inspection, we need a robust method of parsing each word in the EEBO-TCP corpus, extracting a time series from it, and algorithmically identifying words that show the characteristic pattern of rise or fall that we have noticed across orthographic variants if we are to scale up our analysis beyond the anecdotal evidence of individual words. Linguists have long been familiar with the S-shaped curve that is typical of shifts in the usage of particular features (Nevalainen and Raumolin-Brunberg 2003, 53). It represents a slow period of development when the form is first introduced and is atypical, a period characterized by a long and distinctive rise as the form becomes more popular, and finally a flatter period where it gradually achieves complete dominance and stamps out variants to become the accepted standard. The reverse pattern can be noticed in cases where a particular form declines and eventually falls out of usage. Of course, as we have noticed, in the case of orthographic variants, the rise and eventual hegemony of one spelling often comes at the cost of a corresponding variant. The time series plot of one mirrors the other. Interestingly, the decline in a word's usage tends to lag slightly behind the rise of the word that replaces it, which might indicate that even when there exist relatively well accepted standards, older forms tend to linger, perhaps sustained by the practice of certain compositors who stick to their habitual spelling. Compare, for example, the slightly offset pattern of the decline of "giue" and "loue" against the rise of "give" and "love" as the accepted spellings in Figure 9.

---

[6] N-grams are sequences of one or more words extracted from a text corpus and widely used in computational natural language processing applications. For the study of orthographic change, we will only use single words, or unigrams.

To isolate words that undergo interesting orthographic transformations, therefore, we need to be able to identify the characteristic S-shaped or reverse S-shaped curves over a time span of a few decades algorithmically from the time series vectors of all words in the EEBO-TCP corpus. This involves two main tasks: detecting periods of sudden rise or fall in the relative usage of an orthographic form over a sliding window across the time series, and avoiding as far as possible false positives resulting from local spikes or troughs that result from normal variations in a word's usage from year to year. The analysis of patterns in time series data has received widespread attention in recent years in a wide array of information processing contexts. A subfield of signal processing, these techniques find usage in a variety of domains from machine learning to the analysis of economic data. I adapt a subset of these techniques to parse nearly four million time series representing the relative frequencies of individual tokens in the EEBO-TCP corpus in order to find ones that contain the distinctive "S" curve. Once we find a possible match—i.e., we encounter a time series that is a close fit for the pattern we have noticed for grapheme shifts—we can score its closeness of fit to an "ideal" target function. This allows us to build an effective "classifier" for the detection of words that undergo spelling change; we can then rank every word in the corpus, and retain the top 20,000 highest scoring ones. (Readers not interested in the technical details of the model might want to skip the next two paragraphs.)

To identify "S" and reverse-"S" shaped curves we can represent the patterns as target functions f_0 and f_1, denoting the "ideal" forms of the curves (Figure 13). Each curve is represented as a sigmoid function given by the following general equations:

$$f_0(x) = \frac{1}{1 + e^{-x}}$$

$$f_1(x) = \frac{1}{1 + e^{x}}$$

To compute the degree of "fit" between these ideal projections and the actual segments of the time series we gather from the EEBO-TCP corpus, we can calculate the mean-squared error between either of these target functions and a segment of the time series curve for a given number of years, y. Mean squared error is a widely used estimate that quantifies the degree to which the responses for two functions are aligned with each other (James et al. 2013, 29). In essence, given the observed pattern of spelling change for a specific subset of words, we want to be able to build a highly scalable and flexible search function that will identify similar patterns across millions of token time series. We also want our search to be unconstrained by either the specific x-axis-location, or the amplitude within the time series in which the pattern occurs. In other words, our algorithm should detect the pattern regardless of whether the spelling shift occurs in the mid-seventeenth or the late sixteenth century or elsewhere. It should also be unconstrained by the actual frequency of the token (whether the word undergoing transition is a relatively common one such as "goe," or a rarer word such as "whylome"), and refer only to the *shape* of the time series plot. We can achieve the first goal by using a sliding window of y years across the time series, while the second goal is easily achieved by scaling or normalizing the time series values to a range of [0,1] so as to match the range of the target functions in Figure 13.

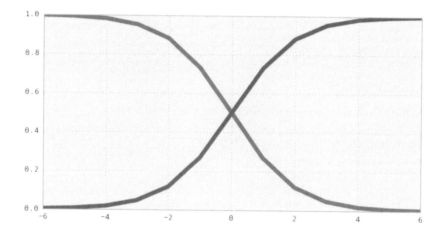

Figure 13. Sigmoid functions f_0 (blue) and f_1 (purple). The range of the functions (the Y axis) is [0, 1], while the domain on the X axis can be offset or multiplied by a constant.

If there are *n* words in the EEBO-TCP corpus, each produces a time series written as a vector X with 228 columns (denoting relative frequencies from 1473 to 1700): {X1, X2, X3, ... Xn}. We want to be able to detect "S" or reverse-"S" shaped patterns ranging over a certain number of years y within each time series. For each vector Xi, we calculate mean squared errors between a sliding window of length y years and the ideal forms of our target sigmoid functions f_0 and f_1. We represent a window of y years in a time series vector

Xi, as Xiy, where y ranges from 0 to |X|-y. We can then compute the mean squared error between any window Xiy and the target functions f_0 and f_1 by the following formula:

$$MSE = \frac{1}{y} \times \Sigma_{n=0}^{y} (X_{iyn} - f(x_n))^2, \text{ where } f \in \{f_0, f_1\}$$

Of course, each vector Xi will return |Xi|-y values for each of our target functions. To calculate the best fit, we need to minimize this function, or in other words take the minimum value from this set of results as our score for the vector. This score represents a degree of fit for the point in the time series curve that best matches either of our target functions. Thus, this algorithm will identify the years over which the time series for any token in the EEBO-TCP best approximates an "S" or reverse-"S" curve. If the mean squared error for a target function f that we calculate for each y-year window over a vector Xi in the time series is denoted MSE_{yf}, we can calculate the minimum score as:

$$MSE_{min} = min(MSE_{yf}), \text{ where } y \in \{y_0, y_1 ... y_{|X\_i|-y}\}; f \in \{f_0, f_1\}$$

We sort this calculated minimum MSE over all tokens in our corpus to isolate the 20,000 words that show the most distinctive patterns associated with grapheme shifts. The years over which each token undergoes the shift is simply given by the value of y at which the minimum value of MSE is returned.

As an example, consider the time series plots for "giue," "give," "loue," and "love" (Figure 9). The algorithm successfully identifies the broad "S" curve in the time series and locates the beginning of the civil war as the period when these two shifts occur. Other words that have grapheme shifts in different periods are identified by their characteristic curve but, of course, return the match over a different window y. Identifying the 20,000 words that show these distinctive S-curve patterns, either upward or downward, gives us a sample large enough for us to explore the general contours of change across a wide range of graphemes in the early modern period. To identify when the most number of words undergo orthographic shift in the early modern period, we can sum up the cumulative transition periods of individual tokens to see if there is a general time pattern to linguistic change. We plot the periods of transition for words that show evidence of an upward "S" curve—i.e., new forms that gain wide acceptance—in Figure 14. A corresponding barplot (Figure 15) visualizes the number of words undergoing a downward transition in a given year—in other words, it plots the rates at which orthographic forms fall out of usage.

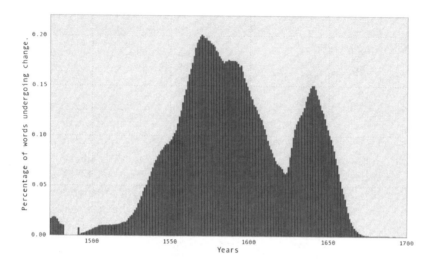

Figure 14. Percentage of orthographic forms per year that are coming into widespread use. This is represented by the slope of the "S" curve, denoting a rapid rise in the usage of individual tokens.

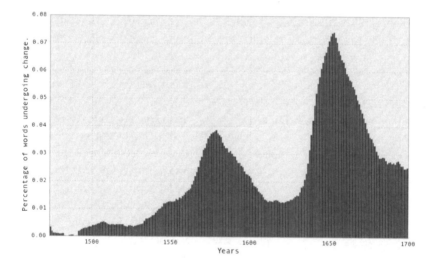

Figure 15. Percentage of orthographic forms undergoing a rapid decline in usage per year.

Together these plots give a broad account of the rate of spelling change across the first two centuries of English print. It is an account that fits in certain limited aspects with a familiar narrative, but also reveals surprising new patterns that contradict the prevalent linear model of orthographic standardization. A comparison of the two plots shows that the rate of decline for graphemes lags slightly behind the rate of acceptance for new forms. Older spellings continue to occur in relatively low numbers even after the new forms have established themselves. This is to be expected because, as we have seen in the time series plots from Figure 7 to Figure 12, variant forms continue to persist, perhaps in isolated print-houses or in the practices of a small number of possibly older typesetters. Sixteenth- and seventeenth-century print certainly doesn't eliminate all variation, and these plots should serve as a reminder that variants continue to exist even when one particular orthographic form dominates overwhelmingly enough to constitute a de facto standard—or, to use the word cautiously, even the "correct" way of spelling a word at a given time. The spike around the mid-seventeenth century in Figure 15 indicates what we might think of as a major "cleanup"—a wave of regularization where a lot of lingering variant forms were finally eliminated, perhaps fueled by the increasing demands of efficiency made on presses by the exploding volume of print during the civil war.

However, it is the distinctive pattern in Figure 14 which should give us the most pause. It indicates that there were two major waves of grapheme shifts followed by periods of relative lull in the rate of orthographic change within the first two centuries of English print. The second wave around the time of the civil war fits our expectation. It is well defined and easily identifiable. But it is the first wave around the mid- to late sixteenth century that is the larger one. This is the period where the process of printing matures in England. The Stationers' Company recieves charter in 1557, and a steadily increasing number of books are printed by a generation of printers and apprentices mostly trained in London. So we should perhaps not be surprised that a huge number of words undergo transformation in this period. The spike is not only higher (a greater number of words) but also broader than the later wave of regularization, and the corresponding spike in word eliminations in Figure 15 is much smaller. This indicates that the few words we have already observed as shifting before and around the mid-sixteenth century are not aberrations, but part of a broad set of innovations that come out of the printing press. However, in spite of the spike, the broadness combined with the relatively low number of words falling completely out of usage indicates that a degree of variation remains. Even when one form establishes itself as the dominant spelling, alternatives are not completely eliminated. This persistence of variant forms should not be construed as evidence of chaos or a complete lack of standardization. On the contrary, the pattern alerts us to a very significant phase of regularization in the mid-sixteenth century that has been almost totally overlooked.

Perhaps the most significant observation, and one that deserves more sustained attention, is the fact that the first great wave of orthographic standardization in English—one that dwarfs the second major wave around the civil war—predates most of the writings on linguistic reform. In fact, the period from around 1570 to 1620, when most orthoepists wrote, turns out to be one of relative orthographic stability. This pattern of change, along with the evidence we have considered of similar highly structured shifts for graphemes such as "ea" in the early sixteenth century, should make us reconsider the current model of orthographic standardization. If the writings of the orthoepists are not the prime movers in the processes of standardization, we should look to the repetitive processes of the printing press and reassess their impact on the evolution of English orthography. Such a model invites us to rethink the simple opposition between accident and intellectual intervention, between practice and theory.

The scale of the EEBO-TCP corpus allows us to sketch the outlines of a *materialist* history of standardization, one that is animated not only by the intellectual and ideological debates that characterize early modern discussions of language but is equally inflected by material practices and technologies, hierarchies and social structures within guild and print-shop, that contribute toward the production of the book as an artifact. These questions need further investigation, and for us to be able to address some of the issues it raises—such as the development and propagation of certain conventions among individual print-houses—we will need to get access to better metadata or other aspects of information on the EEBO-TCP corpus. But the corpus of early modern print is already available and tractable for computation. As it continues to improve, we would do well to reconsider how some of the biological and evolutionary metaphors used in the discussion of linguistic change in recent years (Hope 2000) might make new insights possible or allow us to grapple with the scale of the data we encounter. To a large extent, such evolutionary metaphors of mutation and propagation already structure the ways in which we conceptualize cultural influence and change at scale. However, the central insight of the evolutionary model that seeks to explain how apparently chaotic processes, when reiterated at massive scale, can lead to fundamentally ordered and directed change has often been overlooked in the ways we have translated and applied these metaphors to the study of culture. I hope the scalable approaches that computation makes possible will invite us to rethink not only the underlying mechanisms that drive orthographic change and standardization over the first two centuries of English print, but also the extent to which such models of repetition, mutation, and accumulative change can become novel paradigms of thinking about culture.

WORKS CITED

Bale, John. ca. 1570. *The Image of Both Churches after the Most Wonderfull and Heauenly Reuelation of Sainct Iohn the Euangelist, Contayning a Very Fruitfull Exposition or Paraphrase Vpon the Same. Wherin It Is Conferred Vvith the Other Scriptures, and Most Auctorised Histories. Compyled by Iohn Bale an Exyle Also in Thys Lyfe, for the Faithfull Testimony of Iesu.* London: Thomas East. Early English Books, 1475–1640 / 485:01. http://name.umdl.umich.edu/A02872.0001.001.

Barber, Charles Laurence. 1976. *Early Modern English.* The Language Library 1. London: Deutsch.

Baron, Alistair, Paul Rayson, and Dawn Archer. 2009. "The Extent of Spelling Variation in Early Modern English." *International Computer Archive of Modern and Medieval English* 30. http://eprints.lancs.ac.uk/42530/2/icame_presentation_v3.pdf. Accessed October 1, 2013.

_____ 2012. "Innovators of Early Modern English Spelling Change: Using DICER to Investigate Spelling Variation Trends." http://www.helsinki.fi/varieng/series/volumes/10/baron_rayson_archer/.

Basu, Anupam. 2015. "Early Modern Print: Text Mining Early Printed English." http://earlyprint.wustl.edu/.

Blake, N. F. 1965. "English Versions of Reynard the Fox in the Fifteenth and Sixteenth Centuries." *Studies in Philology* 62 (1): 63–77.

Brengelman, F. H. 1980. "Orthoepists, Printers, and the Rationalization of English Spelling." *Journal of English and Germanic Philology* 79 (3): 332–54.

British Library. *English Civil War: Thomason Tracts.* http://www.bl.uk/reshelp/findhelprestype/news/thomasontracts/. Accessed October 26, 2015.

Bullokar, William. 1580. *Bullokars Booke at Large, for the Amendment of Orthographie for English Speech.* London: Henrie Denham. Early English Books, 1475–1640 / 951:04.

Burnley, J. D. 1989. "Sources of Standardization in Later Middle English." In *Standardizing English: Essays in the History of Language Change in Honor of John Hurt Fisher*, edited by Joseph B. Trahern, Jr., 23–41. Tennessee Studies in Literature 31. Knoxville: University of Tennessee Press.

Burns, Philip R. 2013. *MorphAdorner v2.0.* Northwestern University. http://morphadorner.northwestern.edu/.

Christianson, C. Paul. 1989. "Chancery Standard and the Records of Old London Bridge." In *Standardizing English: Essays in the History of Language Change in Honor of John Hurt Fisher*, edited by Joseph B. Trahern, Jr., 82–112. Tennessee Studies in Literature 31. Knoxville: University of Tennessee Press.

Cusack, Bridget. 1971. "'Not Wreton with Penne and Ynke': Problems of Selection Facing the First English Printer." In *Edinburgh Studies in English and Scots*, edited by A. J. Aitken, Angus McIntosh, and Hermann Pálsson, 29–54. London: Longman.

Fisher, John H. 1996. *The Emergence of Standard English.* Lexington: University Press of Kentucky.

Görlach, Manfred. 1991. *Introduction to Early Modern English.* Cambridge and New York: Cambridge University Press.

Hart, John. 1569. *An Orthographie Conteyning the Due Order and Reason, Howe to Write or Paint Thimage of Mannes Voice, Most like to the Life or Nature. Composed by I.H. Chester Heralt. The Contents Wherof Are next Folowing.* [London: [Henry Denham?] for] William Seres. Early English Books, 1475–1640 / 321:10.

Hope, Jonathan. 2000. "Rats, Bats, Sparrows, and Dogs: Biology, Linguistics, and the Nature of Standard English." In *The Development of Standard English, 1300-1800: Theories, Descriptions, Conflicts*, edited by Laura Wright, 49–56. Studies in English Language. Cambridge and New York: Cambridge University Press.

Howard-Hill, T. H. 2006. "Early Modern Printers and the Standardization of English Spelling." *The Modern Language Review* 101 (1): 16–29.

James, Gareth, Daniela Witten, Trevor Hastie, and Robert Tibshirani. 2013. *An Introduction to Statistical Learning: With Applications in R.* New York: Springer.

Jockers, Matthew L. 2014. *Text Analysis with R for Students of Literature*. Cham, Switzerland: Springer International Publishing.

Kao, Justine, and Dan Jurafsky. 2012. "A Computational Analysis of Style, Affect, and Imagery in Contemporary Poetry." NAACL Workshop on Computational Linguistics for Literature, 8–17.

Milroy, Jim. 2000. "Historical Description and the Ideology of the Standard Language." In *The Development of Standard English, 1300-1800: Theories, Descriptions, Conflicts*, edited by Laura Wright, 11–28. Studies in English Language. Cambridge and New York: Cambridge University Press.

Mueller, Martin. 2009. "Nupos: A Part of Speech Tag Set for Written English from Chaucer to the Present." *WordHoard*. http://wordhoard.northwestern.edu/userman/nupos.pdf.

Mulcaster, Richard. 1582. *The First Part of the Elementarie Vvhich Entreateth Chefelie of the Right Writing of Our English Tung, Set Furth by Richard Mulcaster*. London: Thomas Vautroullier. Early English Books, 1475–1640 / 426:04.

Nevalainen, Terttu, and Helena Raumolin-Brunberg. 2003. *Historical Sociolinguistics: Language Change in Tudor and Stuart England*. Longman Linguistics Library. London and New York: Longman.

Puttenham, George. 1589. *The Arte of English Poesie Contriued into Three Bookes: The First of Poets and Poesie, the Second of Proportion, the Third of Ornament*. London: Richard Field. Early English Books, 1475–1640 / 1839:18; Early English Books, 1475–1640 / 421:12.

Rissanen, Matti. 2000. "Standardization and the Language of Early Statutes." In *The Development of Standard English, 1300-1800: Theories, Descriptions, Conflicts*, edited by Laura Wright, 117–30. Studies in English Language. Cambridge and New York: Cambridge University Press.

Salmon, Vivian. 2000. "Orthography and Punctuation." In *The Cambridge History of the English Language, 1476-1776*, edited by Roger Lass, 13–55. Vol. 3 of *The Cambridge History of the English Language*. New York: Cambridge University Press.

Scragg, D. G. 1974. *A History of English Spelling*. Manchester: Manchester University Press.

Smith, Jeremy J. 2012. "From Middle English to Early Modern English." In *The Oxford History of English*, edited by Lynda Mugglestone, 147–79. Updated edition. Oxford: Oxford University Press.

Vallis, Owen, Jordan Hochenbaum, and Arun Kejariwal. 2014. "A Novel Technique for Long-Term Anomaly Detection in the Cloud." 6th USENIX Workshop on Hot Topics in Cloud Computing (HotCloud '14). Philadelphia, PA: USENIX Association. https://www.usenix.org/system/files/conference/hotcloud14/hotcloud14-vallis.pdf.

Virgil. 1490. *Here Fynyssheth the Boke Yf [sic] Eneydos, Compyled by Vyrgyle, Which Hathe Be Translated Oute of Latyne in to Frenshe, and Oute of Frenshe Reduced in to Englysshe by Me Wyll[ia]m Caxton, the Xxij. Daye of Iuyn. the Yere of Our Lorde. M.iiij.Clxxxx. The Fythe Yere of the Regne of Kynge Henry the Seuenth.* [Westminster: William Caxton.] Early English Books, 1475–1640 / 17:18.

Wright, Laura. 2000. "Introduction." In *The Development of Standard English, 1300-1800: Theories, Descriptions, Conflicts*, edited by Laura Wright, 1–8. Studies in English Language. Cambridge and New York: Cambridge University Press.

Linked Open Data and Semantic Web Technologies in *Emblematica Online*[1]

Timothy W. Cole
University of Illinois
at Urbana-Champaign
t-cole3@illinois.edu

Myung-Ja K. Han
University of Illinois
at Urbana-Champaign
mhan3@illinois.edu

Mara R. Wade
University of Illinois at Urbana-Champaign
mwade@illinois.edu

This paper explores the new affordances for emblem scholarship made possible through the integration of Linked Open Data (LOD)-based services and LOD-compatible authorities into *Emblematica Online*. This corpus and its associated portal[2] provides search and discovery services for digitized emblem books and associated emblem resources aggregated from six international institutions as of December 2015. While this research portal supports traditional scholarly approaches to inquiry, the addition of LOD-based features adds meaningful new functionality, enabling new modes of inquiry.

Traditional emblem research requires the ability to study authors and national traditions at scale. By bringing together 1,388 complete digital facsimiles from collections of emblematica worldwide, *Emblematica Online* supports this research at a greater scale than ever before. As a digital library of emblematica, the aggregate corpus of *Emblematica Online* allows scholars to search for emblem books across geographically remote locations and make efficient use of their time and financial resources. Beyond the very practical features of the reduction of labor and expenditures, *Emblematica Online*

---

[1] The research was completed with a grant from the National Endowment for the Humanities, Preservation and Access: Humanities Collections and Reference Resources, # PW5145413. All images courtesy of *Emblematica Online*: the Rare Book and Manuscript Library of the University of Illinois at Urbana-Champaign; the Herzog August Bibliothek, Wolfenbüttel; Glasgow University; and Utrecht University.
[2] See http://emblematica.library.illinois.edu/.

enables scholars to study books from diverse collections side by side in real time, allowing them to make comparisons that were previously not possible. The collocation of so many emblem books means that scholars no longer study only the handful of books and authors available in facsimile reprints. Thus the portal's extensive corpus enlarges the canon of works studied, and authors and texts previously marginalized because of the lack of access are now available for research and pedagogy. The open digital corpus of *Emblematica Online* is as large as or larger than any single print emblematica collection, with the notable exception of the Stirling Maxwell Collection.[3] Its larger scale, diversity, and increased facility for searching across collections also increase the opportunity for serendipity, the creative coincidence that is a prerequisite for new insights into the material.

Combining both the document-based web of digital surrogates for emblem books and the Semantic Web with its LOD for individual emblems, a practice enabled by new technologies, *Emblematica Online* also makes it possible for scholars to recognize new relationships and ask new questions. The research described in this paper demonstrates how our ability to pursue these research questions is enabled by the affordances of new technologies. Being able to identify clusters of emblems across formerly discrete collections allows researchers to study them in a nuanced manner—i.e., to ask new questions of the images and their associated texts by juxtaposing changes in the textual and pictorial articulation of concepts over time with regard to subtle variations in the iconographic presentation, new linguistic contexts, and a changed historical setting. For example, how do representations of the themes of Peace and War change or remain consistent across the early modern period, a time characterized by the unrelenting strife of conflicts such as the Dutch Revolt (1568–1648) and the Thirty Years' War (1618–1648)? Which pictorial elements are used to represent Peace and War in this period? Do the texts reflect changes in mentalities and attitudes over time? The new capacity to search at the granularity of the emblem (rather than solely at the granularity of the book) reflects enormous advantages in terms of both scale and complexity.

Because emblems themselves operate on the basis of textual and pictorial analogies and parallels, they are well suited to benefit from Semantic Web technologies. The flexible, multilingual searchability on consistent notations with a navigable hierarchy supports the associative thinking required of creating and decoding emblems. Therefore, emblems connected by LOD provide

---

[3] See http://www.gla.ac.uk/services/specialcollections/collectionsa-z/stirlingmaxwellcollection/.

multiple points of access into the innovatory processes integral to emblematica and thus foster research creativity. By using Semantic Web technologies, scholars are now better able to search and discover nuances across broader chronologies, thematic corpora, and iconographical meanings than previously possible. This demonstrates the advantages and enabling role of moving from opaque (to the computer) string data, as found in document-based web resources, to LOD, as confirmed when using *Emblematica Online*. Collating completed projects representing subsets of the early modern corpus into a single portal creates greater scale both quantitatively and qualitatively, facilitating new scholarship. This larger digital corpus can be placed into a much larger discourse by integrating the study of emblems and other contemporary print resources. By employing LOD, *Emblematica Online* also enables researchers to connect emblems to information resources from related domains available on the web—not just popular culture resources like Wikipedia, but also substantial scholarly resources such as those of the *Virtuelles KupferstichKabinett*, or VKK (the Virtual Print Room),[4] and *Festkultur Online* (Festival Culture Online).[5] With these capabilities, *Emblematica Online* fosters vertical and horizontal knowledge management and integration.

LOD-based services maximize the usefulness of rich, hierarchical, multilingual vocabularies such as Iconclass.[6] Iconclass was embraced early on by the international emblem studies community. However, as a practical consequence, this meant that Iconclass was applied in different languages by emblem catalogers working in different locales. This practice became an interoperability limitation when integrating digitized emblem resources from institutions in different countries. As described below, LOD-based services have helped to overcome this interoperability limitation. More importantly, Iconclass LOD services are also helping scholars discover new connections between emblems and other kinds of resources used in the study of Renaissance culture, such as the graphic prints of the *Virtuelles KupferstichKabinett* and the illustrated festival books of *Festkultur Online*. Using Iconclass-based links to connect disparate resources in different discipline repositories reveals new insights and avenues for exploration that to date have escaped notice. Beyond the use of Iconclass as illustrated in this paper, the potential offered by LOD generally for discovering new connections is significant.[7]

---

[4] See http://www.virtuelles-kupferstichkabinett.de/.

[5] See http://www.hab.de/de/home/wissenschaft/forschungsprofil-und-projektefestkultur-online.html.

[6] See http://iconclass.org, http://iconclass.org/help/lod.

[7] According to the *State of the LOD Cloud* report (http://lod-cloud.net/state/), the

These are important, concrete ways in which LOD approaches can facilitate scholarly inquiry.

From Documents to Data

The history of the transition in technology as evidenced in the evolution of the web is pertinent to understanding the evolution we are seeing not only in emblem studies, but also more generally in historical humanities studies. Created in 1989 by Tim Berners-Lee, the HTML (HyperText Markup Language) and HTTP (HyperText Transfer Protocol) standards enabled the creation of a document-centric World Wide Web (Raggett et al. 1998, Chapter 2). More recently, best practice guidelines for LOD, in combination with updates to the Resource Description Framework (RDF) first introduced in 1998, and RDF-related protocols, are facilitating the emergence of the Semantic Web, a next-generation enhancement of the World Wide Web that is data-centric rather than simply document-centric.[8] Documents and document-like objects are still to be found on the Semantic Web, of course, but through the use of LOD and RDF, the way they are connected is enhanced. RDF supports the use of properties and classes in ways that enable standard, precise identification and differentiation of objects and the links between them. Coupled with LOD best practices on how to reference and talk about non-web entities (people, places, intellectual works, events), a broader range of linkage types and descriptive granularity can be expressed, allowing scholars to extend traditional research methods and paradigms and conduct their research in the digital environment more efficiently. The availability of sub- and supra-documental items enables scholars to better interrogate collections (e.g., of digitized emblem books) across institutional boundaries and at multiple levels of granularity, and to link resources used in one discipline to resources in other disciplines on the basis of their LOD-described properties, links, and attributes. Using an RDF-compatible ontology, links can now be differentiated as to their role, and entities within web resources can be more precisely labeled—e.g., as a person, a place, an event, a bibliographic entity. The implications and affordances of these technologies for various digital humanities scholarly domains are still being explored, but in the context of emblem

---

number of links in the entirety of the web's LOD cloud had already exceeded 500 million by August 2011.

[8] See, for example, Berners-Lee 1989. His February 2010 presentation at TED also reflects the role of the Semantic Web and Linked Open Data: https://www.ted.com/talks/tim_berners_lee_the_year_open_data_went_worldwide?language=en.

studies, LOD and RDF allow scholars to recognize and answer new questions and better overcome limitations of scale and geographical distribution.

From Printed to Digital Emblem Lexicon

Fifty years ago, in their landmark volume *Emblemata*, Henkel and Schöne demonstrated the research potential of moving from a strictly volume-level bibliographic perspective to a more granular, emblem-level view of the emblem book corpus.[9] By indexing, in some cases partially, 47 volumes, Henkel and Schöne enabled scholars to recognize new connections among a limited set of individual emblems. Scholars benefited from the access to information, and *Emblemata* became the standard reference work in the field of emblem studies. There were, however, drawbacks to *Emblemata* that can now be overcome in a LOD-based digital environment. For example, of the 47 volumes indexed by Henkel and Schöne, 26 of those were "almost completely analyzed," while the remaining 21 volumes were "only selectively" indexed. On the one hand, by disassociating the emblem from its book, Henkel and Schöne enabled users to search thematically across a select corpus of emblematica. On the other hand, the corpus was very narrow; it focused primarily on books from the sixteenth century and contained no emblem book after 1640, even though emblems were very popular well into the eighteenth century (Henkel and Schöne 1996, xxv).

Of the 3,713 emblems organized thematically, only 2,428 were "completely classified," and 901 classified only "to various abbreviated degrees," while 384 entries presented only *picturae* (xxv). The emblems' classification is significant to the present discussion as the avenue for entry to emblem-level data. Limited by scale and technology, Henkel and Schöne present emblems according to a single relevant image from the *pictura*. Thus, Alciato's emblem of Peace with the bees using a discarded helmet as a hive can be found only under the heading of an "implement of war," "Kriegsgerät" (Henkel and Schöne 1996, cols. 1489–1490), and does not occur in the insect world with the bees (Henkel and Schöne 1996, cols. 918–930)—although the indices do cross-reference "bee hive" and "peace" to this emblem.[10] The privileging of a single pictorial element in the classification process distorts the results, offering only an incomplete view of the string of associations for any given emblem. Because the helmet, and not the theme of Peace central to the meaning of Alciato's emblem, is indexed, the emblem's association with other emblems of Peace, such as ones featuring the elephant, cannot be discovered

---

[9] Originally published in 1967; the third edition of 1996 will be cited here.
[10] See "Ex bello pax" in Alciato 1531: http://www.emblems.arts.gla.ac.uk/alciato/emblem.php?id=A34a046.

(Henkel and Schöne 1996, cols. 418 and 409). Henkel and Schöne's lexicon is also a German-language reference resource, thus limiting access for modern researchers (although some of the indices are in Latin and Greek). Nonetheless, in spite of these limitations, the unequivocal success of *Emblemata* at transforming the discipline by creating pathways into the content across a select corpus does suggest the even greater potential for impact of a larger, web-based resource with LOD-structured metadata.

At nearly 30 times the scale of Henkel and Schöne, and with the technical wherewithal interactively to tailor views of the emblem literature to user requirements, we believe that LOD-based services and authorities will facilitate and in some instances enable new scholarship involving emblems. Ultimately, rich, RDF-based descriptions will allow scholars to discover and consider emblems beyond a bibliocentric perspective—e.g., in event-centric, geo-spatial-centric, social network-centric perspectives. In further unbinding the emblems from the book, *Emblematica Online* becomes a powerful new resource that enables research in a variety of contexts and from multiple points of view. By simultaneously offering scholars the emblem disbound and the emblem within the context of its book, we continue to support traditional research of books and authors, while concurrently providing a previously unimagined flexibility for research.

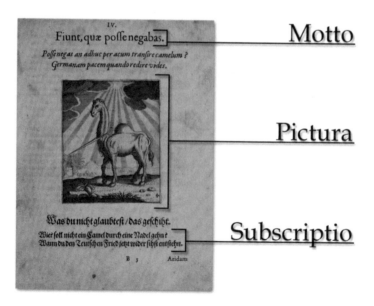

Figure 1. What is an emblem?

What is an Emblem?

The emblem expresses "highly complex ideas in compact and compelling forms" (Wade 2012a, 401). A bimedial genre, it combines texts and images in a typically tripartite structure consisting of a brief motto (*inscriptio*) in Latin or a European vernacular language, an enigmatic picture (*pictura*), and an epigram (*subscriptio*) (Manning 2002, 18–24) (Figure 1). The emblem is more than the sum of its parts; the motto, *pictura*, and *subscriptio* all work together to produce a greater meaning, the goal of which is to reorient the reader's thinking and produce new knowledge. Emblems were not only printed in books, they also pervaded the decorative arts, and are therefore integral to the understanding of both sacred and secular art of the period. The emblem is a critical genre to the understanding of Renaissance and Baroque Europe, owing both to its wide geographic spread and to the window it opens on the mentalities and attitudes of the period concerning nearly every aspect of life, ranging from love and politics to war and peace and religion. Emblems suggest a sophisticated strategy for repurposing, reorganizing, and reading texts and images through a system of parallels and analogies that narrow, or open, the meaning to impart a new perspective or idea. Because emblems embody both a rhetorical structure and a process, they are ideally suited to digital presentation in an LOD environment that can reflect semantic patterns of associative thought. LOD-based approaches enhance cataloging precision and interoperability, thereby opening a new breadth of exploration for emblem scholars and simultaneously making emblems published in books more visible to scholars in related disciplines: art historians, historians of Renaissance and Baroque cultures, comparative literary scholars, and scholars interested in the wider relationship between literature and the visual arts, theories of representation, and iconography. Because emblems arose concurrently to the development of European literatures in the vernacular and were also widespread in Latin, they became one of the primary vehicles for the transfer of cultural knowledge across Europe and beyond.

Major Emblem Collections and Emblem Digitization Projects

Digital emblem studies began in the late 1990s and early 2000s. Two early projects worth noting are: (1) the *Biblioteca Digital de Emblemática Hispánica* at the Universidade da Coruña, Spain (1995), which was later complemented by a database of emblem books translated into Spanish (1999),[11] and (2) *Digital-*

---

[11] See *Biblioteca Digital Siglo de Oro* for both projects, http://www.bidiso.es/estaticas/ver.htm?id=4.

isierung von ausgewählten Emblembüchern der frühen Neuzeit, a joint enterprise of the Ludwig Maximilians Universität and the Bayrische Staatsbibliothek, Munich, Germany.[12] The Spanish emblem project digitized and described 27 emblem books that were later supplemented by five additional emblem books translated into Spanish. The Munich project analyzes and classifies ~12,000 individual emblems from 139 books. Each of these early projects developed independent standards for digitization, metadata, and presentation on the web. These projects conducted pioneering work in emblem book digitization, and it is an urgent desideratum of research to transform the idiosyncratic data from these silos of information and aggregate them with other projects.[13]

A leader in coordinating the digitization of the European emblem is Alison Adams, Professor Emerita of French and former Director of the Center for Emblem Studies, Glasgow University, who together with her research team convened a series of meetings of international emblem scholars focused on digitizing emblem books and presenting them on the web beginning in the early 2000s. These ambitious, forward-looking meetings brought together an international cohort of scholars, who over time agreed to standards and norms that paved the way for future collaborative work (Graham 2004, 14). These projects are the first to use common standards and open practices in Emblem Studies and resulted in the completion of *French Emblems at Glasgow*[14] and *Alciato at Glasgow*.[15] *The Italian Project* at Glasgow is ongoing.[16] The Glasgow projects digitized and fully indexed 27, 22, and 7 emblem books, respectively. Concurrently, at the University of Utrecht, Els Stronks led a research group for the Emblem Project Utrecht that digitized and indexed 27 Dutch love emblem books from the seventeenth century.[17] The latter project continues with a web edition of Herman Hugo's *Pia desideria* (1624), and thus continues to explore scholarly communication for the emblem genre.[18] These projects,

---

[12] See http://mdz1.bib-bvb.de/~emblem/.

[13] See, for example, the comment by Professor Dr. Dietmar Peil on the Munich website: "Zur Zeit ist nicht abzusehen, ob sich eine weitere Förderungsmöglichkeit ergeben könnte, die eine Ergänzung der Datenbank um weitere relevante Titel sowie die äußerst wünschenswerte systematische Vernetzung mit anderen vergleichbaren Projekten (Utrecht, Wolfenbüttel, University of Illinois) ermöglichen würde."

[14] See http://www.emblems.arts.gla.ac.uk/french/.

[15] See http://www.emblems.arts.gla.ac.uk/#alciato.

[16] See http://www.emblems.arts.gla.ac.uk/#italian.

[17] See http://emblems.let.uu.nl/index.html.

[18] See http://emblems.let.uu.nl/hu1624.html.

in particular their common use of Iconclass as a classification system, provide the emblem corpora which were later ingested into *Emblematica Online*. As early as 2001, before any of the emblem projects were launched, scholars at the University of Illinois at Urbana-Champaign (UIUC) were discussing an Emblem Portal (Wade 2004b).

The Iconclass Thesaurus and Emblem Studies

Iconclass is a multilingual classification system for describing cultural heritage art and images. It contains more than 28,000 basic definitions representing a broad range of concepts, categories of art, people, events, and abstract ideas. Iconclass is hierarchical, with 10 top-level divisions, approximately 60 second-level subdivisions, nearly 400 third-level subdivisions, and so on through additional levels. It is also extensible. Through the use of the mechanisms described below, its database has grown to more than one million terms as of 2015.[19] Art historians, museum curators, emblem scholars, librarians, and others use Iconclass to describe and classify the content and meaning of curated images in a variety of formats, including paintings, emblems, drawings, and photographs (Graham 2004, 13–16). This allows scholars to search using Iconclass terms, keywords, or notations to discover images of interest.

Although it is being used in a digital context, Iconclass actually predates the web. Henri van de Waal (University of Leiden) began work on what was to become Iconclass in the 1950s. After his death, the first complete edition of Iconclass was published over multiple years between 1973 and 1985. The current online edition of Iconclass is maintained by the Rijksbureau voor Kunsthistorische Documentatie (RKD) in the Netherlands.

Iconclass has four features that are especially relevant in the context of digitized emblem literature and LOD.

(1) First is its hierarchical and extensible nature. *Picturae* can be described in great detail, i.e., to the level of a specific individual, event, locale, etc. Each base Iconclass definition is assigned an alphanumeric notation (classification code), a textual correlate (human readable label), and keywords. Iconclass notations begin with a digit (0–9) corresponding to the main divisions of Iconclass. The second character of the notation is also a digit, this one corresponding to the subdivision (to a maximum of nine subdivisions for any main division). Upper case letters are used to indicate the next level of hierarchy, with digits used for subsequent levels. Iconclass extensibility mechanisms,

---

[19] The numbers quoted here were taken from: http://iconclass.org/help/lod.

called *auxiliaries*, allow for more precise classification. Some Iconclass headings allow for *bracketed text*. The general Iconclass notation for literary proverbs, sayings, etc., is "86." A subdivision "86(...)" is provided to allow the inclusion of the specific text of a saying or proverb, e.g., "86(LIGHTLY COME, LIGHTLY GO)." Bracketed text is also used to assign headings for specific historical persons, places, and events; for example, starting from the base notation "11H(with NAME)," the Iconclass extended term "11H(FRANCIS)" can be constructed as the notation for classifying images of St. Francis of Assisi. Additional constructs, *structural digits* and *keys*, are provided to further refine meaning.

(2) Second, Iconclass works well for classifying emblem *picturae* because it is multilingual. Textual correlates and keywords are currently available in four languages: English, French, German, and Italian. Most emblem literature was published in Latin and contemporary vernacular languages (Renaissance variants of today's French, English, German, etc.). Emblem scholars today often specialize in emblems of one language or another. Iconclass notation, however, is invariant across languages. In an LOD context, this means that a librarian most comfortable in German can search or browse Iconclass in German to find the best terms to assign to an image, and a scholar working in English will still be able to find these images when searching in English. The language-invariant notation is used to bridge the gap between languages.

(3) A third benefit of Iconclass is the ability to discover emblem resources in multiple languages. For example, an early modern historian might be searching for "Turks" for an article on early modern European encounters with the Ottoman Empire. The Iconclass notation "32B33(TURKS) Asiatic races and peoples: Turks" in English reveals 41 hits for emblems treating Turks in French, Italian, Spanish, German, and Latin, thus amplifying the quantity of search results that could lead to qualitatively more significant understanding of early modern portrayals of Turks in both texts and images. Owing to the alphanumeric notations in Iconclass, better and more comprehensive searches are possible because scholars who search only in a single language get results from all of the languages. The serendipity of searches is also significantly enhanced. Thus scholars can build the portal independent of the languages used to encode the emblems, and other scholars can discover emblems independent of the language in which they search. This is one of the elegant solutions for a technical hurdle that reaps great benefits for research.

(4) Finally, Iconclass is LOD compatible: each Iconclass term (heading) has been assigned a persistent http-scheme Uniform Resource Identifier (URI), and the entire Iconclass database has been made available under an Open Database license.[20] (These URIs derive in part from Iconclass notation values.) Iconclass URIs can be used to retrieve RDF descriptions of Iconclass headings via Web Services at iconclass.org. This includes headings with extensions. (If a heading with extension is not found in the Iconclass database, information about the parent, i.e., base, term will be returned.) LOD descriptions are also available serialized as JavaScript Object Notation (JSON) to improve performance of consuming applications that rely on JavaScript, and term descriptions can be returned for more than one term at a time. The Iconclass. org site also supports Representational State Transfer (RESTful) searches of the Iconclass database. The return from the RESTful search is not RDF (and therefore not technically LOD), but rather HTML—which, helpfully, includes the URIs of Iconclass terms that match the search query in whichever language. Because LOD-compatible descriptions are used to build the index on which *Emblematica Online* relies, Iconclass term-level URIs are used as index keys in the *Emblematica Online* database rather than textual correlate or keywords in a particular language. This allows us to delegate Iconclass keyword and textual correlate searches to Iconclass.org—i.e., search algorithms parse the HTML returned from a search for term-level Iconclass URIs, then check the index for matches on URI (rather than language-specific strings). This represents a new, LOD-specific form of distributed searching functionality.

Emblematica Online: A Pragmatic Implementation of LOD

Emblematica Online's browse, search, and discovery services depend on descriptive metadata, and all books indexed for addition to the digital corpus are described using the SPINE metadata schema as developed by Rawles specifically to describe an emblem book, its emblems, and its emblem components (Rawles 2004). Thomas Stäcker at the Herzog August Bibliothek (HAB), with contributions from others in the emblem studies community, developed an XML schema based on the outline proposed by Rawles (Stäcker 2012, 1–26).

The SPINE schema allows three child components: book description, copy description (<copyDesc>), and emblem description(s). The book description component is used to express the bibliographic attributes of a specific

---

[20] See http://opendatacommons.org/licenses/odbl/1.0/. Any rights in individual contents of the database are licensed under the Database Contents License: http://opendatacommons.org/licenses/dbcl/1.0/.

edition or imprint of an emblem book. Copy description elements are used to describe "the copies on which the new digital works are based," and record information including ownership of the physical copy or copies digitized as representative of the printed edition (Rawles 2004, 21). A sub-element, <digDesc>, describes digitization details for digital instances created from the specific physical copy, and includes links (URLs) to each book-level digital instance available from a digitization. (It is not uncommon to have multiple digital instances derived from a single digitization.) The <emblem> element provides emblem-specific information, including information about emblem subcomponents, such as motto, *pictura*, and *subscriptio*. Controlled vocabulary descriptors, including Iconclass headings, are included here. Each <emblem> node also includes a unique and persistent handle system URI registered with the UIUC.[21] Consistent with LOD best practices, this facilitates emblem identification and linking.

Not all components of the schema are required. *Emblematica Online* provides independent search and discovery services at different levels of granularity based on the information available or not available in a SPINE metadata record. This flexibility allows *Emblematica Online* to provide search and discovery over emblem book collections that have no descriptions, or less than comprehensive descriptions, of individual emblems contained in their books. The minimum requirement for inclusion is a book-level bibliographic description and URL of the digital facsimile.

Multi-granular Discovery Services and the Integration of LOD

SPINE metadata descriptions serialized as XML were ingested into the data store supporting *Emblematica Online*, then subsequently enriched in accord with LOD best practices. (A provisional export as RDF using schema.org semantics has been implemented and is being tested.) Services were then implemented that take advantage of these enrichments and the range of descriptive granularity inherent in the original SPINE records, both to promote discovery and to provide context for and additional pathways from resources discovered.

Emblematica Online users can enter queries to find either whole emblem books or individual emblems. Keyword searching (searches of all fields) and field-specific searching are both supported. Thus, when searching for books, users can search by entering keywords, title words, names (authors or contributors), and/or publication information (date, publisher's name, or place of

---

[21] See http://www.handle.net/.

publication). For example, searching for books having the keyword "pace" (peace) anywhere in their descriptions retrieves five books from three different institutions (Figure 2). Note that the collection to which each book belongs is identified by the institutional logo: here Duke University, the Getty Research Institute Library, and the UIUC. Clicking on the title of any book in the results listing takes users to a page with more information about the book selected.

Figure 2. A keyword search of book descriptions for the word "pace" (peace) returns five results.

Similarly, searching emblem motto transcriptions for the word "pace" returns the *picturae* and mottos of 39 individual emblems (Figure 3). Users can click on a given *pictura* from the results listing to see more information about a specific emblem.

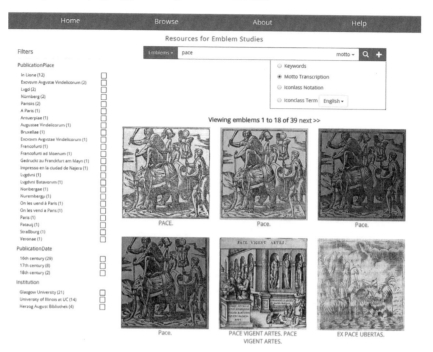

Figure 3. A search of emblem motto transcriptions for "pace" (peace) returns 39 results.

The implications for the scholar of being able to do this kind of search are significant both in terms of quantity and efficiency, but most significantly also qualitatively in the sense that scholars can study disparate resources together. Now able to survey a relatively broad cross-section of the emblem corpus, they can discover, juxtapose, compare, and contrast both books and individual emblems from different books; they can analyze a corpus across time and space, interpret in broader context and across domains, thereby deepening insights through new quantitative and qualitative approaches.

The details page for an individual emblem book, such as Johann Vogel's *Meditationes emblematicae de restaurata pace Germaniae = Sinnebilder von dem widergebrachten Teutschen Frieden* (Nürnberg, 1649), provides all of the book's bibliographic details, including a link(s) to view the complete digitized volume. For books that have been fully analyzed, another link ("View emblems from this book") is provided to display the *picturae* and mottos for emblems from the book. When viewing the Book Detail page for a book for which emblem descriptions are not yet available, a note is provided: "Emblem level searching for this book is not yet available." Where emblem book author names have been linked to LOD authority services, e.g., the LOD authority services of the Deutsche Nationalbibliothek (DNB),[22] a "More info" link is added to the Book Detail page, providing users with a way to see information gleaned from these LOD services about the author (e.g., birth/death dates, place of birth, gender, occupation) and links to further information (e.g., to the German-language Wikipedia[23] page about the author). Because emblem authors generally published in a range of contemporary genres, these features enable scholars to situate a given author in a broad discursive system. For example, Lutheran pastors who wrote emblem books also wrote sermons. Alciato himself was a jurist, and wrote learned tomes on various aspects of law. Relying on LOD-based web services helps to establish the context of emblem authors and thereby lead to insights about their habits and mentalities. For example, such information can reveal their literary and artistic networks. This is particularly useful because emblems are a pan-European genre, while modern scholars generally are trained within a national tradition, such as French engravers, German poets, and the like.

To enable this functionality, personal name information contained in the book-level metadata was analyzed to discover links to the Virtual International Authority File (VIAF) maintained by the Online Computer Library Center.[24] After normalizing for variations in punctuation, our current, unsophisticated algorithm identified 666 unique personal name strings associated with the 1,388 emblem books indexed in *Emblematica Online*. This number included some publishers and printers as well as book authors (which constituted the majority of names identified). VIAF, which itself aggregates a wide range of descriptive information about individuals, provides a number of LOD services. Through these services we can link users of *Emblematica Online* to additional context about emblem book authors, publishers, and printers. VIAF is also a link to more information about select individuals in resources such as WorldCat Identities[25] and Wikipedia.

---

[22] See http://www.dnb.de/EN/Service/DigitaleDienste/LinkedData/linkeddata_node.html.

[23] See https://de.wikipedia.org.

[24] The VIAF (https://viaf.org/) provides access to name authority files from 34 agencies in 29 countries as of July 2014, including 24 national libraries. The VIAF offers information related to each name, such as variant names, published work titles, major collaborators, publishers, nationality, and more.

[25] See http://www.worldcat.org/identities/.

Continuing to a more detailed level of granularity by clicking on a given *pictura* in an emblem search results listing or in a list of emblems from a specific book provides a preview image of the emblem accompanied by additional metadata describing the emblem (Figure 4). Iconclass descriptors assigned to the emblem, based on pictorial elements and topoi from the emblem *pictura*, are also shown when available, along with alternate transcriptions and translations of the motto as appropriate. As illustrated in Figure 4, the default language of the Emblem Detail page is English. Iconclass heading labels are fetched in real time from the Netherlands using the Iconclass LOD services. Users can switch the language of Iconclass headings displayed to German, French, or Italian by clicking on the appropriate item in the pull-down language menu. Additionally, as discussed in further detail below, the Iconclass headings attached to an emblem can be used to find additional emblems and images that have been assigned the same Iconclass descriptors.

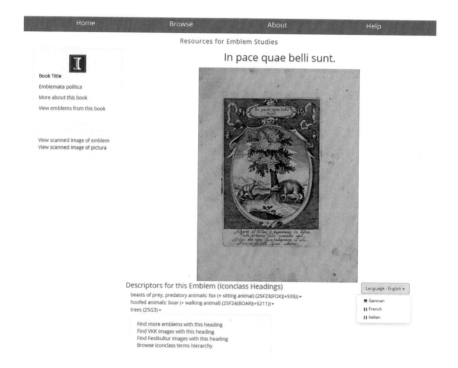

Figure 4. By clicking on the *pictura* thumbnail, the user can view a digital surrogate of the emblem image with rich descriptive metadata.

Browsing and Searching Using Iconclass LOD Services

The display and use of Iconclass in *Emblematica Online* exploit Iconclass LOD services. *Emblematica Online* uses Iconclass LOD services in lieu of maintaining a local copy of the full vocabulary, which would be wastefully duplicative and impossible to keep synchronized. Multiple modes of interacting with Iconclass are currently supported. In *Emblematica Online*, users can enter and search for an Iconclass notation (alpha-numeric code) directly, or as is more commonly done, enter and search for Iconclass terms using keywords or phrases. The user can specify an Iconclass term search in any of four languages. When this approach is taken, the search queries Iconclass LOD services in the Netherlands in order to find the Iconclass notations most relevant to the keyword or phrase entered (Figure 5). The Iconclass notation codes (expressed as URIs) are then used to identify emblems indexed with these same Iconclass notations.

Figure 5. A search for emblems having "Pace" (Peace) in the emblem motto and the Iconclass descriptor 45C221, "Helmet" (or "Casque" in French).

In addition, by using the pull-down menu for a particular Iconclass heading in an Emblem Detail display, users can link to where this Iconclass term fits in the Iconclass hierarchy (Figure 6). While browsing the Iconclass thesaurus, users can select the language of the labels shown and navigate up and down the hierarchy. The Iconclass browse and language-switching functionality is also supported by real-time calls to Iconclass LOD services. Having browsed to an Iconclass heading of interest, users can then search for other emblems indexed in *Emblematica Online* that have been classified with the same Iconclass heading.

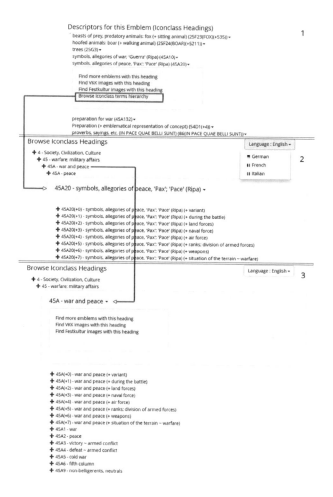

Figure 6. Users can view the Iconclass thesaurus structure and browse the hierarchy for Iconclass terms of interest.

Because Iconclass headings are used to describe resources found in other repositories of digitized images from the early modern period, Emblem Detail displays of Iconclass headings and Browse Iconclass displays of headings also provide links by which users can discover images in other collections. This functionality is illustrated in Figure 7. Here a user has browsed the Iconclass hierarchy to the heading "45A20 symbols, allegories of peace, 'Pax'; 'Pace' (Ripa)," and selected the pull-down menu next to the term. From this menu a user can initiate a search of either repository—"*Festkultur Online*" or "*Virtuelles Kupferstichkabinett*"—for images to which the Iconclass heading "45A20" also has been assigned. The interoperability and openness of Iconclass LOD services makes linking directly from *Emblematica Online* into these other significant repositories of Renaissance culture possible. One needs only to think of the Thirty Years' War to realize that scholars can aggregate here a substantial corpus of texts and images from multiple sources to reveal European mentalities about their social and political condition.

Figure 7. "Browse Iconclass" displays also provide links to the *Virtuelles Kupferstichkabinett* and *Festkultur Online*.

As a further illustration, consider a user looking for both emblem and non-emblem images pertaining to Peace. Searching from *Emblematica Online* into *Festkultur Online* on the Iconclass term "45A20 symbols, allegories of peace, 'Pax'; 'Pace' (Ripa)" reveals, for example, a pageant for the entry of King Augustus II of Poland into Danzig in 1697 (Curicke 1698).[26] The illustration reached through LOD depicts a fireworks architecture on the theme of Peace. Pyrotechnic displays were the prerogatives of royalty, and generally fireworks were divided into those fireworks for peacetime entertainments and those for waging war.[27] Since the Saxon Elector August's coronation as king of Poland was contested, the print and the emblem imagery elucidates arguments about the role of the festival welcoming the new monarch into Danzig. Searching the same term from the *Emblematica Online* to VKK retrieves 42 hits, including an image of Fortuna on the theme of Peace.[28] This print bears the inscription "Pace viget Fortuna favens, terraque marique Iam blandiuntur omnia" ("Through peace a benevolent fortune flourishes, on land as well as at sea. Then everything delights"), thereby suggesting emblematic strategies for this graphic and the deep intellectual connections across printed works in different genres. This expert research proceeds compactly and efficiently, contributing to a more complex understanding of European print heritage.

The features for linking to external projects still need to be refined, although we have convincingly demonstrated the feasibility of the concept. For example, fine-tuning how hierarchical authorities are searched needs to be addressed. A search in *Emblematica Online* one step up the Iconclass hierarchy on "45A2 Peace" produces more results, here 243, as one would expect. While in *Festkultur Online* and VKK the search on "45A20 symbols, allegories of peace, 'Pax'; 'Pace' (Ripa)" produces a strong set of results, the broader search on "45A2, peace" retrieves few or no results in these portals. These are among the problems to be addressed in future projects.

Leveraging Iconclass for Innovative Scholarly Inquiry

The LOD-based services afforded by Iconclass and exploited in *Emblematica Online* can be understood on the basis of the case studies discussed below. The emblem "Ex bello pax" from Andrea Alciato's *Emblematum Liber* (1531) illustrates one of the ways that Iconclass descriptors can support advanced scholarship. A search of *Emblematica Online* for emblems described with Iconclass

---

[26] See http://diglib.hab.de/drucke/gm-4f-256/start.htm.
[27] See http://diglib.hab.de/drucke/gm-4f-256/start.htm?image=00072.
[28] See http://www.virtuelles-kupferstichkabinett.de/index.php?selTab=3&currentWerk=2665&.

headings containing the word "bees" retrieves 369 emblems containing bee motifs, among them Alciato's emblem with bees swarming a helmet used as a hive, a powerful symbol of Peace. Clicking on this emblem brings up an Emblem Detail page, which in turn links to bibliographic information about the book in which this emblem is contained. From this page a user can learn that this particular emblem book was digitized from the collections of Glasgow University. The digital surrogate for the complete book is viewable on that institution's Alciato website.[29] Links to related information are also provided on the Book Detail page, including links to VIAF and DNB name authority information for authors and contributors. Users interested in the other 103 emblems in this book can click on links provided on both the Book Detail and the Emblem Detail pages to retrieve the emblem *picturae* from this book. Links back to the original search for "bees" are also provided.

The Emblem Detail page presents comprehensive information about Alciato's "Ex bello pax," beginning at the top of the page, in descending order, with the collection (the insignia of Glasgow University as a branding device); a persistent link to the digital facsimile of the book at Glasgow also is provided. The Emblem Detail page presents the cropped *pictura*, the transcribed emblem motto, and links to the entire emblem and to the *pictura*, concluding with the Iconclass headings. Flags indicate the multilingual thesauri available for users to view Iconclass terms in English, Italian, German, or French. The headings consist of an alphanumeric notation as a hotlink into the Iconclass hierarchy, followed by the Iconclass labels in the selected language. For example, the notation, here "25F711" for bees, remains constant, while the label changes for "bee" (English), "ape" (Italian), "Biene" (German), and "abeille" (French). The linguistic flexibility of Iconclass allows users to search for items in any language in its thesaurus, regardless of the language in which they were encoded—a critical component for international co-operation in both the creation of and research conducted in the *Emblematica Online*.

As described above, users can navigate the Iconclass thesaurus within *Emblematica Online* to discover related resources in *Emblematica Online*. For example, Alciato's "Ex bello pax" signifies the abstract concept of Peace, one of the Iconclass headings for this emblem. By clicking on the notation for Peace, "45A20 symbols, allegories of peace, 'Pax'; 'Pace' (Ripa)," users enter the Iconclass hierarchy and can search for related emblem resources in *Emblematica Online*, in this case retrieving 107 emblems on diverse aspects of Peace, including ones with mottos such as "Que tras la guerra viene la paz," "Paix,"

---

[29] See http://www.emblems.arts.gla.ac.uk/#alciato.

"De Guerre Paix," "Che dalla guerra procede la pace," and "Ex bello pax. Frid auß krieg," among the many hits. The results are conceptually related to the allegory of the emblem and not just tied to elements from the emblem *pictura*, such as the bees or the helmet. These examples also demonstrate the usefulness of Iconclass headings as URIs rather than strings, because this single click retrieves emblems regardless of language in which the emblem was described.

Because Iconclass allows collocating of emblems based on concept, not just on string descriptions of image features, the search reveals another famous motif from Alciato, the elephant drawing a triumphal chariot as an emblem of Peace. In several variations on this motif, an elephant pulls a chariot that crushes the implements of war under its wheels, in another the elephant itself treads on the shields and spears, and in yet another an elephant bears torches on its back, alluding to the chess piece and the animal as a living tank, a vehicle of war. As symbols of Peace the emblems of the bees and the elephant are remarkably consistent across the corpus both chronologically and linguistically. Identifying these emblems allows researchers to study them in a nuanced manner and to ask new questions of the images and their associated texts. Scholars and their students can recognize and envision trends that were previously hidden. This new capacity reflects both enormous advantages in terms of scale as well as of complexity. The flexible, multilingual searchability on consistent notations with a navigable hierarchy supports the associative thinking required of creating and decoding emblems. Emblems connected by LOD provide multiple points of access into thought processes integral to emblematica, and thus foster research creativity.

Significance of LOD Services for Historical Humanities Research

With the rise of the printed book and most especially of the printed image, new forms of visual and literary communication saturated early modern Europe: the emblem, the festival book, and graphic prints. New technologies made these works more affordable and thus more widespread. By employing the widely recognized classification standard of Iconclass as LOD, *Emblematica Online* not only makes these extremely rich collections of Renaissance texts and images available on the web, but it also makes them more usable. Through the multilingual classification system Iconclass, librarians can create more powerful metadata and scholars can return improved quantitative and qualitative search results for their research and teaching. The canon of emblem literature is widening to include more authors and texts. And it is now possible to study emblems more easily in the broader discourses of the early modern period in conjunction with printed graphics and festival books. Because scholars can search and discover at increased levels of granularity, their research results can incorporate various granularities of analysis and interpretation.

Emblematica Online aggregates geographically dispersed special collections into a scholarly space to explore primary materials through a single point of access. Writing for historians, Brent Rogers (2008) suggests that carefully curated digital collections have the potential to facilitate new, more complex and connective work. Because of LOD, emblem scholars can now conduct true digital research, asking and answering entirely new research questions, reading, grouping, and analyzing emblems in a manner productive to their individual line of scholarly inquiry. Individual emblems can be studied "connectively" in the context of their book, their author, their national tradition, European emblems, and related resources in a compact and efficient manner, employing valuable resources of time and energy for the work of interpretation.

Iconclass headings as LOD have allowed us to use resources and technology to good effect in *Emblematica Online* to overcome the barriers of project silos. The use of LOD makes it possible for *Emblematica Online* to integrate its emblem resources with those from its founding partner at the HAB, Wolfenbüttel.[30] The HAB encoded its emblems with Foto Marburg, whose encoding language is German. UIUC's encoding was accomplished with Arkyves in the Netherlands,[31] whose encoding language is English. Nor does the language of the emblem book present a hurdle to indexing. Both UIUC and HAB encode emblem literature from Latin and all vernacular European languages. The potential linguistic barriers posed no problems, however, owing to Iconclass's established vocabulary and notations within a multilingual thesaurus, as discussed above. Moreover, this feature also made the project more cost effective, as each agent was able to encode in the most practical and efficient manner for its own institution.

The use of Iconclass as a classification system also stimulated further international cooperation and enabled us to aggregate existing emblem metadata from the *Emblem Project Utrecht* and *Glasgow University Emblems Website*, both completed in about 2006, with those from the HAB and UIUC in a second

---

[30] See http://www.hab.de/en/home/research/projects/emblematica-online.html.
[31] See http://arkyves.org/. Hans Brandhorst, Arkyves, did the encoding for UIUC.

phase of *Emblematica Online*.³² A striking feature of this international research community, the so-called "OpenEmblem Scholars" composed of a loose association of researchers at Glasgow, HAB, Utrecht, and UIUC, is their early adaptation of Iconclass in anticipation of later international cooperation. They also developed a flexible schema to accommodate their different projects and goals. By working in loose consultation and not under the direction of a single design authority, the OpenEmblem Scholars initiated, in the words of John Unsworth (2014), an "evolvable system" for the presentation of emblems on the web. As Don Waters (2012) argues, many projects stall or fail because they have designed independent systems and the resulting hurdles to cooperation are significant, sometimes even insuperable, because of attitudes, national funding policies, and the realities of workflows, among other causes. Owing to the agreement on using Iconclass and a flexible schema, emblem researchers evolved a pretty good platform that worked for most of the projects. As Unsworth continues: "Evolvable systems begin partially working right away and then grow [...]" (Unsworth 2014, [4]). While *Emblematica Online* has created new data from emblem resources at UIUC and HAB, it has also evolved to aggregate existing data at all levels with Glasgow and Utrecht and at the book level from Duke University Library and the Getty Research Institute Library by ingesting emblem books from the latter collections that were digitized with the Internet Archive.³³ The future-oriented perspective of the "OpenEmblem Scholars" has allowed *Emblematica Online* to converge silos of previously unconnected emblem resources in a single portal by employing LOD.

We have delivered proof of concept for several key aspects of international cooperation in digital humanities. Beyond the important goal of linking internationally distributed emblem resources in *Emblematica Online*, we have demonstrated the vast potential of Iconclass indexing as LOD by linking it to key non-emblem resources. Owing to the alphanumeric notations, Iconclass can be linked to virtually any other project employing Iconclass, and we confirm this by linking outward from *Emblematica Online* to two significant portals for Renaissance Studies, the *Virtuelles KupferstichKabinett* (VKK) and *Festkultur Online*. *Festkultur Online* contains 567 completely digitized and fully indexed Renaissance festival books from the HAB and the British Library, while the VKK currently provides full access to more than 50,000 fully

---

³² See National Endowment for the Humanities (NEH), Humanities Collections and Reference Resources, https://securegrants.neh.gov/publicquery/main.aspx?f=1&gn=PW-51454-13.

³³ See https://archive.org/.

indexed printed graphics from the holdings of the HAB and the Herzog Anton Ulrich Museum, Braunschweig, and is growing. For research and pedagogy, scholars can now search across significant repositories of visual and textual resources, including emblematica, festival books, and printed graphics, at the heart of Renaissance culture.

On average the Iconclass indexing depth per emblem is 11, resulting in an index of ~233,000 Iconclass points of access for all of the emblems so far described. This index, encoded according to a consistent classification system in a multilingual thesaurus, offers unprecedented access to emblem literature and is by traditional humanities standards a large resource. Taking into account Iconclass heading extensions used, i.e., headings containing Iconclass *Auxiliaries*, *bracketed text*, and *keys* features, our Iconclass index includes a total of 32,752 distinct Iconclass notations. The resulting corpus is available as LOD. The data and the infrastructure accommodate the linguistic and cultural diversity of the emblem as well as that of the researchers making the emblems available on the web.

Expanding the LOD Graph Through Annotation

A report completed in 2012 as part of the Open Annotation Collaboration research project identified several scholarly use cases for annotating digitized emblematica resources.[34] Implementing support for such annotation services into *Emblematica Online* in a way consistent with Semantic Web and LOD best practices is a challenge. For scalable, practical implementation it is crucial, for example, that it be possible to store annotations separate from the resource being annotated. This requires persistent identification of book and emblem resources (which is already integral to the portal design) as well as robust methods for targeting segments of a resource (e.g., a part of an emblem *pictura*). The World Wide Web Consortium (W3C) recently chartered a new Web Annotation Working Group to develop specifications for doing just this.[35] Experimentation to add annotation functionality to *Emblematica Online* in accord with the work emerging from this Working Group is underway.

---

[34] See http://www.openannotation.org/documents/ProjectReports/EmblematicaFinalReport.pdf.
[35] See http://www.w3.org/annotation/.

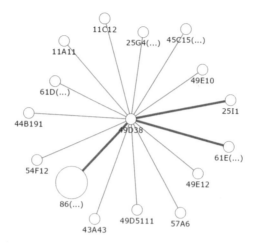

Figure 8. Visualization will allow scholars to browse related Iconclass terms.

Visualization of Iconclass Co-occurrence

As described above, on average 11 Iconclass headings are assigned to each emblem. These terms describe concepts as well as specific symbols and iconography used. Questions naturally arise about the frequency with which specific combinations of symbols or juxtapositions of concepts and iconography were used. Elephants, for example, often appear in emblems about peace; do they also appear with similar frequency in emblems about other concepts? To suggest avenues for further research, we plan to create a visualization of how frequently Iconclass headings appear together in emblem descriptions. It would work as follows: the view of term co-occurrence data could be centered on any Iconclass term, and the user would be able to see all other Iconclass headings with which that Iconclass term co-occurred. To provide more information about the co-occurrences, the size of the nodes would indicate the number of emblems with that particular Iconclass term, and the thickness of the connectors would indicate the number of co-occurrences between the two Iconclass terms (Figure 8).

Expert Crowdsourcing

Transcribing mottos onto a spreadsheet and associating them with the individual emblems registers the individual emblems in the portal. Hand-tagging individual elements of the emblem for Iconclass indexing requires considerable linguistic, textual, and iconographic skills. It is also expensive. A new path to explore would be to open up transcription and Iconclass indexing, for example, to interested members of the Society for Emblem Studies[36] or the Renaissance Society of America.[37]

Corpus Development

The development of an automated means to ingest high quality digital facsimiles of emblem books into *Emblematica Online* is a desideratum. We have provided a proof of concept with books digitized for viewing on the Internet Archive. Streamlining the ingestion of digital emblem books in the HathiTrust,[38] *Early English Books Online* (EEBO),[39] the *Bibliography of Books Printed in the German Speaking Countries from 1601 to 1700* (VD17),[40] Europeana,[41] and other resources would extend the corpus exponentially and exploit already digitized resources worldwide.

Working to ingest the existing emblem-level metadata from the projects at da Coruña and Munich, developed under independent designs particular to each project, would both add to the corpus of digitized emblematica and provide insights on how researchers in related fields might overcome hurdles to sharing resources.

Assessment

Assessment is central to the development of a flexible scholarly instrument for researchers, teachers, and students. How well does a resource support the research and pedagogical needs of scholars? As described throughout this paper, our studies of emblem literature have been greatly facilitated and in some instances enabled by the extensive corpus and functionality of *Emblematica Online*. Integral to developing the scope and functionality of *Emblematica Online* has been an ongoing assessment of scholars' use of emblem resources. A core of use cases, from which we have built (and continue to build) our collections and new services, has been a foundation for development. A first round of formal assessment of interface design, functionality, and features, and testing with scholars not connected to the project, has just been completed. Through the results of this testing we have gained a better understanding of what functionality, facilitated or enabled by the emerging

---

[36] See http://www.emblemstudies.org/.
[37] See http://www.rsa.org.
[38] See http://www.hathitrust.org/.
[39] See http://eebo.chadwyck.com/home.
[40] See http://www.vd17.de/.
[41] See http://www.europeana.eu/.

linked data services, is perceived as having the most potential to benefit scholars and support their research. Some early results have already been incorporated into the latest search and display design. As in earlier stages of the development of this resource, input from emblem scholars will continue to be critical to taking the best advantage of available LOD-based services.

Conclusion

Through implementation of LOD-based services at various levels of emblem metadata, *Emblematica Online* addresses key research requirements for emblem studies:

1. how to create access to remote, rare, and even unique resources;
2. how to afford greater access to the individual emblems in the books;
3. how to enable the study of emblems from many books across collections; and
4. how to study emblems within the context of related resources, such as festival books and graphic prints, from the early modern period.

In a cooperative effort by scholars from various research libraries and academic disciplines, the project developed technical solutions that serve scholarly needs. These new research capabilities are creating broader audiences for emblematica, as scholars from many disciplines of early modern Europe increasingly recognize the importance of emblems for their research and teaching.

Emblematica Online collections represent a research resource that serves all fields of Renaissance studies well, and potentially numerous other fields of text-image research. Scholars of the European vernacular literatures, classical philologists, art historians, musicologists, theologians, and social, religious, and political historians—even scholars of comics and graphic novels—are already using *Emblematica Online* collections in their research and pedagogy; the number of users is expected to grow as the portal matures. *Emblematica Online* demonstrates that emblematic metadata can be used to link other genres of digital humanities projects in Renaissance studies that previously were not linked. Freely available information and technological innovation itself does little to advance the creation of new knowledge (Waters 2012, 2), but by mirroring the complex processes of collaboration and creativity originally required to produce a Renaissance emblem, *Emblematica*

Online accommodates a wide array of learning styles, research questions, and pedagogical methods.

Development of *Emblematica Online* has been facilitated through close collaboration among domain scholars, librarians, and library developers following the principles of user-centered design and "agile" development practices. As compared to more traditional application development approaches, agile software development methodologies support incremental release and the ongoing, iterative identification of priority application features and requirements. While it is at present a fully functional robust resource for research and study, *Emblematica Online* could potentially become an "evolvable system" (Unsworth 2014, [4]). We have executed pilot studies demonstrating proof of concept for annotation tools and practices, the integration of visualizations into the interface, and the potential of using VIAF linkages to provide more context and background about individuals involved in the creation of emblem resources. Even as we explore new portal features and functionalities, we also want to continue to build the corpus by identifying and cooperating with institutions worldwide to make their emblem books discoverable.

Emancipating the emblem from its bound context created hundreds of thousands of new pathways into the emblem corpus, mimicking the processes of analogy, parallels, and even serendipity that foster creativity in research and teaching. It allows scholars to identify intersections in topoi and motifs across the geographically dispersed collections in several genres, while the emblem remains anchored to the full digital facsimile of its book. The principles of collaboration and sharing as inherent to LOD and Semantic Web technologies are key to building this significant scholarly resource, and are manifested at all levels of its development. Our research has convincingly demonstrated that the efforts and resources required to develop a large corpus of metadata enriched with linked data produce an authoritative foundation for new research, and that the use of LOD services can more efficiently facilitate advances in Renaissance scholarship.

WORKS CITED

Alciato, Andrea. 1531 (April 6). *Emblematum Liber*. Augsburg: Steyner. http://www.emblems.arts.gla.ac.uk/alciato/books.php?id=A31b&o=.

Berners-Lee, Tim. 2006. "Linked Data." Last modified June 18, 2009. http://www.w3.org/DesignIssues/LinkedData.html.

Curicke, Georg Reinhold. 1698. *Freuden-Bezeugung Der Stadt Dantzig....* Dantzig: Jansson von Waesberge. http://diglib.hab.de/drucke/gm-4f-256/start.htm.

Graham, David. 2004. "Three Phases of Emblem Digitization: The First Twenty Years, the Next Five." In *Digital Collections and the Management of Knowledge: Renaissance Emblem Literature as a Case Study for the Digitization of Rare Texts and Images*, edited by Mara R. Wade, 13–18. Salzburg: DigiCULT.

Henkel, Arthur, and Albrecht Schöne. 1996. *Emblemata: Handbuch zur Sinnbildkunst des XVI. und XVII. Jahrhunderts.* 3rd ed. Stuttgart: J.B. Metzler.

Manning, John. 2002. *The Emblem.* London: Reaktion.

Raggett, Dave, Jenny Lam, Ian Alexander, and Michael Kmiec. 1998. *Raggett on HTML 4.* New York: Addison Wesley Longman.

Rawles, Stephen. 2004. "A SPINE of Information Headings for Emblem-Related Electronic Resources." In *Digital Collections and the Management of Knowledge: Renaissance Emblem Literature as a Case Study for the Digitization of Rare Texts and Images*, edited by Mara R. Wade, 19–28. Salzburg: DigiCULT.

Rogers, Brent M. 2008. "The Historical Community and the Digital Future." Presented at the Third Annual James A. Rawley Graduate Conference in the Humanities, *Imagining Communities: People, Places, Meanings.* Lincoln: University of Nebraska-Lincoln, April 12. http://digitalcommons.unl.edu/historyrawleyconference/26/.

Stäcker, Thomas. 2012. "Practical Issues of the Wolfenbüttel Emblem Schema." In *Emblem Digitization: Conducting Digital Research with Renaissance Texts and Images*, edited by Mara R. Wade. Special issue 20, *Early Modern Literary Studies.* http://extra.shu.ac.uk/emls/si-20/WADE_Staecker_EMLS_Schema.htm.

Unsworth, John. 2014. "Sanity Check." Remarks at the American Geophysical Union, Washington, DC, May 15–16. http://people.brandeis.edu/~unsworth/SanityCheck.pdf.

Vogel, Johann. 1649. *Meditationes emblematicae de restaurata pace Germaniae = Sinnebilder von dem widergebrachten Teutschen Frieden*. Nürnberg: Zunner.

Wade, Mara R., ed. 2004a. *Digital Collections and the Management of Knowledge: Renaissance Emblem Literature as a Case Study for the Digitization of Rare Texts and Images*. Salzburg: DigiCULT. http://www.digicult.info/downloads/dc_emblemsbook_lowres.pdf.

——— 2004b. "Toward an Emblem Portal: Local and Global Portal Construction." In *Digital Collections and the Management of Knowledge: Renaissance Emblem Literature as a Case Study for the Digitization of Rare Texts and Images*, edited by Mara R. Wade, 115–20. Salzburg: DigiCULT. http://www.digicult.info/downloads/dc_emblemsbook_highres.pdf.

——— "Emblem." 2012a. In *The Princeton Encyclopedia of Poetry and Poetics*, 4th ed., edited by Roland Greene et al., 401–2. Princeton, NJ: Princeton University Press.

——— ed. 2012b. *Emblem Digitization: Conducting Digital Research with Renaissance Texts and Images. Early Modern Literary Studies*, Special Issue 20. http://extra.shu.ac.uk/emls/si-20/si-20toc.htm

Waters, Donald J. 2012. "Digital Humanities and the Changing Ecology of Scholarly Communications." Opening Keynote, TELDAP International Conference, Taipei, Taiwan, February 21. http://msc.mellon.org/staff-papers/waters_teldap.docx.

GLOSSARY

Term	Definition / Notes
Emblem	A bimedial early modern genre combining textual and graphical components, typically in a tripartite structure.
Emblem – Motto	An emblem component. A brief text in Latin or a European vernacular language, sometimes referred to as the emblem *inscriptio*.

Emblem – *Pictura*	An emblem component. The *pictura*, typically an enigmatic depiction, is the graphical element of the emblem.
Emblem – *Subscriptio*	An emblem component. Typically a short verse in Latin or a European vernacular language. The epigram of the emblem.
Emblem SPINE Schema	An XML document that declares and defines the names of and relationships between XML markup components (e.g., elements and attributes) used to describe an emblem book and each of the emblems that the book contains.
eXtensible Markup Language (XML)	An open standard used to serialize structured data and descriptions (metadata) to aggregate, organize, and share information about digitized emblems and emblem books. XML facilitates the automation of these tasks and is a foundational technology for *Emblematica Online*. XML documents are textual, and, as with HTML, angle bracket characters ('<,' '>') delineate markup from content.
eXtensible Markup Language – Attribute	A component of XML markup to associate additional properties with an XML element. Attributes elaborate and refine the role of XML elements appearing in metadata records.
eXtensible Markup Language – Element	A core component of XML markup to delineate the structure and hierarchy of an XML metadata record.
eXtensible Markup Language – Schema	A special class of XML document to enumerate and describe the names of and relationships between XML markup components (e.g., elements and attributes) used in metadata records describing a specific kind of resource (e.g., emblems).
HyperText Markup Language (HTML)	An open standard that defines the rules for creating web pages. Like XML, HTML is a text-based approach that uses markup (elements and attributes) to describe how the content of a web page is to be interpreted and displayed.

HyperText Transfer Protocol (HTTP)	An open application protocol for distributed, collaborative, hypermedia information systems. HTTP defines the rules that web browsers and web servers use to communicate.
Iconclass	A classification system designed for art and iconography. See http://www.iconclass.nl/home.
Iconclass Classification Code (Notation)	The alphanumeric code assigned to each of the more than 28,000 Iconclass Descriptors in the Iconclass classification system hierarchy. To maximize expressiveness, Iconclass notations are extensible.
Iconclass Definition (a.k.a. Term, Descriptor, Heading)	Iconclass Definitions are used to index, catalog, and describe the subjects, themes, and motifs of images found in works of art, photographs, and other sources, including emblems. In emblem studies, Iconclass terms help scholars discover connections and relationships between emblems.
JavaScript Object Notation (JSON)	An open standard for serializing information, including descriptive metadata. Similar to XML in purpose but different in syntax, and optimized for web applications that make use of the JavaScript programming language.
Linked Open Data (LOD)	An approach that uses web technologies like RDF to link related data that was not previously linked, or was linked in a less machine-actionable way. LOD lowers the barriers for linking related resources on the web. See http://linkeddata.org/.
Representational State Transfer (REST)	A web software architectural design approach for application-to-application interactions that relies on the HTTP protocol. In this approach all transactions are represented with URLs. See https://en.wikipedia.org/wiki/Representational_state_transfer.
Resource Description Framework (RDF)	A family of World Wide Web Consortium (W3C) specifications that provide a framework for describing and linking information resources on the web. Facilitates machine interactions with these resources. Based largely on a subset of first-order logic. See http://www.w3.org/RDF/.

Resource Description Framework – Class	Used to divide or classify resources on the web into types or categories that share common attributes; can be defined as an extension or an intension in an ontology. Classes may classify individuals, other classes, or a combination of both. Instances of a Class serve as Subjects and Objects in RDF Statements.
Resource Description Framework – Object	The final component of an RDF statement. Similar meaning in the context of RDF to the traditional meaning of "Object" in the context of English (or another language's) grammar. In RDF serializations, the Object is always represented by its URI.
Resource Description Framework – Ontology	A formal naming and definition of the RDF classes, properties, and interrelationships to be used when describing a particular domain of information resources found on the web. See https://en.wikipedia.org/wiki/Ontology_(information_science).
Resource Description Framework – Property (Predicate)	A directed binary relation for use in expressing the characteristics of an instance of a Class. The role of a property in an RDF statement is similar to the role of a predicate in grammar or predicate logic, but less nuanced.
Resource Description Framework– Statement	In RDF an information resource (e.g., an emblem or emblem book) is described by a set of RDF Statements. Each RDF Statement consists of a Subject, a Property, and an Object and conveys one facet of description. It can take dozens or even hundreds of RDF statements to adequately describe a resource.
Resource Description Framework – Subject	The first component of an RDF Statement, identifying unambiguously the resource or entity being described in the Statement. As with Objects, Subjects are represented in RDF Statements by their URI.
schema.org	An initiative sponsored by Bing (Microsoft), Google, Yahoo, and Yandex to create and maintain a shared set of classes and properties to be used in describing the content of web pages and web resources. The initiative also deals with strategies for serializing resource descriptions in HTML. See http://schema.org/.

Semantic Web	An initiative of the World Wide Web Consortium to promote standards-based data formats and exchange protocols on the web. A goal is to be able to treat a portion of the Internet as a web of linked data, organized and described in a manner that will facilitate machine-mediate reasoning and inference. See http://www.w3.org/standards/semanticweb/.
Semantic Web technologies	The foundational technologies for the Semantic Web, most fundamentally HTTP, the Resource Description Framework (RDF), and Linked Open Data (LOD).
Serialization	The process of transforming intellectually-realized metadata and other forms of structured data into a format that can be stored by a computer, exchanged between computers, and acted on by software applications.
Uniform Resource Identifier (URI)	A string of characters used to unambiguously identify a resource (e.g., an emblem). URIs have the same syntax as URLs, but may not resolve to the resource itself when entered into your web browser. For example, entering the URI for the Eiffel Tower into your web browser will not result in the Eiffel Tower appearing on your desk. It is a LOD best practice that when a URI is de-referenced, either the resource or its description will be returned. See https://en.wikipedia.org/wiki/Uniform_Resource_Identifier.
Uniform Resource Locator (URL)	A reference (an address) to a resource on the web. See https://en.wikipedia.org/wiki/Uniform_Resource_Locator.
Virtual International Authority File (VIAF)	An international service designed to provide convenient access to the world's major name authority files. VIAF provides a unique, persistent URI for authors and contributors involved in publishing emblems and emblem books, and links to descriptions of these individuals at the Library of Congress, the *Deutsche Nationalbibliothek*, etc. See http://www.oclc.org/viaf.en.html.

WorldCat Identities   A service that provides information about personal, corporate, and subject-based entities (writers, authors, characters, corporations, horses, ships, etc.) based on information in WorldCat, a catalog database. See http://www.oclc.org/research/themes/data-science/identities.html.

Mapping Toponyms in Early Modern Plays with the *Map of Early Modern London* and *Internet Shakespeare Editions* Projects: A Case Study in Interoperability

Janelle Jenstad
University of Victoria
jenstad@uvic.ca

Diane K. Jakacki
Bucknell University
dkj004@bucknell.edu

Over the past two years the *Map of Early Modern London* project (*MoEML*) has been collaborating with the *Internet Shakespeare Editions* (*ISE*) and its sibling sites, the *Queen's Men Editions* (*QME*) and *Digital Renaissance Editions* (*DRE*), to link data across sites and enrich the data on all sites. While *MoEML* aims to harvest toponyms[1] from a wide range of texts for its digital gazetteer of early modern London, the *ISE/DRE/QME* editors need a way to identify location and annotate toponyms in their editions. *MoEML* director Janelle Jenstad and *ISE* editor Diane Jakacki share an interest in how dramatists, poets, and chroniclers envisioned sixteenth- and seventeenth-century London, and in how this vision in turn encouraged a sense of identity for contemporary London audiences and readers. Our understanding of that topographical identity provides modern scholars with an important perspective on how London is a subject in, as well as a setting for, the plays upon which the *ISE/DRE/QME* editors focus. The development and expansion of *MoEML* offers an invaluable opportunity to foreground the idea of London as locale and referent in these plays.

For this essay, we will focus on a case study: the *Internet Shakespeare Editions* text of *King Henry VIII*, which aims to display the toponyms from *Henry VIII* in both the *MoEML* and the *ISE* environments using a single underlying XML file. This case study anticipates the development of the *Linked Early Modern Drama Online* (*LEMDO*) platform, a major research partnership that will bring together well-established projects and institutions. *LEMDO*'s work addresses one of the most pressing questions in digital scholarship: the need to link data and projects in order to facilitate interoperability and build a global

---

[1] The *Oxford English Dictionary* defines toponym as "A place-name; a name given to a person or thing marking its place of origin."

research environment—in this case for early modern drama. Our work suggests that interoperability achieves more than remobilization of data in new environments; it also generates new research questions on all sides and gives us greater critical purchase on our respective data sets and texts. This additional benefit more than justifies the technological challenges of sharing data.

Interoperability affords editors across the sites the opportunity to identify and amplify textual intersections regarding place and time, and it models ways in which these editorial intersections can be leveraged to emphasize the importance of place in early modern dramatic, literary, and chronicle texts. After an overview of *MoEML*'s and the *ISE*'s independent histories and an explanation of the tagging protocols, this chapter works through four challenges that have ultimately helped both projects formulate better understandings of how place signifies in early modern drama:

1. Implied place. While *MoEML* tags only toponyms (named places), play texts often imply places that an editor needs to identify. Balancing the needs of *MoEML* with those of the imagined *ISE* reader has focused our attention on the many ways that texts establish a sense of place.
2. Uncertainty. Plays written for a bare, unlocalized stage use dialogue to indicate location, but the unlocalized nature of that stage also means that a play does not always need a specific location. The editor wishing to perform a platial analysis must contend with stage conventions that are comfortable with uncertainty.
3. Temporality. Plays and stories are time machines. They tell stories from the past, set in historic Londons where names of places may well have been different. At the same time, historical drama is not necessarily historically accurate. Tagging *MoEML* toponyms and implied locations in an *ISE* text means thinking both about platial change over time and about (mis)representations of the past.
4. Mechanics of tag placement. With one eye on the eventual ingestion of tagged text into the *MoEML* environment, the *ISE* editor must think carefully about where to place a tag and what (if anything) to include in the tag.

MoEML: *History*

Beginning in 1999 with a digitized version of the 1560s woodcut Agas map of London, *MoEML* has grown into a significant research project comprising four distinct and interoperable projects connected through the map platform. In addition to the map (freshly digitized and stitched for the OpenLayers platform launched in 2015) with its accompanying descriptive gazetteer, *MoEML* offers an encyclopedia of London people, places, topics, and terms; a library of marked-up texts; and a versioned edition of John Stow's *Survey of London*.[2] The expansion of *MoEML*, rooted in humanistic questions about the importance of London as a literary place, required an approach to spatial analysis that differs from the more commonly applied geographic information systems (GIS) methods that have marked many digital humanities projects associated with historical and literary place. *MoEML* has grown as a result of its decision to treat the Agas map not just as a graphical user interface to a data set but also as a map-like text deserving its own editorial treatment. The fact that the Agas map—partly bird's-eye view and partly perspective landscape—does not lend itself to georeferencing and georectification, combined with the project director's scholarly inclinations toward textual editing, have moved *MoEML* in the direction of text mark-up and toponym (placeName) tagging, which has in turn fed a dynamically generated gazetteer of place names and variants.

MoEML's research questions point in two directions. On the one hand, *MoEML* investigates toponymy, the linguistic practice of naming places by referring strings (Radding and Western 2010). *MoEML* encyclopedia entries have an onomastic bent in that they tend to discuss the etymology and history of place names. The texts in its library have been chosen for the density of their toponyms, the four editions of Stow's *Survey of London* being the main source of *MoEML* place name variants. On the other hand, *MoEML* has always been concerned with which London spaces are prominent in early modern literature, when they appear, and in what kind of texts. Most of *MoEML*'s users and contributors have been scholars of early modern drama. Although the dramatic toponym density is relatively low even in city comedies, *MoEML* does offer dramatic extracts, mostly compiled in the early days of the project ("Dramatic Extracts" 2015).

---

[2] Stow's *Survey of London* was first published in quarto in 1598. Stow revised it for a 1603 second edition. Anthony Munday continued the work of adding interstitial commentary for the 1618 edition. The much longer 1633 edition announced itself as "complete." *MoEML* is editing all four editions and building a versioning environment that will make it possible to find and visualize the successive additions to the text, on the grounds that the additions offer a purchase on how London changed between 1598 and 1633.

To increase the number of toponyms drawn from dramatic sources, *MoEML* programmer Martin Holmes applied Named Entity Recognition (NER, a Natural Language Processing technique) to four plays in the *ISE* corpus. Even after using training sets and running the text through the *Stanford Named Entity Recognizer* (Stanford), accuracy peaked at about 65% (Jenstad and Holmes). The NER tagger stumbled over spelling variations (a problem that *MoEML*'s new gazetteer may resolve) and place names that double as character names. We see toponymy and prosopography converge, for example, in the stage direction "Enter Yorke, Salisbury, and Warwick" in *2 Henry VI* (2010, TLN 959). NER has no way of distinguishing the place from the person whose title is rooted in hereditary rights to the named territory. Given that *MoEML* had already called upon Jakacki's expertise to confirm NER tagging of toponyms in *Henry VIII*, Jenstad and Holmes began to wonder if it would be reasonable to ask *ISE*/*QME*/*DRE* editors to tag toponyms manually in the process of tagging their texts for publication on the *ISE* platform.[3]

ISE/QME/DRE: *History*

The *ISE* platform, on which the *QME* and *DRE* projects are also published, is complex. All the sibling sites have an in-built performance database, facsimile database, and static pages in addition to the XML editions at the core of each site.[4] The *ISE* requires each editor to correct and tag diplomatic transcriptions of the early texts, then prepare a modern text from one or more of those early texts.[5] Having been developed before XML and its dialects (such as the Text Encoding Initiative tags used by *MoEML*), the *ISE* uses a custom tag set that *ISE* encoders and programmers call ISGML (for *ISE* Standard Generalized Markup Language), or "SMGLish" (in recognition of its evolution toward complexity and extensibility).[6] The tags, published openly along with the

---

[3] An example of the result of the combined hand-tagging and NER can be seen on the *MoEML* encyclopedia entry for "The Tower," which aggregates references harvested from a subset of *ISE* texts. Once harvested, these references are housed in *MoEML*'s database. One goal of our present collaboration is to imagine a more dynamic way of generating these references and extending their value for the *ISE* site.

[4] We will confine our discussion to the *ISE*, but any tools and functionalities on the *ISE* sites also operate on the *DRE* and *QME* sites, and the editorial tagging procedures are common to all three projects.

[5] Bevington offers three modernized editions, each one based on a different copytext (Q1, Q2, and F), as well as an "Editor's Version" modern text (Bevington).

[6] These ISGML tags are now converted to XML via a complex tool chain that separates out, validates, and then recombines (with thousands of join elements) overlapping

rest of the Editorial Guidelines (Best and Jenstad 2015c) are human-readable and range from structural elements like <ACT> to typographical elements like <HW> (for "hung word") to interpretive elements like <PROP>. Editors are also expected to make extensive use of the multi-level <NOTE> tags, thereby enriching the texts in robust ways and setting *ISE* editions apart from traditional editions that are constrained because of print space. Digital editions have much greater capacity for extensive and intensive editorial engagements. These notes are keyed to TLNs (Through Line Numbers), which are the essential piece of information that connects all the files in an edition, including digital images of the early texts and eighteenth-century editions from various libraries.

When Jenstad joined the *ISE* as assistant coordinating editor and Jakacki came on board as the editor of *Henry VIII*, the *ISE* had neither a protocol nor any dedicated tags for marking up toponyms. The <SD> element (i.e., stage direction) does have an @t attribute (i.e., type) that allows one to specify location as a value, but that type is applied to theatrical as opposed to geographical space. Currently the only mechanism for glossing place names is within the stand-off annotations. And yet <SD> offers a place for editors to explore complex and sometimes problematic understandings of place. For example, in *Henry VIII* 2.1, Jakacki first marked up the stage direction type as <SD t="entrance, location">Enter two Gentlemen at several doors.</SD>. As will be examined below, that <SD t="location"> could be used to interpret where in London this scene might take place as well as—or instead of—two stage locations. The Editorial Guidelines invite the editor to provide "Explanations of reference to customs, events, etc.," including "references that may be familiar in Britain [but] might not be clear to readers overseas, e.g. districts in London" (Best and Jenstad 2015a, 5.2.4). Standard editorial practice is to gloss place names on their first appearance in a text and assume that readers will carry the knowledge forward. However, such a practice does not take into account the critical value of identifying the frequency and context of toponyms, nor of the value of working through the evidence for implied locations. In other words, editorial practice has not yet caught up with the intense interest in space and place generated by the spatial turn in literary studies (see Bodenhammer 2010).

---

hierarchies. The individual play editor is generally unaware of the processing performed on her or his SGML-tagged file.

Unlike other London GIS projects,[7] *MoEML* is invested in the theory and practice of textual editing as well as geohumanities considerations about the meaning of place. The *ISE*, with its long-standing commitment to producing editions "native to the medium of the Internet,"[8] is well positioned to model how a digital edition might annotate and then visualize markers of space and place. Indeed, at precisely the same time that Holmes and Jenstad were harvesting toponyms from the *ISE* using NER techniques, Jakacki was preparing a seminar paper in which she mocked up *ISE* editions with a map link in the Toolbox that accompanies each edition (Jakacki 2013; Jenstad and Holmes 2013). While the needs of the projects—and their underlying technologies—remain distinct, it is mutually beneficial to collaborate and plan for interoperability. Only by working together can we fully address questions about how and to what degree Shakespeare invoked London spaces. Through these types of editorial collaborations, we can begin to consider how characters move through the urban environment in particular ways. We can raise questions about the relationship between London and "the court"—especially when the court moves with the monarch across London over time. And we begin to consider how this vision compares to other playwrights, and how that vision might have been informed by historians.

Basic Linking Protocols

The *ISE* has developed a new tagging protocol that we are testing with Jakacki's edition of *Henry VIII*. In its current prototypical form, the tagging protocol distinguishes between primary and secondary *ISE* texts, anticipates more extensive use of linked data in the future, and allows for various forms of processing. For the first prototype, *ISE* lead programmer Maxwell Terpstra suggested an adaptation of the *ISE*'s existing <ilink> tag, which is used to link any string in editorial paratexts (secondary texts, consisting of annotations and critical essays) to a page inside or outside the *ISE* site. Normally, the <ilink> tag is not used within primary texts, but our first prototype used <ilink> tags both in critical paratexts and in primary texts. In working with *Henry VIII*, we recognized that this prototypical <ilink> tag was prescriptive rather than descriptive markup (in the sense that it prescribed a link rather than described the text) and was thus suitable only for the editorial paratexts.

---

[7] We are thinking specifically of the incredibly rich historical GIS project, *Locating London's Past*, and its forthcoming prequel, the *Mapping London* project. Respectively, they georeference and georectify the Rocque and Morgan maps of London for use as surfaces to display georeferenced data sets.

[8] See Best (2006). This commitment was first articulated in Best (1998).

For *MoEML* to harvest London references from the *ISE* texts, the *ISE* will need to add a new descriptive tag to its customized markup language. According to Michael Best, ISGML began as "a simplified version of the tag set devised by Ian Lancashire for his Renaissance Electronic Texts, and suggested in his paper on 'The Public Domain Shakespeare'" (Lancashire 1992, quoted in Best 2010). This tag set has evolved from its SGML origins into a more complex critical tag set that is now being converted into a TEI-compliant tag set. To identify London references in primary texts, the next step for the *ISE* is to develop a <PLACENAME> tag cognate with the Text Encoding Initiative's <placeName> element (TEI Consortium 2015, <placeName>).

We propose to continue using the <ilink> tag for paratexts and a new <PLACENAME> tag for primary texts. The *ISE* wants to build this tagging protocol so that it can be repurposed for places outside London as well. The model for tagging a toponym in a primary text therefore includes a pointer and a unique place id, as in this example:

<PLACENAME n="mol:CHEA2">Cheapside</PLACENAME>

The value of the @n attribute includes the prefix "mol" to indicate that CHEA2 is a *MoEML* value (the project's unique xml:id for Cheapside). Using the "mol" prefix ensures that the value will be unique in the *ISE*'s system (even if it partners with other projects using alphanumeric ids consisting of four letters and an integer). This model is extensible to allow the *ISE* editor to point to digital gazetteers/maps of England, France, Europe, or the world, as necessary, provided those projects have place ids (a private URI scheme, unique authority names, or codes). A further requirement for paratextual tags is the ability to distinguish location links from other types of links, in order to call up the maps that the *ISE* plans to include in the suite of tools linked from the Toolbox (exactly as Jakacki envisioned). The basic model for a location <ilink> tag thus has two attributes, as in this tagged instance of Cheapside in an *ISE* paratext:

<ilink component="geo" href="mol:CHEA2">Cheapside</ilink>

The @component attribute indicates that the <ilink> is a specific type; the value "geo" says that this link is a geospatial link and can be displayed on a map. The familiar @href attribute allows one to point to a specific target; in this case, the value is not the URL for *MoEML*'s page on Cheapside (http://mapoflondon.uvic.ca/CHEA2.htm), but simply the unique part of a *MoEML* URL from which the full URL can be dynamically generated. Once the text and paratext are marked up, they can be processed in various ways by either project. *MoEML* will harvest the text inside the <PLACENAME> tag. The *ISE*'s processing instructions can permit multiple ways of treating the <ilink> and <PLACENAME> tags; for example, the *ISE* might generate a pointer to a map of London on the *ISE* site, or a custom layer within the *MoEML* site. The *ISE* will collaborate with another project to mark up and visualize English locations; that project's ids will be prefixed with an identifying string cognate with "mol." For non-English locations, the *ISE* will probably have to create its own gazetteer based on Sugden's *Topographical Dictionary* of 1925 (with geo-coordinates from a world gazetteer). These tagging protocols and processing instructions will extend to *DRE* and *QME* texts as well.

Henry VIII: *History Play and City Play*

The complexities of *Henry VIII* as both a history play and a city play present a robust challenge to these initial tagging protocols. At the most basic level, this kind of tagging allows *MoEML* to ingest toponyms identified by the *ISE* editor and the editor to gloss place names efficiently by linking each one to *MoEML* resources. However, if this tagging were simply an act of identifying toponyms within the play, the *ISE* editor might not be so committed to participating in the collaboration between *ISE* and *MoEML*. Using *MoEML* ids for scene locations is neither technically or intellectually challenging, except insofar as scene locations are often ambiguous.[9] From a usability perspective, inclusion of a static wayfinding map with markers and a legend would suffice for identifying named places and implied scene locations. However, the act of tagging, harvesting, and visualizing all toponyms to places in London has permitted us to find more extensive correspondences among *MoEML* and the *ISE*, *DRE*, and *QME*. Attempting to achieve interoperability has been critically advantageous on both sides.

The challenge with *Henry VIII* is that the articulation of place is heavily encoded in terms of shifts in political status, the relationships between characters, and the proximity to particular places and types of places. For example, Buckingham is sent away from Westminster to the Tower to be executed; Anne Boleyn comes from the Tower to Westminster to be crowned. Henry uses Wolsey's York Place as the location to meet and woo Anne; he later takes over the residence, renames it Whitehall, and uses it as the location to celebrate his marriage to Anne. Furthermore, different classes and character types

---

[9] On the bare, unlocalized early modern stage, scene location is indicated in the dialogue if location matters. Modern editions are far more invested in pinning down scene location than early modern theatrical practices were. We offer no comment here on the appropriateness of adding location to scenes in early modern plays.

have distinct relationships to places: Katherine and Henry appear only in interior spaces associated with royalty and legality. Middle- and lower-class characters (with the exception of royal servants) are presented only outside of these same buildings. Aristocrats and ecclesiasts are the only characters to traverse exteriors and interior spaces, moving across the London landscape in their implied movements and in their range of dialogic references. The very act of tagging for *MoEML* has produced these observations, which in turn feed the *ISE* editor's critical introduction and commentary notes.

Jakacki's interest in authoring content in both *ISE* and *MoEML* has revealed issues about the complementary—and sometimes competing—needs of the two research platforms and those who tag within them. The first revelation, which came only after work on *Henry VIII* had begun, was that issues of location for a play editor went far beyond the toponyms that *MoEML* wished to harvest. *MoEML* had anticipated processing only the toponyms that occur within dialogue segments; while this works very well in terms of early modern play dialogue that serves to signpost place, it does not take into account questions of place (both explicit and implied) within paratext and stage directions. *MoEML* was specifically interested in augmenting its gazetteer of place name variants with additions from the old spelling texts, whereas the *ISE* editorial guidelines stipulate that interpretive tagging belongs in the modern texts; location is so frequently a matter of interpretation that the *ISE* editorial board wanted to limit location tagging to the modern texts. It became increasingly important for the engaged *ISE* editor to work with the *MoEML* programmer to find ways of tagging toponyms at relative values within Level One and Level Two annotations,[10] as well as within stage directions and other forms of editorial paratext. Editions often indicate scene location in order to orient readers; in such cases, editors infer location from dialogic evidence in the scene itself, from stage directions, and from proleptic information provided in previous scenes. Editorial inferences belong in the *ISE* edition, but not in *MoEML*'s database. Furthermore, even places named in the dialogue are subject to interpretation. "Blackfriars" is a toponym that might point to any one of several places within or associated with the Blackfriars precinct. Ultimately, we recognized the need to record locational certainty for both the *ISE* reader and the *MoEML* processor. As the *ISE* and its sibling projects move toward a full TEI implementation, we will be able to add @certainty attributes and certainty values to the <placeName> elements within <l> (line) elements.

---

[10] The *ISE* Editorial Guidelines identify three levels of annotation: basic (Level One), advanced (Level Two), and discursive (Level Three) (Best and Jenstad 2015a).

Therefore, the next phase of interoperable markup between the *ISE* edition and *MoEML* involves unexpected but potentially vital challenges for the development of the *LEMDO* partnership that arise from Jakacki's attempt to work through issues of space and place in *Henry VIII*. These challenges require consideration at a programmatic level, but also point to opportunities for expanding critical and editorial content that spans and further links both projects.

Challenge 1: Implied Place and Extratextual Information. The *ISE* editor and *MoEML* project have different approaches to implied place. While *MoEML* never identifies implied places in texts and deliberately avoids tagging deictics (such as "that place," "there"),[11] the *ISE* editor is bound to explicate implied locations as part of the general project of making these texts comprehensible to the user. Location tagging for the *ISE* sometimes demands identification of London places that are not named but can be identified with a high degree of editorial confidence. If there is no toponym to be tagged, the *ISE* editor can indicate place with the @location attribute in the <SD> (stage direction) tag, add a Level One annotation in which the place is named and then tag the editorial toponym, or use an empty <ilink> element at the point where a deictic occurs or where the editor infers something about place from the text. So, for example, it is certain that Katherine's trial in 2.4 takes place in Blackfriars, although the place is not named in the scene and there is no toponym to tag. It seems appropriate to use the value "mol:BLAC1" in the stage direction for 2.4 to indicate implied place. The editor's high degree of confidence comes from extratextual information, the 1577 edition of *Holinshed's Chronicles*,[12] as well as from a proleptic reference in the text itself. In 2.2, the

---

[11] *MoEML* does not tag deictics because its tagging is meant for data harvesting rather than textual annotation. *MoEML* does not want to ingest referring strings like "this place" and "that church" into the Gazetteer. Generally, deictics are clear from context because the place has already been identified toponymically. When a *MoEML* text gestures toward an implied and unnamed location, the identification is usually provided in an editorial footnote. Interestingly, the biggest challenges arise in *MoEML*'s complete edition of the mayoral shows, performed in the streets but frustratingly silent on the place of performance.

[12] "The place where the Cardinals ſhould ſit to heare the cauſe of Matrimonie betwixt the king and the Quene, was ordeined to be at the blacke Friers in London, where in the greate Hall was preparation made of ſeates, tables, & other furniture, accordyng to ſuche a ſolemne Seſſion and apparāce. The king & the Queene were aſcited by Doctor Sampſon to appeare before the Argates at the forenamed place, the xxviij. of May being the morrow after yᵉ feaſt of Corpus Chriſti" (Holinshed 1577, 4.1464).

King announces the location for the trial: "The most convenient place that I can think of / For such receipt of learning is Blackfriars. / There ye shall meet about this weighty business" (2.2.140–142). Here the toponym itself is tagged with the <PLACENAME> element and the @n value of "mol:BLAC1" because the place is named. Implied London locations can be identified with an empty, self-closing <ilink> tag, and so the implied reference to Blackfriars in 2.4 would be tagged thus: <ilink component="geo" href="mol:BLAC1"/>. The *ISE* and *MoEML* will each make their own decisions about how to process the tags and the content thereof, including how to treat empty tags.

The editor could also complicate the importance of Blackfriars to the play in a note or in supplemental material within the edition by considering the shift in place type between 1530 and 1613, when the space in which the play was performed could be identified with a high degree of certainty as Blackfriars Theatre, the indoor home of the King's Men and probable venue for performances of the play. Therefore, the place could be tagged as both "mol:BLAC1" (Blackfriars in 1530, the monastery) and "mol:BLAC6" (Blackfriars in 1613, the theatre).

Several examples of implied location in *Henry VIII* require a good deal of extratextual research to identify. One of these is the masque in 1.4, which takes place at Cardinal Wolsey's palace. In 1.3 the Chamberlain asks Lovell where he is going:

LOVELL. To the cardinal's
    Your lordship is a guest too.
CHAMBERLAIN. O 'tis true!
This night he makes a supper, and a great one,
To many lords and ladies. (1.3.50–54)

The masque to which Lovell and the Chamberlain are invited, and at which the King is introduced to Anne Boleyn, is based upon one that took place at York Place in 1527 and was described at length in the 1587 edition of *Holinshed's Chronicles* (Holinshed 1587, 6.921–22). York Place would become Whitehall after Wolsey's fall. Clearly, place is implied, but several assumptions have to be spelled out in order to equate "the cardinal's" with York Place.

More complex considerations of place within *Henry VIII* call for additional enhancements to the ISGML tagging system that will benefit other editors. The complexities relate to ambiguous locations, and to changes in a place's identification over time within the play. Challenges 2 and 3 therefore concern uncertainty and temporality.

Challenge 2: Uncertainty. Some locations are more difficult to identify but should be included in the editor's locational analysis. For example, we have a high degree of confidence that the action in 2.2 takes place within the Westminster palace complex because (a) while not identified by name, it involves business between the king and various aristocratic and ecclesiastic counselors and petitioners, and (b) the King appears on stage by means of a stage direction in which he "draws the curtain and sits reading pensively" (2.2.53.1).[13] This direction clearly demands that the stage accommodate the king's public and his private spaces, and therefore must take place in one of his London palaces, which Jakacki contends is Westminster.

Examples of implied location occur in scenes 2.1 and 4.1, in which three Gentlemen supply exposition by means of congregating at a place in order to forecast an event. Information about location comes from dialogic clues, combined with contextual information from other primary sources. In 2.1, the first Gentleman stops the second Gentleman, who is hurrying past. When asked where he is going, the second Gentleman replies, "Even to the Hall to hear what shall become / Of the great Duke of Buckingham" (2.1.2-3). The first Gentleman describes the trial at which Buckingham was found guilty of treason. At 2.1.53 Buckingham enters, accompanied by Sir Thomas Lovell, Sir Nicholas Vaux, and Lord Sandys—presumably under guard, as he is described as having, "Tipstaves before him, the axe with the edge toward him, Halberds on each side." Lovell later interrupts Buckingham's speech by telling him, "To the waterside I must conduct your grace, / Then give my charge up to Sir Nicholas Vaux, / Who undertakes you to your end" (2.1.95-97). Holinshed reports that the trial took place at Westminster Hall and that Buckingham was beheaded at the Tower (Holinshed 1587, 6.864 and 6.865). So it is possible for Jakacki to identify the scene as an exterior one between Westminster Hall and a quay or landing, and thus give it the tag "mol:WEST2/."[14] Lovell's indication that he must conduct Buckingham to the Thames ("the waterside") reinforces the idea that Buckingham will be conveyed by water to the site of his beheading. Therefore "to your end" could be tagged with a high level

---

[13] The *ISE* Editorial Guidelines state that, "Where a quarto or modern edition [...] add material the numbers will be added using a decimal (<TLN n="1033.1"> etc.)" (Best and Jenstad 2015a, 2.7.1).

[14] A tag for Westminster Palace is currently not included in the *MoEML* gazetteer; Jakacki has proposed to the *MoEML* editorial board that it be added as WEST2.

of certainty as "mol:TOWE5/"—that is, the Tower of London—with a self-closing empty tag because there is no toponymic content to tag.

Act 4, scene 1 also deploys the gentlemen to convey dialogic information about Anne Boleyn's coronation procession and celebration. Like 2.1, this scene requires interpretive work to tease out the implications of place. The two Gentlemen are now joined by a third, who gather to narrate the events and identify the participants in the procession as they cross the stage. The scene is in an exterior area that Jakacki identifies as Westminster, near the Abbey. Her editorial inference is based on dialogue between the first two Gentlemen, in which the first Gentleman asks the second: "You come to take your stand here and behold / The Lady Anne pass from her coronation" (4.1.2-3). The degree of certainty here is based on the greeting exchanged by the Gentlemen, in which the second Gentleman says, "At our last encounter, / The Duke of Buckingham came from his trial." (4.1.4-5). After the Coronation procession crosses the stage, the third Gentleman enters, tells the other two that he has been "Among the crowd i' th' Abbey" (4.1.57), and goes on to describe the coronation spectacle. At the end of his account, he announces that the new queen, "parted, / And with the same full state paced back again / To York Place, where the feast is held" (4.1.92-94). The slippage of time in the scene must be addressed more fully elsewhere, but it is interesting to note that Anne is first identified as "Lady Anne," which would suggest that the Coronation has not yet taken place when the Gentlemen anticipate the procession. Stage time being elastic, Anne has already made her way to the feast when the third Gentleman describes the events in the past tense only forty lines later.

Challenge 3: Temporality. The elasticity of stage time such as the one described above is a particularly intriguing challenge for the editor considering references to place that are encoded within the play text. As 2.1 and 4.1 demonstrate, *Henry VIII*—like other early modern plays—deploys various dramaturgical techniques to dramatize the story: enacting, reporting (through dialogue), and narrating. In this same sequence, the first Gentleman chastises the third about his politically dangerous identification of York Place:

Sir,
You must no more call it 'York Place' – that's past;
For since the Cardinal fell, that title's lost.
'Tis now the King's, and called 'Whitehall.' (4.1.94-97)

The passage points to a problem that bedevils *MoEML*: what to do about physical spaces that change name (a common process to which this passage from *Henry VIII* bears witness), change function (thereby requiring a different location type in the *MoEML* Gazetteer), or even undergo division, reconstruction, or amalgamation.[15] The preliminary editorial decision about this sequence was to identify the location of the scene being described as "mol:WEST2" with reference to Westminster Abbey (even though events occur offstage) as "mol:WEST1". York Place was initially treated as a variant of Whitehall, and thus "mol:WHIT5". But upon further reflection and consideration of Henry's repossession of York Place from Cardinal Wolsey—a site that had been occupied by high-ranking ecclesiasts since the thirteenth century—certainty about toponymic reference has become problematized.[16] Relying upon the reference in Holinshed, it becomes valuable to the spirit of *MoEML*'s reflection of place over time to identify the place as both "mol:YORK2"[17] and "mol:WHIT5", both as named places. And yet, this duality of place name dependent upon ownership presents an unexpected challenge in terms of negotiating toponymical certainty in the *MoEML* tagging structure. It also demonstrates the value of ongoing discourse between editors and programmers. Suddenly uncertain about the transition from York Place to Whitehall in 4.1 and its implications for references to "Court" prior to 4.1, Jakacki contacted *MoEML* lead programmer Martin Holmes and asked if there might be a way to mark up a degree of certainty for a toponym reference with an implied place name. In his answer, Holmes reflected upon the changing nature of

---

[15] Full explanation of *MoEML*'s solutions to these challenges is beyond the scope of this paper. In general, *MoEML* has one xml:id for a space. As the place occupying that space changes, *MoEML* uses the TEI's <when> and <timeline> elements to chart changes to the name. For wholesale destruction and repurposing (e.g., the building of the Royal Exchange), *MoEML* creates a new xml:id for prior uses of the space.

[16] In the 1587 edition of Holinshed's *Chronicles*, the transformation of York Place into Whitehall is recounted thus: "The king hauing purcha|sed of the cardinall after his attendure in the premu|nire his house at Westminster, Yorke place or white Hall now the palace of Westmin|ster. called Yorke place, and got a confirmation of the cardinals feoffement thereof, made of the chapter of the cathedrall church of Yorke, purchased this yeare also all the medows about saint Iames, and there made a faire mansion and a parke for his greater commoditie & pleasure. And bicause he had a great affection to the said house at Westminster, he bestowed great cost in going forward with the building thereof, and changed the name, so that it was after called the kings palace of Westminster. (Holinshed 1587, 6.928).

[17] YORK2 has yet to be added to the tag set.

this approach to editorial interoperability.[18] Holmes spoke to the importance of collaboration at the highest levels of diplomatic digital editing, teasing out technical questions about certainty, raising again the issues of interoperability at a platform-to-platform level in order to anticipate a broader collaborative venture across multiple research projects, and pointing to ways in which the editor might pursue thorny questions about certainty in terms of place. In the case of the transition from York Place to Whitehall in terms of location as a place for Court events in the play, Jakacki might tag later references to Court in the play as <ilink component="geo" href="mol:WHIT5" cert="medium" ana="not before 4.2">Court</ilink>.

Challenge 4: Placing the <ilink> Tag. The final challenge of marking up these sequences is determining where the <ilink> tag should be placed in the text. While the <PLACENAME> tag must enwrap a toponym, the <ilink> tag is more flexible since it points to *MoEML*. On the surface, as demonstrated above, it would seem to make sense to put the Blackfriars tag at both 2.2.141 and in the stage directions for 2.4, using "mol:BLAC1". But to do so would root both scenes in Blackfriars, when in fact we have established that 2.2 should be tagged as taking place at Westminster Palace (which is in itself an editorial decision because its location is inferred by the action of the scene rather than the dialogue). One might include an empty <ilink> after "To the Cardinal's"

---

[18] Holmes wrote: "We can make use of the <certainty> element, as you do above, where we're uncertain. The question is how we work with those uncertainties in the interoperation between the two sites. Right now, *MoEML* just harvests pointers to its places from the *ISE* code. Where there's uncertainty, we may have to harvest the uncertainty too. [...] We have talked a lot about temporal bounds for toponyms, and there's no reason not to provide them if they're significant. Again, we have to be principled about when we do and don't supply them. Our 'location' files contain a <listPlace> element, so the same 'location' can have multiple <place> elements associated with it. Each <place> can have one or more <placeName> elements, and <placeName>s are datable; so the same *MoEML* id could conceivably have multiple <place>s, each with multiple names, and also with multiple <location>s (just to make things more confusing), and each of these things could be dated. [...] The situation you describe here, though, sounds more like two completely different places which happen to share the same location; so I would suggest that they need separate ids and write-ups, but obviously those write-ups should refer to each other. Their geographical coincidence would be evident from their position on the Agas and GIS maps. But if we imagine some sort of timeline feature being deployed on the maps, we would want some way to specify that York Place didn't exist after a certain date, and Whitehall didn't exist before that date" (2014).

for the purposes of plotting places mentioned in the text on a map of London, but the work of explaining the multi-step logic belongs in an extratextual Level Two annotation, with place names tagged therein as necessary. MoEML's interest is in harvesting place names from the *ISE*'s text. At the *MoEML* end, the programmer would harvest only the content of the <PLACENAME> tag wrapped around the toponym "Blackfriars" in 2.2. *MoEML* would ignore "mol" <ilink>s in *ISE* annotations and locations. The editor, however, is interested in encouraging *MoEML* to expand on its interest in toponyms in order to enrich the editorial markup and reflect the forecasting of place between scenes.

New Research Questions

Henry VIII's setting in and around specific locations in London provides abundant editorial opportunities to identify and examine place, not just as a dramatic setting but also as an entity rich with significance and history known to playgoers. Both explicit and implicit references to London places suggest that Shakespeare and Fletcher were highly intentional in their use of chronicle histories. They knew that an audience watching *Henry VIII* at the Blackfriars theatre would be affected by the space's multiple locational identities as a playhouse that only a generation before had been a venue for legal events and that two generations before had been the location of Catherine of Aragon's divorce trial. Dramatic invocation of such places would have triggered in theatregoers a kind of cultural memory tied up in religion, dynastic politics, law, and class identity. The editorial act of engaging in close reading of place is important because it challenges us as editors and readers to think about how these places can be performed—and also because it inspires us to consider why certain places (Blackfriars, Whitehall, Westminster) had such resonance for seventeenth-century theatregoers. These historical and cultural resonances are hard to capture within the current *MoEML* environment, which is still wrestling with how to encode changes of name, ownership, and significance over time.[19]

The experience of trying to make a dramatic text from a different environment interoperable with *MoEML* challenged both the *ISE*'s and *MoEML*'s

---

[19] The *MoEML* Placeography assigns one file to a single space (defined by geocoordinates). The project has developed TEI tagging solutions for changes to the place occupying that space (such as change of name, demolition and rebuilding, and shifts in function), using the TEI's guidelines on dates. But a full exploration of *how* place signified is best performed by the *ISE/DRE/QME* editors, who can draw on *MoEML*'s digital record of changes to place.

conceptions of how and what toponyms signify. Working toward the goal of interoperability helped us recognize that toponyms perform a variety of functions beyond establishing the location of the current scene, such as identifying later scenes proleptically and serving to georeference narrative forms of storytelling and history-making. As the roles of ISE editor and MoEML editor converged and diverged, we recognized a fundamental difference between drawing *MoEML* toponyms and related content into an *ISE* edition and creating *MoEML* content layers that draw textual material from the *ISE/DRE/QME* corpus. The *ISE/DRE/QME* plays extend *MoEML*'s vertical temporal axis further into the past, which means that we have more places occupying particular geospaces. We need to disambiguate various historical places that share GIS coordinates and yet account for the cultural significance of memory. Hand tagging, in an attempt to disambiguate titles and toponyms, revealed the close relationship between toponymy and prosopography.

These salutary effects on our respective and shared critical processes demonstrate the richness of the interoperability model. Already, we have identified several scholarly approaches to analyzing early modern dramatic literature and performance culture using linked data. One straightforward example involves an analysis of ways London locations in the Henriad reflect Hal's changing relationship with his father and with his incipient attainment of the crown. Such a project would requiring analysis of data in *Internet Shakespeare Editions* plays edited by Rosemary Gaby (*1* and *2 Henry IV*) and James Mardock (*Henry V*). More wide-ranging comparisons between the history plays and chronicle documentation of royal associations with the city might reveal to what extent Shakespeare relied upon these works and whether there is a way to further consider the work of scholars such as Annabel Patterson and projects like The Holinshed Project to situate these plays alongside chronicle narratives.

More broadly still, we see opportunities to consider London places within and across genres in plays by tagging a wider cross-section of early modern playwrights—work that would involve not only the creation of specific content layers but also associated scholarly essays. So, for example, an author might develop an argument about how Ben Jonson's use of neighborhoods in *Bartholomew Fair* and *The Alchemist* demonstrates an osmotic intersection among classes that is predicated upon amoral social negotiations, or how city comedies such as Thomas Middleton's *A Chaste Maid in Cheapside* or Middleton and Dekker's *The Roaring Girl* invoke London (Gray's Inn Fields, Holborn, the Strand, the Thames) to reinforce representation of appropriate and inappropriate social behaviors.

Ever more complex arguments about early modern performance culture and reception can play out within the *MoEML* environment. Someone might map and analyze John Holles's anxious letters to the Earl of Somerset regarding his experience crossing the Thames and interacting with the crowd at the Globe watching *A Game at Chess*. We foresee further development of the question about early modern theatrical publics (and counter-publics) first raised by the co-investigators and collaborators of the Making Publics research project (Yachnin 2010), offering an opportunity to tease out post-Habermasian ideas of pluralistic public spheres within the physical sphere and environs of London theatres.

Of course, these are merely suggestions about how the *ISE/QME/DRE* platforms and *MoEML* might serve as a place for scholarly reflection, experimentation, and demonstration of research related to early modern drama that is revealed more fully through spatial analysis. This ambitious process—linking data across four related but distinct early modern digital humanities projects—has been supported by institutional and editorial links among the projects. It has required careful and thoughtful experimentation with ideas of contextualized London spaces. At many points it has demanded the development of new approaches to TEI-compliant XML tagging and the incorporation of Natural Language Processing (NLP). It has been made possible because of the interest expressed by many of the *ISE*, *DRE*, and *QME* editors to enhance their editions of plays with specific relationships between textual and visual identifications of place in dialogue and stage direction.

Ultimately, the critical questions that have emerged are refining our prototype for interoperable tagging between *ISE* and *MoEML*. This type of spatial analysis requires more expansive programming support from *ISE* and *MoEML* to extend existing place name entries in the *MoEML* gazetteer along with enriched notation within the modern *ISE* edition.[20] What has yet to be considered, and will almost certainly be part of the move from prototype to fully implemented network, is how such research projects might be managed. There exists, we believe, a residual echo of the old maxim "information just wants to be free" from the early days of the Internet to current imaginings of digital humanities research. While it is realistic to imagine how scholars

---

[20] The *ISE*'s principal programmer, Maxwell Terpstra, and *MoEML*'s lead programmer, Martin Holmes, have been invaluable in helping us to think through these issues and in devising programmatic solutions to tagging and display challenges. We would like to acknowledge their ongoing intellectual contribution to our work, as well as their technical ingenuity.

only loosely associated with one another and with scholarly digital platforms such as *MoEML* and *ISE* can *use* data in their own work, it is more complicated to put in place best practices for public-facing research reliant upon and published within the framework of these platforms. As the editorial mode shifts from a one editor-one digital text model to something more expansive and complex such as one author working with multiple texts within one or more content layers of *MoEML*, or several authors collaborating across projects, the demands upon the managing editorial boards of these platforms must be taken into account. Questions about attribution, credit, support for data, and metadata access—not to mention sustainability—take on profound importance.

Conclusion

As we have demonstrated in this paper, this prototype phase is crucial to the development of interoperability between just two research platforms. While we strive to establish a supporting linked data structure that provides new and richer ways for editors to consider London places and times within dramatic texts, we have to be careful not to be precocious about establishing best practices. It is important for us to remember that a challenge pertaining to a particular history play might ultimately be unique to that play. While we wish to create and reflect new approaches to research regarding the importance of place in early modern texts, we need to continue to be flexible and responsive to the needs and challenges proposed by an ever-growing cadre of editors across a larger set of research platforms. For now, we must content ourselves with establishing an editorial link between two hubs in what will ultimately be a much more expansive network. The creation of this first spoke between *MoEML* and *ISE* becomes the prototype for the creation of other spokes, which would in turn attach to other hubs. The creation of this particular spoke helps us to see the kinds of issues we will need to address. As envisioned, the *LEMDO* network will comprise a series of hubs that allow a scholar to approach early modern dramatic texts from a variety of research approaches. *Linked Early Modern Drama Online* (and other collaborative endeavours) can thrive only if care is taken in early phases to ensure that linked data can build up from a foundation in praxis rather than be predetermined by theoretical models.

WORKS CITED

Best, Michael. 1998. "Foreword." *The Internet Shakespeare: Opportunities in a New Medium*. Special issues, *Early Modern Literary Studies* 3 (3): 1.1–4. http://extra.shu.ac.uk/emls/03-3/foreword.html.

———. 2010. "Internet Editions of Shakespeare Principles of Tagging." *Internet Shakespeare Editions*. Victoria, BC: ISE, http://internetshakespeare.uvic.ca/Foyer/Tagging.html.

———. 2006. "What we are." *Internet Shakespeare Editions*. Victoria, BC: ISE, http://internetshakespeare.uvic.ca/Annex/Articles/uvic/uvic1.html.

Best, Michael, and Janelle Jenstad. 2015a. "Editorial Guidelines." *Internet Shakespeare Editions*. Victoria, BC: ISE, http://internetshakespeare.uvic.ca/Foyer/guidelines/EdGuidelinesTOC/.

———, coords. 2015b. *ISE (Internet Shakespeare Editions)*. Victoria, BC: ISE, http://internetshakespeare.uvic.ca.

———. 2015c. "Summary of Tags (listed alphabetically)." *Internet Shakespeare Editions*. Victoria, BC: ISE, http://internetshakespeare.uvic.ca/Foyer/guidelines/appendixTagSummary/.

Bevington, David, ed. 2015. *Hamlet*. By William Shakespeare. *Internet Shakespeare Editions*. Victoria, BC: ISE, http://internetshakespeare.uvic.ca/Library/Texts/Ham/.

Bodenhamer, David J. 2010. "The Potential of Spatial Humanities." *The Spatial Humanities: GIS and the Future of Humanities Scholarship*. Bloomington: Indiana University Press. 14–30.

Davies, Matthew, dir. 2015. *Mapping London: A GIS Platform for the History of Early Modern London*. Project in progress at the Centre for Metropolitan History.

"Dramatic Extracts." 2015. *The Map of Early Modern London*. Dir. Janelle Jenstad. Victoria, BC: University of Victoria. http://mapoflondon.uvic.ca/mdtPrimarySourceLibraryDrama.htm.

Henry VI Part Two. (1623) 2010. By William Shakespeare. *Internet Shakespeare Editions*. Victoria, BC: ISE, http://internetshakespeare.uvic.ca/Library/Texts/2H6/.

Hirsch, Brett, ed. 2015. *DRE (Digital Renaissance Editions)*. Victoria, BC: Internet Shakespeare Editions. http://digitalrenaissance.uvic.ca.

Holinshed, Raphael. 1577. *Chronicles of England, Scotland, and Ireland*. London. Published on *The Holinshed Project* website. http://www.cems.ox.ac.uk/holinshed/.

_____ 1587. *Chronicles of England, Scotland, and Ireland*. London. Published on *The Holinshed Project* website. http://www.cems.ox.ac.uk/holinshed/.

Holmes, Martin. E-mail message to Diane Jakacki. October 28, 2014.

Jakacki, Diane. 2013. "Visualization of Geospatial/Textual Relationships in *King Henry VIII*." Shakespeare Association of America 2013 Annual Meeting. Toronto, ON.

_____ ed. 2014. *Henry VIII*. By William Shakespeare. *Internet Shakespeare Editions*. Victoria, BC: ISE, http://internetshakespeare.uvic.ca/Library/Texts/H8/.

Jenstad, Janelle, and Martin Holmes. 2013. "Practical Interoperability: *The Map of Early Modern London* and the Internet Shakespeare Editions." In *Digital Humanities 2013 Conference Abstracts*, 218–21. Lincoln: Center for Digital Research in the Humanities, University of Nebraska-Lincoln. http://dh2013.unl.edu/abstracts/ab-180.html.

Jenstad, Janelle, dir. 2015. *MoEML (The Map of Early Modern London)*. Victoria, BC. http://mapoflondon.uvic.ca.

Lancashire, Ian. 1992. "The Public-Domain Shakespeare." *Renaissance Electronic Texts Supplementary Studies* 2: n.p. http://www.library.utoronto.ca/utel/ret/mla1292.html.

Locating London's Past. 2011. Joint Information Systems Committee eContent Programme. http://www.locatinglondon.org/.

OpenLayers 3. 2014. API. http://openlayers.org/.

Ostovich, Helen, ed. 2015. *QME (Queen's Men Editions)*. Victoria, BC: Internet Shakespeare Editions. http://qme.internetshakespeare.uvic.ca.

Radding, Lisa, and John Western. 2010. "What's in a Name? Linguistics, Geography, and Toponyms." *Geographical Review* 100 (3): 394–412. DOI 10.1111/j.1931-0846.2010.00043.x.

Stanford Natural Language Processing Group. *Stanford Named Entity Recognizer.* http://nlp.stanford.edu/software/CRF-NER.shtml. Software.

Stow, John. 1598. *A Survey of London.* London: John Wolfe. STC 23341.

_____ 1603. *A Survey of London.* London: John Windet. STC 23343.

Stow, John, [and Anthony Munday]. 1618. *The Survey of London.* London: George Purslowe. STC 23343.

Stow, John, [Anthony Munday, and Humphrey Dyson]. 1633. *The Survey of London.* London: Nicholas Bourne. STC 23345.

Sugden, Edward H. 1925. *A Topographical Dictionary to the Worlds of Shakespeare and his Fellow Dramatists.* Manchester: University of Manchester Press.

TEI Consortium. 2015. "P5: Guidelines for Electronic Text Encoding and Interchange." http://www.tei-c.org/release/doc/tei-p5-doc/en/html/index.html.

"Tower of London." 2015. *The Map of Early Modern London.* Dir. Janelle Jenstad. Victoria, BC: University of Victoria. http://mapoflondon.uvic.ca/TOWE5.htm.

Yachnin, Paul, dir. 2010. *Making Publics.* http://www.makingpublics.org/.

Microstoria 2.0:
Geo-locating Renaissance Spatial and Architectural History

Fabrizio Nevola
University of Exeter
F.Nevola@exeter.ac.uk

> "There is still one of which you never speak."
> Marco Polo bowed his head.
> "Venice," the Khan said.
> Marco smiled. "What else do you believe I have been talking to you about?"
> The emperor did not turn a hair. "And yet I have never heard you mention that name."
> And Polo said: "Every time I describe a city I am saying something about Venice."
> "When I ask you about other cities, I want to hear about them. And about Venice, when I ask you about Venice."
> "To distinguish the other cities' qualities, I must speak of a first city that remains implicit. For me it is Venice."
> (Calvino 1974, 69)[1]

Italo Calvino's fictional account of Marco Polo's dialogue with the Emperor of Cathay, Kublai Khan, has become a *topos* among urban theorists and historians for its characterization of the infinite variety of urban form and identity. In *Invisible Cities*, Calvino offers a series of short descriptions of cities and their inhabitants, each of them filtered through a particular frame: emotion and the senses, color and texture, materials, ethnography. When

---

[1] The examples selected for discussion in the text and notes are no more than selective and have been chosen as much as possible to focus on research and projects that address early modern material. My thanks to Alex Butterworth and Jo Reid for discussions about apps, mapping, and geo-location over the past couple of years, and to David Rosenthal, who has worked with me on the *Hidden Florence* project, discussed below.

challenged by the Emperor about the fact that he has made no account of his hometown, Venice, Marco responds that each and every one of the descriptions he has provided was of that city. Through Marco's voice Calvino engages in a postmodern writing exercise; the multiplicity of interconnected evocations of place conjures varieties of forms and styles resolved in a single physical place. It is also a richly evocative articulation of the idea of the city as a palimpsest—a physical text layered with multiple meanings and interpretations that supersede one another on the lived surface of any one era. Long before the idea of the Anthropocene, Calvino captures the physical sense of distinct, deposited, and intelligible layers of the human past interwoven into the physical and sensory fabric of the current urban environment.

Calvino's text has inevitably caught the imagination of many academics and urban theorists, and is introduced here to provide a sense of the power of digital tools to access and order multiple and overlapping narratives of the cities of the past. In particular, this essay considers the ways in which GIS (geo-information systems), GPS (global positioning systems), and more broadly geo-locative technologies have been adopted by both historians and art and architectural historians of the early modern European city to reconfigure research questions and present new findings, often to wider audiences than have traditionally been the principal beneficiaries of academic research. Specifically, the focus here is on how mapping and other geo-spatial approaches have addressed urban history, and how a variety of new technologies have been harnessed to deliver innovative research. The first section of this essay offers a broad overview and selective sampling of projects that adopt GPS and GIS technologies in relation to early modern urban and architectural history. This is followed by a specific case example, delivered as a result of a research project led by the author: the creation of an immersive, location-sensitive audio tour of the Renaissance city of Florence as a smartphone app called *Hidden Florence*. In the final section, this project, which explores the affordances of locative digital tools and participates in the wider technology-enabled spatial turn, will be discussed in relation to the experimental historical methodology of microhistory (or *microstoria*) that emerged in the late 1970s. It will be suggested that technology-led experiments such as *Hidden Florence* do more than simply reconfigure existing historical practices to the digital medium: rather, they also have the potential to propose new approaches and methods in the practice of history.

Digital technologies and the rapid and pervasive adoption of everyday mapping technologies delivered through GPS-enabled devices have renewed and invigorated the "spatial turn" in humanities that cultural geographer Denis Cosgrove identified around a decade ago (Nevola 2013). Digital humanities are rapidly reconfiguring the field of Renaissance and early modern studies, with the spatial turn particularly evident in the application of GIS technology to mapping aspects of past urban environments (Ullyot 2013). Meanwhile, the seemingly unstoppable uptake of smart handheld devices offers an opportunity for exploration of how interactive map interfaces might be deployed on location and applied to urban history research. That so many European cities preserve extensive portions of historic fabric (and street layouts) makes these environments uniquely suited to reinscribing past meanings onto places through tagging of information to them—the city understood as text and palimpsest is one that is now more widely open to interpretation through digital tools that enable spatial visualization and located sensory experiences (Presner et al. 2014).

In fact, there is a growing tradition of historic maps being adapted to digital platforms by researchers to explore the social and architectural history of pre-modern cities (Favro 1999). Such maps obviously allow the historian to engage with the physical fabric and urban layout of cities prior to the widespread major changes that have resulted from the town planner's zeal to accommodate wider boulevards from the nineteenth century onwards. By deploying "urban markup" on historic maps, digital humanities projects operate on past layers of the city, offering a more authentic representation of period-specific information (Farman 2013b, 529–32).[2]

A first generation of the digital application of historic maps as interactive and enhanced environments can be observed in two early projects that used Giovanni Battista Nolli's remarkable figure-ground map of Rome, created in 1748 and one of the most accurate depictions of the city in the pre-modern period (Pinto 1976). Already in the later 1990s at Princeton, John Pinto was exploring the potential of a CD-ROM support for a database of information relating to the city and monuments of Rome and linked to highlighted areas on Nolli's map (Connors 2011; Waldron 1999). In turn, the interactive Nolli Map Website, launched in 2005 and based at the University of Oregon, was an innovative attempt to make this stunning and highly detailed map into a web-based "dynamic, interactive, hands-on tool."[3] Less than a decade on, it is easy to overlook the significance of a project in which the interactivity of

---

[2] Farman (2013b, 532) discusses the term "urban markup," citing analog equivalents described by Malcolm McCullough.
[3] Jim Tice and Erik Steiner, http://nolli.uoregon.edu/preface.html.

the Nolli Map Engine 1.0 does little more than offer the user clickable layers (as in Photoshop) that highlight such landmarks and sites as walls, gates, gardens, and the River Tiber. Nonetheless, it is an important stepping-stone in the development of interactive mapping of an early modern city, and suggests how powerful the combination of historic basemap with contemporary historical interpretative interventions can be. It is also primarily designed in relation to user experience, with a clear aesthetic that draws attention to the historic map itself as an artifact worthy of study.

Considerably more ambitious is the *Aquae Urbis Romae: the Waters of the City of Rome* project, developed by Katherine Wentworth Rinne, initiated in the 1990s and updated in 2007, and evolving with the emerging opportunities of online presentation.[4] Here, the basemap interface is not a historic one, but rather the 1992 Cadastral Plan of Rome; while perhaps less visually compelling than that of Nolli, it offers a clear map onto which the complex website plots not only key elements in the hydraulic infrastructure of the Eternal City, but does so in chronological sequence through the use of a slidable timeline. This enables the user to track the construction, use, and decline phases of particular features from classical antiquity to the sixteenth century. Moreover, by hovering the mouse over a particular object marked on the map as a pin (e.g., fountains, sewers, aqueducts, and other features), the user is able to draw up additional text-based information about each element. A prodigious amount of material has been assembled in the website, which in turn underpins a traditional monographic study of Rome's Baroque water supply system (Rinne 2011). It is also clear that the project's ambitions extended beyond the specific research publications of its initiator and sought a dialogue with "historians, classicists, archaeologists, hydrologists, engineers, and geographers" by offering a long chronological coverage and proposing that other scholars might contribute to the site contents, as well as to an online journal also hosted on the site.[5]

More recently, the *Orbis Urbis* project, developed through the Biblioteca Hertziana Max-Planck-Institut in Rome, has gone further still in terms of proposing the city map as a means of accessing and interrogating a vast mixed-media database.[6] Here, the large corpus of library holdings—including early printed books, etchings, and photographs—has been provided

---

[4] See http://www3.iath.virginia.edu/waters/first.html; the project owes a great deal to the personal research of its director, published as a book (Rinne 2011).
[5] *Aquae Urbis Romae*, accessible at http://www3.iath.virginia.edu/waters/first.html.
[6] See http://db.biblhertz.it/orbisurbis/html/ou/frameset.htm.

with location coordinates and linked to a map or satellite view of the city of Rome. This enables users to identify a monument (e.g., the Coliseum) and draw down the items in the collection pertaining to it. This is a massive undertaking, and is designed with the potential to be extensible to other library holdings and collections—in essence adding a spatial geo-reference code to the more traditional library codes and search terms usually used to classify objects.

While ambitious in its scope, projects such as *Orbis Urbis* tend to focus on the process of tagging documents (be these historical events or material objects) to location coordinates on a map. Concurrently, over the last three or four years, various web-based resources have begun to redefine spatial and mapping practices, making urban markup into a simple practice embedded in social media platforms and tools, so that both academics and the general public can deploy powerful instruments to tag material to place. Such sites as Pinterest and Historypin enable mapping of any subject through the tagging of objects to place on easy-to-use map interfaces; Historypin in particular has been taken up by museums, local community groups, and individuals to assemble visual archives of impressive proportions.[7] Indeed, the popularity of Historypin as a social media platform makes it a powerful tool in the mobilization of crowdsourced citizen-science-style projects, where the general public can be drafted to assist scholars in such tasks as locating the subjects of paintings or the particular viewpoint chosen by an artist.[8] The Tate Gallery's ArtMap project, for example, sets out to locate the subject of artworks in the museum's collections from 1500 to 1900, while the British Library's geo-referencing project has tagged over 3,000 maps to their modern-day locations; both rely on crowdsourced contributions.[9] There is of course some debate about how such collaborative projects are checked for quality, although a growing body of evidence from the sciences as well as humanities suggest that crowdsourcing can make a valuable contribution while also extending the research process outside the narrow pool of professional researchers (Dunn and Hedges 2012).

While the volumes of geo-tagged historical data are growing, it is nevertheless worth observing that such processes of cataloguing do not significantly

---

[7] See https://www.pinterest.com/ and https://www.historypin.org/en/ (both were founded in 2010).
[8] See, for example, the Imperial War Museum project, *Putting Art on the Map*, www.historypin.org/project/41-putting-art-on-the-map/#!mysteries/index/.
[9] See artmaps.tate.org.uk/ and www.bl.uk/maps/georeferencingmap.html.

alter interpretative paradigms unless they are interrogated in new ways. Indeed, it is, I think, helpful to consider such efforts in the light of a dichotomy observed in the context of archaeology, where field survey practices ("broken pots and meaningless dots") have become quite detached from theoretical and interpretative frameworks applied in other areas of the discipline (Witcher 2006). In the instance of GIS mapping and tagging projects, the risk would seem to be that while the amounts of things and events that can be tagged is almost limitless, critical and interpretative frameworks are required to put such information to good use. In this respect, mapping projects that layer multiple data sets—and offer the potential for external contributions and extensions of the material they support over time—offer quite a different degree of potential.

Michael Ullyot has recently written about a number of mapping projects that use the visual, pictorial interface of historic maps as the key public-facing point of access for complex underlying databases in relation to *The Map of Early Modern London* (MoEML: University of Victoria), which uses the 1560s "Agas Map" of London, and *Locating London's Past* (Universities of Hertfordshire and Sheffield), which is structured on John Rocque's 1746 map of the city (Ullyot 2013, 240–41; Nevola and Rosenthal 2011).[10] Geo-information systems applied to historical research (H-GIS) underpin such projects that are built on GIS-compliant historic maps; location coordinates are added to the other database archiving coordinates (such as date, unique ID number, etc.), so that items can be precisely plotted on the maps that have been rendered machine-readable as map tiles (Gregory and Geddes 2014; Bodenhamer 2010). To this small family of projects, the recently-published Digitally Encoded Census Information and Mapping Archive (DECIMA)—based at the University of Toronto and developed by a research team led by Nicholas Terpstra—has begun to provide a remarkably fine-grained spatial-historical account of Florence in the mid-sixteenth century, using census data overlaid on Stefano Bonsignori's 1584 map of the city (Terpstra 2016).[11] The DECIMA project distinguishes itself from the outset as being "open" in that its historical basemap is structured so as to be able to accommodate layers of additional data following the first input of the census material. As such, it has the ambition to be "a collaborative resource built by and for Early Modernists of

---

[10] See https://mapoflondon.uvic.ca/ and www.locatinglondon.org/.

[11] See http://decima.chass.utoronto.ca/. For another earlier use of the Bonsignori map, in addition to the *Hidden Florence* project discussed below, see Robert Burr Litchfield's searchable Online Gazetteer of Sixteenth Century Florence (2006): http://cds.library.brown.edu/projects/florentine_gazetteer/.

all stripes," in which future participants might add new documentary data sets that would combine to offer a growing body of searchable information that can be cross-referenced on screen.[12] This opens up huge possibilities for researchers, as well as providing a remarkable teaching resource that visually frames database queries through the important relationship between the archival evidence and the physical place where events were played out, plotted on a map.

In all these cases, and there are of course many others, the basemap serves as a tool for the user to visualize locations of historical events, ranging from census data to architectural monuments or events that occurred at particular locations. The use of the historical map functions as a visual strategy that gives a historical "look and feel" to the interface, while also providing the user with a more chronologically accurate rendering of the urban environment than satellite-delivered basemaps such as Google Earth. With growing computing power and large research teams working in digital humanities projects comes the potential for layering more data on maps, and the opportunity for users to dynamically interact with the underlying mapping systems. Looking for a parallel to another area of digital humanities, we can see how the groundbreaking *Mapping the Republic of Letters* (Stanford University's network visualization project), which explored the intellectual and correspondence networks of the Enlightenment, has led to the development of Palladio, a platform for editing and visualizing any sort of comparable data set (Ullyot 2013, 939–40).[13] What is exciting here is to see how an initially "closed" project for analysis of a particular set of historical information has been turned into a platform that might be used by students and researchers alike; it seems likely that the fine-grained spatial detail of city-map projects such as DECIMA or MoEML will similarly expand in unanticipated ways as they become more collaborative and open.

It is of course also worth noting that significant innovations in mapping outside humanities point the way for future directions in historical digital research. Thus, as researchers working in the contemporary fields of political and environmental studies, sociological analysis, or urban planning and design have long realized, smartphones create prodigious quantities of data relating to users' consumption practices in their everyday interactions, a form of real-time data that can be aggregated and analyzed through maps

---

[12] See the home page statement at http://decima.chass.utoronto.ca/.
[13] See http://republicofletters.stanford.edu/ and http://hdlab.stanford.edu/projects/palladio/.

(Bauman and Lyon 2012; Silva and Frith 2012; Presner et al. 2014). Such data ceases to be static; real-time maps can trace both physical and temporal movement in dynamic ways through new forms of visualization in which the two-dimensional vectors of the map, latitude and longitude, are enhanced through the action of time. In the work of MIT's Sense*able* City Lab, for instance, real-time movement in an entire city can be viewed to understand such factors as commuting and traffic patterns, infrastructure capacity, or crowd behaviors at occasional events (e.g., sports events and concerts).[14] As early as 2008, a project called *Real Time Rome* analyzed the city, revealing how people favored particular streets over others in their pedestrian movement, or gathered and socialized in specific places at particular times (Ratti and Calabrese 2008; Girardin et al. 2008). Referring to the Nolli map discussed above, Carlo Ratti and Francesco Calabrese note that "Nolli didn't think that what happened in the spaces of the city was any of his cartographic business," while "we can now model the city as a dynamic system in which social, architectural and technological aspects cohabit" (Ratti and Calabrese 2008, 43).

Naturally, no sufficient body of real-time data can help us visualize movement and experience in past urban environments. As we have seen in the opening examples above, historical maps have primarily served historians as a static base for digital humanities projects that layer historical evidence. Nonetheless, the dynamics of everyday use and movement might be at least in part recovered through digital mediation. The potential of real-time geo-location enabled by smartphones through apps that adopt the affordances of geo-location technologies have yet fully to be explored by historians. Apps enable the research process to be locative, embodied, and kinetic, and lead in potentially quite different directions to the desk-based screen projects discussed in the opening section of the paper. At their most basic level, portable handheld devices enable research content to be taken out "into the field" to interact directly with the urban environment; this has obvious benefits for research dealing with urban history. So, to take an example, the Museum of London's *Streetmuseum* app allows users to interact on-site with historic images (prints, paintings, and photographs) of London overlaid onto the contemporary street view, and to select various enhanced reality historical trails through the city following particular themes.[15] GPS and compass features in

---

[14] See http://senseable.mit.edu/. For *Real Time Rome*, see http://senseable.mit.edu/realtimerome/.

[15] See http://www.museumoflondon.org.uk/Resources/app/you-are-here-app/home.html.

the handheld device help the user reach specified locations and facilitate quite accurate image overlays. A similar strategy underpinned the *Artist's View* app produced in 2012 to allow the public to overlay painted views of Florence onto extant sites in the city.[16] In both cases—and these are no more than examples of an emerging genre—the portable screen allows the user to compare the real urban environment with its representation in earlier periods.

Other apps focus on more traditional forms of content delivery on location, using GPS to pinpoint sites in the city at which various types of information are delivered on screen as text, photographs, or film. This is a standard format for the repurposing of guidebooks as apps, and again, as with the desktop-based projects discussed earlier, social media tools such as Historypin also provide an app-based service that enables users to access tagged information in real time through GPS-enabled devices. The format is also starting to be adopted in the context of early modern research as a means of presenting findings to a peer community and the general public, usually in the form of themed itineraries.[17] So, for example, ShaLT (*Shakespearean London Theatres*) is a research project based at De Montfort University (Leicester, UK) that set out to trace and map theatres active in the capital at the time of Shakespeare; a simple app offers five walking itineraries that invite the user to explore these theatre locations and to understand better their connections and proximity to other sites such as ale-houses, courtrooms, and churches, delivered on site in the form of short texts and images.[18] While most of the original buildings no longer survive, their proximity to modern-day sites of sociability are richly evocative, and reward the user with a valuable site-specific experiential encounter.

---

[16] Produced to coincide with the exhibition *Americans in Florence* (2012).

[17] This is a fast-growing area, especially when viewed as a focus for audience engagement (what in UK research funding is usually termed as "impact"). Public funding from the UK's Arts and Humanities Research Council (AHRC), for example, has contributed to the development of a range of apps; examples developed at the University of Exeter are discussed at https://humanities.exeter.ac.uk/classics/news/title_422730_en.html.

[18] See http://shalt.dmu.ac.uk/. Another approach that has been adopted for "locating" history is through the use of QR codes, literally attached to sites and locations of significance (a physical tagging of real objects to virtual content held on a website). I am not discussing this approach, however, as it is somewhat mechanical and misses many of the innovative affordances of technology discussed here.

Apps functioning on smartphones can immediately set up quite a different relationship between the information they draw on and the urban fabric with which they engage, as the screen serves as a direct interface between the physical environment and the enhanced or augmented reality that can be triggered through the user's location-aware device. Moreover, immersive experiences can be designed whereby the user does not manually access information through the phone, but through GPS triggers (such as geo-fences or i-beacons) that can automate the release of content on the go (Farman 2012; Farman 2013a).[19] Here, significant innovations have been made primarily through the medium of audio, whereby curated experiences can be designed that use movement, GPS, and audio content to offer a seamless narrative experience through the urban realm, a form of audio guide such as those that have long been used in museums, extended through GPS into the streets of the city. The museum comparison is pertinent here especially for the disciplinary areas of art and architectural history, as an app design focus centered on audio allows the end user to concentrate on looking at the urban environment rather than gazing at a screen.

Hidden Florence, a project I led and recently published for Apple and Android phones, allows us to explore more closely the potential of apps, and more specifically app-enabled walking trails as a medium for delivering an innovative research practice and process that directly engages with the urban fabric (Figure 1).[20] We developed *Hidden Florence* in 2013 during an intensive six-month research collaboration between two academics, an IT developer specializing in the creation of location-based apps, and an audio producer, and it was published in 2014.[21]

---

[19] Farman discusses the potential for such technologies in relation to narrative and storytelling.

[20] Available from the App Store and GooglePlay in July 2014, this is a collaboration with the developers Calvium Ltd., funded by the AHRC and published by the University of Exeter. For the full project team see: http://www.hiddenflorence.org/credits/.

[21] Further details can be read at http://hiddenflorence.org/about/ and on the project blog.

Microstoria 2.0: Geo-locating Renaissance Spatial and Architectural History   269

Figure 1. Screenshot of home screen of the *Hidden Florence* app (© University of Exeter).

While a number of audio-guide apps have been published that engage with European historical urban spaces, these have to date tended to relate to quite recent periods no earlier than the nineteenth century, and one of our objectives was to explore how we might evoke the more remote past of early modern Italy.[22] This project, designed as an experiment, proposed to use smartphone technology to provide an on-site audio guide to the architectural and socio-cultural features of Renaissance-era Florence.

Hidden Florence adopts a GPS-triggered audio-guide format which enables an engagement strategy that seeks to combine elements of specialist information delivered in a standard museum audio guide with a more immersive experience provided by a narrative-driven character, on site in front of actual places and objects scattered around cities. Working with app-developers at Calvium, we benefited from their guidance in user experience design, and as a result created a semi-fictional character—Giovanni di Marco, a late-fifteenth-century wool-worker—to lead the user on a context-specific themed walking tour through the city of Florence.[23] Giovanni was scripted and written by historians and voiced by actors (English and Italian language options are provided).

Furthermore, in order to add a heightened level to the immersive experience created by the audio, the modern-day user navigates the city through the interface of a sixteenth-century map of the city (Figure 2). Stefano Bonsignori's 1584 map of Florence was etched in spectacular detail (it is also used as the basemap for the DECIMA project discussed above), and the relatively precise cartography enabled it to be overlaid on the modern street map.[24]

---

[22] We chose to approach Calvium because they had developed a number of projects, including audio App-trails for the National Trust and the Guardian newspapers; see http://calvium.com/services/app-trails/. Among other interesting projects created around the concept of GPS-triggered audio walks, see the Wellcome Trust's "Magic in Modern London" (developed by Amblr): http://wellcomecollection.org/magic-modern-london.

[23] For Calvium's user experience design, see http://calvium.com/research/, as well as Farman 2012 and Farman 2013a. The "Giovanni" character was developed using firm historical evidence, much of which is made available through the project website in articles written by Fabrizio Nevola and David Rosenthal.

[24] For a summary account of the process applied in transferring the map and overlaying it to the satellite route-map, see http://hiddenflorence.org/about/about-bonsignori-map/.

Figure 2. Screenshot of a navigation screen from the *Hidden Florence* app, which uses Stefano Bonsignori's 1584 map of Florence (© University of Exeter).

Bonsignori's remarkable piece of mapmaking combines a more traditional axonometric (bird's-eye view) approach with the innovative ichnographic (accurate figure ground) survey to offer the most accurate contemporary map of Florence in the sixteenth century (Frangenberg 1994; Else 2009; Ballon and Friedman 2007). Thus the key network of streets stands out (and is indeed revealed to be quite similar to the modern layout in most areas), while landmarks are also prominent. The use of the historic map as a navigational tool both historicizes the experience of moving through the city and estranges the user from their surroundings, and by so doing highlights ways in which the Florence of today is different from its Renaissance self. As far as we know, *Hidden Florence* is the first geo-located app to adopt a pre-modern map interface; it offers a remarkable experience akin to walking in the city using what we might loosely term a sixteenth-century street view interface.

The app is composed of two guided walks, one that takes the user to the neighborhood where the narrator lives (the eastern parish of Sant'Ambrogio), while the second takes him to the city center and the world of work. In so doing, of course, this choice sets up a comparison between the nature of spaces on the working-class periphery of the Renaissance city and the densely packed urban core.[25] Each walk is made up of a series of locations plotted along a route; each location offers a significant building-type or site in the city (a street corner, an apothecary's shop, a tavern, a bridge), and this provides an opportunity to consider its social, cultural, and visual history. As will be discussed further below, the choices made for each location are fundamental to the structure of the app and the nature of the information encoded in it. In addition to the fictional guide, further content is provided in the form of brief audio-interviews with academics on subjects directly pertinent to each site. The app is tightly integrated to a website compatible with the app's design and layout; each site page is linked to a page on the *Hidden Florence* website, where additional information and interpretation is provided in the form of short articles, bibliography, and web links to further resources. The app and website also include Twitter feeds to enable full integration with social media used by target user groups.[26]

---

[25] I have chosen not to provide bibliography references in this article for the underpinning historical research (which is far too extensive to be cited adequately), as this is not the main substance of the discussion.

[26] While this article does not discuss the technical details of the app, it is worth noting that it has been designed to work using basic GPS services and does not rely on roaming, 3G, or other higher level location services. This has been done to avoid charges for users who will in many cases be tourists or students using phones with roaming turned off.

Microstoria 2.0: Geo-locating Renaissance Spatial and Architectural History 273

Figure 3. Guided tour through Florence showing the OpenStreetMap underlying map in the *Hidden Florence* app (photo: Ross Gill).

The app functions as a publication that integrates a range of different types of content. Users taking a tour of the city then move conceptually from the physical fabric of the city, through the screen interface of the sixteenth-century map, to hear an immersive audio "located" in the Renaissance past, from which they can then link out from the app to more traditional text content delivered on the website. They can then interact with that content through social media, potentially offering further information, or simply commenting on their experience. As such, the project participates in the wider actions of social media communities, although the content is carefully scripted and curated. Furthermore, the app has been designed so that it can also be used remotely in "armchair mode," and initial analytics evidence suggest that to date it has been used as much remotely as on site. We also decided to make all the audio content and map routes available through the website as downloadable files so that it can be accessed without a smartphone.

To give a sample of the user experience, I will discuss the example of one site, the Bargello. Users navigate to the site using the Bonsignori map—a sort of immersive Renaissance street view experience in which a Renaissance-styled avatar character plots their movement until they are in sight of the object they are to look for. At this point the geo-fence triggers a new screen showing a photographic cue to identify a window on the building, with the prompt: "Look for the small window on the corner of the first floor—the hanging window."

The user then presses "found it" and the Giovanni audio begins:

> Now you're gonna have to just believe me when I tell you that sometimes, just sometimes when you come by this way, what do you get when you look up at that window? A hanging body!
> You do, honest!!
> This is the Bargello. It's the fortified home of the chief of police in this city.
> You don't want to end up in there. If you do, you're in trouble, believe me …
> The cops—the *birri*—are based there—and there's a law court, torture chambers, and a prison—and for the worst of crimes, you don't get to languish in the jail down the road. Oh, no, it's a public execution straight out of the window up there. That's what happened to the Pazzi conspirators who plotted to kill the Medici a few years ago. Everyone came to see them—and that just added to their family's shame.
> … Y'see, shame is almost worse than death in this city. That wall there, below the windows, it's usually covered in portraits. *Pitture infamanti*, we call them—shaming portraits. It's a rogues' gallery of traitors, turncoat fighters, and also of people you can't trust your money with—debtors and people that have been declared bankrupt.
> Some say that the risk of getting your portrait painted up there is the main reason people avoid getting into trouble. Well, yes … it's certainly quite a warning.

This short sample offers a taste of a block of the full two-minute audio script for the Giovanni piece delivered at the site. Should users wish to pursue their interests further, they would then press "hear more" to launch the additional two-minute interview with a historian (in this instance Fabrizio Nevola) talking about the Bargello in the exercise of policing and justice in the city. They could then follow a link further to the website article, or indeed listen to Giovanni in Italian.

Readers of the above text—and historians who might use the app—will note a number of things here. There is no intention to tell a full history of the Bargello as a monument (as a guide book might try to do); rather, a clear set of decisions have been made by the researcher—to explore a series of themes specifically associated to the object of inquiry. From the window on the building façade, we move to the Bargello as police headquarters, to touch on the phenomena of portraits used as a form of punishment, to capital punishment, to the place of reputation (*fama*) in Renaissance urban society. These issues are woven into a brief narrative delivered in front of a physical object and are all the more memorable for this; they can of course be explored further by both curious tourists and students through reading the linked website article and its associated further readings and links. Place, and the user's embodied experience in the spaces of the city, underpins the narrative (Freedberg 2009; Crowther 2001).

As we have already seen, the spatial turn and interest in street life in the early modern city is a fast-growing area in Renaissance and early modern studies, and this approach is both at the center of my own research and also of the questions that the *Hidden Florence* app sought to address and present to users (Nevola 2013). A recent collection of essays places sensory experience (sound, speech, performance, vision) at the heart of an analysis of urban space (Nevola and Clarke 2013a). Overarching and innovative research areas that emerge from such an approach include, but are not limited to: the performative nature of everyday actions and exceptional rituals played out in public space (including crime, punishment, and violence); the agency of public space and strategic locations for information and knowledge exchange (street-corners or piazzas); the interplay of public and private spaces occurring on the interface of the street (property, sounds, exchange of goods) (Nevola and Clarke 2013b). It is themes such as these that have informed the app design, in which a close reading and focus on sites, locations, and objects still extant in the city fabric is proposed. The research process serves to conceptualize the meaning embedded in the material culture of public space in relation to the fluid and dynamic nature of urban experience. As such, I would suggest that the medium of the app shapes the message—and perhaps more significantly the research questions and methods—that we can explore in relation to the urban environment of cities of the past. This more direct engagement with the embodied and everyday experiences of living, and walking, in the city of the past extends the potential of traditional text-based research (books and articles; static web supports), without of course replacing the significance of these.

Figure 4. Screenshot of the "found it" screen from the *Hidden Florence* app (photo: Ross Gill).

Hidden Florence is a stand-alone app produced with the support of UK research council funding as an experiment to explore the potential and affordances of the medium, as applied to Florence in the early modern period. We have built an analytics package into the app to monitor how it is being used, and plan to report on this after a year has elapsed from its publication. We are also seeking funding opportunities in the context of further academic research projects to develop an editing platform that will enable us to simplify the process by which new content can be implemented for Florence, and so that new cities might also be more easily added. As will be readily imagined, the app format is extensible in a multitude of ways; it is easy to imagine how new stories might be developed to present the research of other academics, and indeed also how the basic concept could be repurposed for any urban setting. There is also considerable potential—which I am beginning to explore—for adopting this format in a simplified form in the context of teaching.[27] At present, audio-guide apps such as *Hidden Florence* tend to be custom-built products, but like some of the mapping projects discussed above, the movement is toward the creation of editing platforms and technologies that would result in the production of such work being both easier and less expensive. Calvium, the developers of *Hidden Florence*, offer a service called *AppFurnace* that enables relatively easy editing and creation of geo-located apps.[28] Such offerings suggest that a service allowing users to create geo-located audio walks (like Historypin for tagging and WordPress for blogs and websites) may be only a short way off.

In conclusion, the *Hidden Florence* project raises questions about the nature of the work in relation to the methods and practice of history. It is evident, as this article has shown, that it contributes to an evolving field of digital humanities applied to the early modern period; likewise, I have also suggested that it is particularly well-suited to exploring specific concerns of the spatial turn as applied to studies of the material culture of the early modern city. Rather than simply an output that presents research on space using digital tools, however, I would argue that geo-located historical narratives of the sort described here can be understood to constitute an innovative methodology in their own right (Nevola and Rosenthal 2016). In seeking parallels with existing approaches, perhaps the strongest affinity can be found with *microstoria*—microhistory, or a practice of history that is often told from "below," from the perspective of the everyday, and through close analysis and description of almost ethnographic precision.

Microstoria was defined in part by a book series Giovanni Levi and Carlo Ginzburg edited from the late 1970s for the Turin-based publisher Einaudi, in which the choice of a narrative mode for historical writing owed something to the literary family background of both historians and to the experimental

---

[27] In January–March 2015, I experimented with writing for geo-location with a group of students using a platform (http://www.placeify.co.uk) developed by colleagues at the University of Exeter.

[28] Additional information can be found at http://www.appfurnace.com/home, with links to the educational (school) package they have also launched. Announced in 2014 was a service called Detour, which provides "tools for anyone to create" location-aware audio walks, although none have yet been published; see http://www.detour.com.

writing (as Ginzburg has often noted) of Raymond Queneau's *Les fleurs bleues*, translated into Italian by Italo Calvino in 1967 (Ginzburg 1993, 14–16, 30–32). Microhistory affords special attention to the narrative mode, with detailed stories of individual actors in the historical process presented to illuminate a wider picture, or an overlooked perspective. Giovanni Levi and Carlo Ginzburg, the founder-practitioners of *microstoria* as it developed in the 1970s in Italy, never set down a set of hard-and-fast methodological rules. And yet, if we turn to Levi's identification of their key concerns as being for "the reduction of scale […] the small clue as scientific paradigm, the role of the particular (not, however, in opposition to the social), the attention to reception and narrative, a specific definition of context and the rejection of relativism," I think there is clear resonance with the spatial and geo-located research questions that underpin *Hidden Florence* (Levi 1991, 113, 117–19). Likewise, Ginzburg has described microhistory as:

> based on the definite awareness that all phases through which research unfolds are constructed and not given: the identification of the object and its importance; the elaboration of the categories through which it is analyzed; the criteria of proof; the stylistic and narrative forms by which the results are transmitted to the reader […] the insistence on context. (Ginzburg 1993, 32–33)

In the *Hidden Florence* project, the choice to propose an oblique perspective on a well-known period through directing the spotlight on non-canonical or peripheral sites (street-corners, street-shrines, taverns and apothecaries' shops, minor neighbourhood piazzas) or non-elite protagonist perspectives is also informed by the objectives of *microstoria*. While critical of the deployment of microhistory in some cases as a practice for revealing no more than "human interest stories" from the archives of the past, Peter Burke has spoken of it as a "'strategy of knowledge' which keeps close to human experience" (Levi 1991, 116). Moreover, it is also worth noting that microhistory was not uniquely confined to social histories of "minor" events and individuals from social history—Ginzburg wrote a well-known book on the painter Piero dell Francesca, while the influential architectural historian Manfredo Tafuri contributed to the series with a book about the church of San Francesco della Vigna, by Palladio, in Venice (Ginzburg 1981; Foscari and Tafuri 1983).

It might then be suggested that *Hidden Florence* coincides in a number of ways with how historical evidence is deployed in the practice of microhistory, albeit that it does so through digital means, and that its experimental approach still needs to be tested, challenged, and refined. Moreover, by engaging both the researcher and the user in a dialogue between real and virtual environments, the geo-located narrative audio guide format could even be said to conform to web 2.0 characteristics of interactivity. Rather than merely tagging information to place, the audio-guide app format proposed has inquiry and interpretation at its heart—it is intentionally a work of history, not of information delivery.[29] In the final chapter of his magisterial overview of historical methodologies, John Burrow discusses the place that the TV documentary has taken as a new form of "making" history, with special reference to Ken Burns's sophisticated use of images to convey dynamic narratives about eras that predate the moving image (Burrow 2009, 518). While it is all too easy to see such work as public-facing outreach, I would like to suggest that easily accessible geo-location services delivered through smartphones offer an opportunity for methodological innovation, creating a means by which history is not only presented, but researched. "Altering the scale of observation" and putting research—physically locating it as practice and process—back in the actual spaces where events were played out allow us to explore and reveal the powerful layers and traces of the past that still populate the public realm of the present (Levi 1991, 102).

WORKS CITED

Ballon, Hillary, and David Friedman. 2007. "Portraying the City in Early Modern Europe: Measurement, Representation, and Planning." In *The History of Cartography*, vol. 3: *Cartography in the European Renaissance*, Part 1, edited by David Woodward, 680–705. Chicago: University of Chicago Press.

Bauman, Zygmunt, and David Lyon. 2012. *Liquid Surveillance: A Conversation*. Cambridge: Polity.

Bodenhamer, David J. 2010. "The Potential of Spatial Humanities." *The Spatial Humanities: GIS and the Future of Humanities Scholarship*, 14–30. Bloomington and Indianapolis: Indiana University Press.

---

[29] The first iteration of *Hidden Florence* includes links to Twitter, and we anticipate that future versions would develop further the potential for user interactivity through social media-enabled participation, comment, and perhaps even content creation.

Burrow, John. 2009. *A History of Histories: Epics, Chronicles, Romances and Inquiries from Herodotus and Thucydides to the Twentieth Century.* Harmondsworth, UK: Penguin.

Calvino, Italo. 1974. *Invisible Cities.* Translated by William Weaver. London: Picador.

Connors, Joseph. 2011. *Piranesi and the Campus Martius: The Missing Corso—Topography and Archaeology in Eighteenth-Century Rome/Piranesi e il Campo Marzio: Il corso che non c'era—Topografia e archeologia nella Roma del XVIII secolo.* Rome: Jaca Book.

Crowther, Paul. 2001. *Art and Embodiment.* Oxford: Oxford University Press.

Dunn, Stuart, and Mark Hedges. 2012. "Crowd-Sourcing Scoping Study. Engaging the Crowd with Humanities Research." Centre for e-Research, Department of Digital Humanities, King's College London. Available for download at http://crowds.cerch.kcl.ac.uk/wp-content/uploads/2012/12/Crowdsourcing-connected-communities.pdf.

Else, Felicia. 2009. "Controlling the Waters of Granducal Florence: A New Look at Stefano Bonsignori's View of the City (1584)." *Imago Mundi* 61 (3): 168–85.

Farman, Jason. 2012. *Mobile Interface: Embodied Space and Locative Media.* London: Routledge.

——— ed. 2013a. *The Mobile Story: Narrative Practices with Locative Technologies.* London: Routledge.

——— 2013b. "Storytelling with mobile media." In *The Routledge Companion to Mobile Media*, edited by Gerard Goggin and Larissa Hjorth, 528–37. London: Routledge.

Favro, Diane. 1999. "Meaning and Experience: Urban History from Antiquity to the Early Modern History." *Journal of the Society of Architectural Historians* 58 (3): 364–73.

Foscari, Antonio, and Manfredo Tafuri. 1983. *L'armonia e i conflitti: La Chiesa di San Francesco della Vigna nella Venezia del '500.* Turin: Einaudi.

Frangenberg, Thomas. 1994. "Chorographies of Florence: The Use of City Views and City Plans in the Sixteenth Century." *Imago Mundi* 46 (1): 41–64.

Freedberg, David. 2009. "Movement, Embodiment, Emotion." In *Histoire de l'art et anthropologie*. Paris: INHA and Musée du Quai Branly («Les actes»). http://actesbranly.revues.org/330.

Ginzburg, Carlo. 1981. *Indagini su Piero: Il battesimo, il ciclo di Arezzo, la flagellazione*. Turin: Einaudi. (Translated by Martin Ryle and Kate Soper as *The Enigma of Piero della Francesca: The Baptism, the Arezzo Cycle, the Flagellation*. London: Verso, 1985.)

―――― 1993. "Microhistory: Two or Three Things That I Know about It." *Critical Inquiry* 20 (1): 10–35.

Girardin, Fabien, Josep Blat, Francesco Calabrese, Filippo Dal Fiore, and Carlo Ratti. 2008. "Digital Footprinting: Uncovering Tourists with User-Generated Content." *IEEE Pervasive Computing*, 7 (4): 36–43.

Gregory, Ian N., and Alistair Geddes. 2014. *Towards Spatial Humanities: Historical GIS and Spatial History*. Bloomington and Indianapolis: Indiana University Press.

Levi, Giovanni. 1991. "On Microhistory." In *New Perspectives on Historical Writing*, edited by Peter Burke, 95–119. University Park: Pennsylvania State University Press.

Nevola, Fabrizio. 2013. "Review Essay: Street Life in Early Modern Europe." *Renaissance Quarterly* 66 (4): 1332–45.

Nevola, Fabrizio, and Georgia Clarke, eds. 2013a. "The Experience of the Street in Early Modern Italy." Special issue, *I Tatti Studies in the Italian Renaissance* 16 (1/2): 47–229.

―――― 2013b. "Introduction: The Experience of the Street in Early Modern Italy." *I Tatti Studies in the Italian Renaissance* 16 (1/2): 47–55.

Nevola, Fabrizio, and David Rosenthal. 2011. *Urban Communities in Early Modern Europe (1400-1700): A Research Review*. Available for download at http://www.earlymoderncommunities.org.

―――― 2016. "Locating Experience in the Renaissance City Using Mobile App Technologies: The 'Hidden Florence' Project." In *Mapping Space, Sense, and Movement in Florence: Historical GIS and the Early Modern City*, edited by Nicholas Terpstra, 187–209. London: Routledge.

Pinto, John A. 1976. "Origins and Development of the Ichnographic City Plan." *Journal of the Society of Architectural Historians* 35 (1): 35–50.

Presner, Todd, David Shepard, and Yoh Kawano. 2014. *HyperCities: Thick Mapping in the Digital Humanities*. Cambridge, MA: Harvard University Press.

Ratti, Carlo, and Francesco Calabrese. 2008. "Nolli 2.0: Or How to Rebuild Rome Without Even Asking Permission from the Historic Preservation Authority." In *Uneternal City: Urbanism Beyond Rome*, edited by Aaron Betsky, 42–47. Venice: Marsilio.

Rinne, Katherine W. 2011. *The Waters of Rome: Aqueducts, Fountains, and the Birth of the Baroque City*. New Haven, CT, and London: Yale University Press.

Silva, Adriana de Souza e, and Jordan Frith. 2012. *Mobile Interfaces in Public Spaces: Locational Privacy, Control, and Urban Sociability*. London: Routledge.

Terpstra, Nicholas. 2016. *Mapping Space, Sense, and Movement in Florence: Historical GIS and the Early Modern City*. London: Routledge.

Ullyot, Michael. 2013. "Digital Humanities Projects." *Renaissance Quarterly* 66 (3): 937–47.

Waldron, Ann. 1999. "The Magic Classroom: Interactive Web Programs are Changing How Art History is Taught." *PAW: Princeton Alumni Weekly*, 7 April. http://www.princeton.edu/~paw/archive_old/PAW98-99/13-0407/0407feat.html.

Witcher, Robert E. 2006. "Broken Pots and Meaningless Dots?: Surveying the Rural Landscapes of Roman Italy." *Papers of the British School at Rome* 74: 39–72.

Gazing into Imaginary Spaces:
Digital Modeling and the Representation of Reality

John N. Wall
North Carolina State University
jnwall@ncsu.edu

Digital modeling provides humanists with powerful tools for integrating large amounts of historical information into a unified and coherent display. Our Virtual Paul's Cross Project (vpcp.chass.ncsu.edu), for example, allows us to bring evidence from paintings, drawings, and engravings about the appearance of Paul's Churchyard, outside St. Paul's Cathedral in London in the early 1600s, together with archaeological evidence about the location, size, and design of the structures within that space and with meteorological evidence about the climate, local weather, and angle of sunlight providing illumination for that space at various times of day throughout the year. We believe, as a result, that we can experience that space in new and more profound ways, and get from it a deeper understanding about early modern urban life in England's largest city.

This essay is, however, about something else, about what we are doing when we believe we have discovered, from our experience with a digital environment, things about past events that are not documented by traditional sources. For example, based on experience with the Paul's Cross acoustic model, I have concluded that the sound of church bells ringing out the eleven o'clock hour at the cathedral and at the many churches surrounding St Paul's Cathedral in the City of London would have rendered inaudible for a minute or more the voice of a preacher—in the case of our model John Donne—delivering a sermon at the Paul's Cross preaching station. The cathedral's clock also sounded out the quarter-hours as well, not an insignificant interruption. Presumably, the preacher would have had to pause in his delivery, until the sound of these bells died away, before continuing with his sermon.

Having reached this conclusion, I believe, therefore, that I now understand more fully what the experience of a Paul's Cross sermon would have been like, both for the preacher and for his congregation. This observation has led me to a series of other questions about the experience of preaching and hearing a Paul's Cross sermon. If, for example, the preacher's voice were interrupted for such an extended period, and with it, his hold on the congregation's attention, he would have had to reengage their attention before continuing with his argument.

But how might he do that? If Donne were able to anticipate the timing of the bell marking the time, he could bring one section of his sermon to a close just before the bell sounded, permit the sound of the bell to signal a pause and underscore the importance of the point he had just made, then resume his sermon with a new point after the sound of the bell died away. Interestingly, from time to time in his sermon for Gunpowder Day, November 5, 1622, Donne pauses in the development of his argument to summarize where he is at that point, then lays out the subject he will address next, before launching into it.

Could these moments represent the points in Donne's sermon at which the bell sounded during his delivery? They do not come, in the sermon, at evenly spaced points. Yet this does in itself not rule out such a practice, since we know that Donne preached from notes, not from a fully-written-out text, and that as a result he could have shortened or expanded his argument in response to the reaction of the crowd or the passage of time. What we have for this sermon is a text written out after the fact, free of the constraints of time, and hence it represents what Donne remembered saying, or wished he had been able to say, not a direct transcript of what he said.

Yet this discussion brings us to a crossroads in interpretation. I believe, on the one hand, that my inference about the sound of bells impeding the progress of Donne's sermon is based on experience of an accurate model of the event. I believe the model is accurate, either precisely accurate because the model is based on the hard data of measurements, or generally accurate because it is based on our interpretation of visual images, or accurate to a degree of approximation because we use images of real early modern buildings of the kind we know were there, if not the specific example of the kind that really was there. I know from other kinds of historic evidence (Stern, 2012) that there were clocks at the cathedral and at London's parish churches, and that they announced the time by the ringing of bells. But as for my specific conclusion about the sound of those bells and their impact on the experience of preaching and hearing a Paul's Cross sermon, I must acknowledge that this conclusion is based entirely on inference from my experience with the Paul's Cross model. I have not been able to find any traditional forms of historical evidence—no references to such a phenomenon in contemporary accounts of preaching, no diary entries, no references in poems or plays or letters, or any of the traditional documentation we customarily use— to support this understanding.

We have thus reached the point described by Willard McCarty concerning the Virtual Paul's Cross Project:

> *Virtual Paul's Cross* takes the observer to the limits of what is securely known, then offers a standpoint from which securely to infer more. (Does knowledge result? If not quite that, then what gives—knowledge or what we mean by "knowledge"?) And by the nature of the medium, the observer, having done that, can then return to the sermon (which is, at this point, what exactly?) to look and listen again, then incorporate what has been inferred into the simulation model. And so it goes, "where no one has gone before." I am tempted to say, if this does not make you nervous and excited all at once, then either you or I are at fault.

Traditional historical data, organized through digital modeling, has brought us to the point where we can hear the sermon in the space for which it was prepared to be delivered, and hear the bells, also in that space, and recognize that—if our data is correct, and our modeling is correct—the bells and the preaching could not coexist, at least for the length of time it takes for the bells to mark the time and fall silent. We are at the moment when we can realize this, and then ask ourselves: where, along McCarty's inferential chain, do we go next? What do we begin to understand about the delivery of sermons at Paul's Cross, in London, in the early seventeenth century?

"Along this trajectory," McCarty continues, "lies the creative scholarship in digital humanities that we must have for the discipline to become strong." I, for one, agree with McCarty, but as a very traditionally trained scholar who has come to the use of digital technology for scholarly purposes rather late in my career, I also believe in moving forward along this inferential chain carefully, thoughtfully, fully aware of how far from traditional forms of evidence in humanities research we have come, and how far we may yet go. I believe in holding in tension the contrast between traditional methodologies and understandings of knowledge and their grounding in the evidence of texts and objects and the methods and evidentiary bases of conclusions we reach by working with digital models. I also believe in holding those conclusions under some kind of provisional status while we use them—and the process of our arriving at them—to help us address the kinds of questions McCarty brings up: "Does knowledge result? If not quite that, then what gives—knowledge or what we mean by 'knowledge'?"

To consider how that "holding under provisional status" might work, let us begin by noting one fundamental fact about the Virtual Paul's Cross Project: however accurate our research, however capable our technology, however detailed our modeling, however careful our reasoning, the event we are creating actually never happened as we model it. The title page of the first printed edition of John Donne's sermon for November 5, 1622, clearly states that this sermon was "Intended for Pauls Crosse, but by reason of the weather, [was] Preached in the Church." No one has been able to figure out exactly *where* "in the Church" it was preached (not in the nave, which was never used to house worship services after the Reformation), but all the evidence points to the conclusion that in fact this sermon—unlike other surviving texts of Paul's Cross sermons preached by Donne or other clergy of the period—was not preached at the Paul's Cross preaching station, but instead was "preached in the Church."

Yet in spite of the fact that Donne did not preach this sermon while standing in the Paul's Cross preaching station, but somewhere else, somewhere inside St Paul's Cathedral,[1] there were still compelling reasons to use it anyway as our model of the Paul's Cross sermon. We will explore some of those reasons in the course of this essay. But I begin our discussion with the acknowledgment that there is, always, built into the experience of the Virtual Paul's Cross Project the distance between our experience of the visual and acoustic models included in the Project and our knowledge that it didn't happen this way. This serves as a constant reminder that the Virtual Paul's Cross Project does not aspire to the status of time travel. It cannot ever claim to "hold the mirror up to nature" in any precise sense, but is better thought of as a special kind of thought experiment. Thus, what we learn from it is always provisional, always at best representational and approximate. We can never say that this is what it would have been like to be there when Donne preached that day in Paul's Churchyard; we can say, on the other hand, that to the extent of our ability to recreate the conditions of the churchyard on that day, we can approximate

---

[1] Presumably in the cathedral's Choir, but since seating in the Choir was limited to fewer than 200 people, other spaces such as the space beneath the Choir designated as St. Faith's parish church, or the Chapter House on the south side of the cathedral, have been suggested as alternatives.

the experience of Donne preaching the sermon as he intended to preach it in the place from which he intended to preach it.

Hence, we are now involved in a conversation about authority, and its limits, in digital modeling. Ultimately, we will return to the question of what we know when we learn it from our experience with digital models. For now, however, I want to review the processes of assembling and evaluating the traditional kinds of evidence we drew on in the creation of the Virtual Paul's Cross Project, the evidence on which any claim to the value of what we learn from these models must rest. I hope to show, ultimately, that the kinds of scholarship we do with digital models are remarkably similar to the kinds we do using traditional modes of research, which also depend for their authority on the assembly and interpretation of traditional kinds of evidence. For, of course, traditional research in the humanities depends—as does the Virtual Paul's Cross Project—on the selection, evaluation, interpretation, and synthesis of materials that survive from the period of time that engages us in our work.

To get us into that conversation, let's review the development of the Virtual Paul's Cross Project, with special attention first to the development of the visual model and second to the acoustic model of Paul's Churchyard. We will then turn to our choices of the sounds one hears when using the website to experience Donne's Paul's Cross sermon for Gunpowder Day, November 5, 1622.

The Virtual Paul's Cross Project was conceived to develop our understanding of the post-Reformation Church of England as a religious body functioning in a culture of speaking and hearing as well as a culture of writing, manuscript, and the printed text (Wall 2014b). We set out with this project to recreate as accurately as the evidence allows the experience of a Paul's Cross sermon, all two hours of it, from 10 a.m. until noon, in the northeast corner of Paul's Churchyard, outside St. Paul's Cathedral in London. At the center of our attention in Paul's Churchyard was a small building that had been used for decades as a site for preachers chosen by the Bishop of London to deliver a series of two-hour-long public sermons every Sunday, as well as on other days of particular significance (Morrissey 2011, 1–34).[2] The phenomenon of the Paul's Cross sermon is widely regarded by scholars as a significant public event in the unfolding of English social, religious, and political history in the early modern period. At Paul's Cross, during this period, crowds of up to five or six thousand people gathered to hear religious controversies of the Reformation period debated and to have the developing policies of the Crown in matters of religion proclaimed and defended.

---

[2] Such as Gunpowder Day, which in 1622 was on a Tuesday.

My original goal for this project was to explore how the Paul's Cross sermon functioned as an occasion for communication, inspiration, and, perhaps, the development of a sense of urban public identity. I was concerned especially with the audibility of open-air preaching, with just how well the preacher at Paul's Cross could be heard in a large open space and in the presence of large, though variable, crowds of people. With funding from the National Endowment for the Humanities, I was able to assemble a team of scholars, historians, architects, actors, and acoustic engineers who used both visual and acoustic digital modeling technology to recreate as fully and accurately as possible the experience of a Paul's Cross sermon, making it available to us as an experience unfolding in real time as it would have sounded in Paul's Churchyard, not just as a text on the printed page of a scholarly edition. The digital tools we employed to bring this sermon to life once more are tools customarily used by architects and designers to anticipate the visual and acoustic properties of spaces that are not yet constructed. Here, we use them in the Virtual Paul's Cross Project to recreate the visual and acoustic properties of spaces that have not existed for hundreds of years. Digital modeling technology used in the Virtual Paul's Cross Project enables us to integrate the physical traces of pre-Fire St. Paul's Cathedral—measurements of the foundations and study of the surviving stones—with the historic visual record of the cathedral and its surroundings into a highly accurate visual model of the cathedral and its churchyard. Acoustic modeling technology starts with the visual model and transforms it into an acoustic space through which we can hear recorded ambient sounds that invoke the experience of that space in 1622, as well as the actor Ben Crystal's performance of Donne's sermon text.

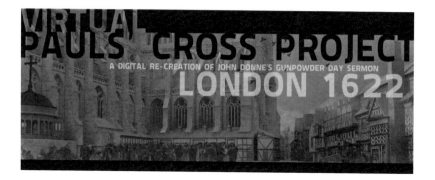

Figure 1. Paul's Churchyard, the Cross Yard, looking west. From the visual model, constructed by Joshua Stephens, rendered by Jordan Gray. Banner design by David Hill.

The visual model, shown from various angles in the figures that accompany this article, depicts the northeast corner of Paul's Churchyard, including the Choir and north transept of the Cathedral, the Paul's Cross Preaching Station, and the buildings surrounding the churchyard, chiefly mixed-use houses with retail book shops on the ground floor and living accommodations above. The model also includes the buildings along the streets that run alongside the northeast corner of the churchyard, specifically Paternoster Row to the north and the Old Change Street to the east, as well as their intersection at the west end of Cheapside.

The acoustic model enables us to experience how sound behaves in the space depicted in the visual model under different conditions of weather and crowd size. This acoustic model offers us an opportunity to join the crowd gathered at Paul's Cross, the preaching station in the churchyard of St Paul's Cathedral, as it might have come together on the morning of November 5. Surrounded by the visual world of the cathedral and its adjacent buildings, we can hear Donne's sermon performed inside the acoustic model of this space and thus experience it as an event unfolding in real time, moment by moment, taking our place at an interactive occasion heard in the physical space for which Donne composed it. We can experience this event from multiple angles, hearing Donne's sermon for Gunpowder Day in its full two-hour version from two different locations in the churchyard. We can also hear portions of the sermon from eight different listening positions and in the company of four different sizes of crowd, from five hundred people to five thousand people, modeling the range of contemporary estimates of attendance at these sermons.

Our choice of Donne's sermon for Gunpowder Day, November 5, 1622, was mandated by the same commitment to accuracy as the rest of the project. Donne preached frequently at Paul's Cross; the sermon text that survives from that day shows every hallmark of his Paul's Cross sermons, even though it was—"by reason of the weather"—actually delivered inside the cathedral instead of in Paul's Churchyard. Even better for the sake of accuracy, King James, who asked Donne to preach this sermon, also asked for a copy to be delivered to him shortly after the event. Donne prepared this copy, which survives in the British Library as MS Royal. 17.B. XX; it is thus, uniquely, the surviving Paul's Cross sermon text that was written out the closest to the day of delivery, hence representing the most likely of all these sermons to represent the words actually spoken by the preacher on the day of delivery.

As a result of using this capability to restore multimedia and real-time experience to our interactions with traces of the past, we raise the possibility of transforming our concern for the priority of the printed or displayed word into a deeper concern with the word-as-spoken. We can reintroduce into our consideration of past sermon texts—and from a fresh perspective—questions of their political and social as well as their physical and liturgical settings. In the case of the Paul's Cross sermon, for example, we can begin to reconceptualize it as an event taking place within a customary practice of gathering, attending, delivering, responding, and concluding, all within a specific set of circumstances, as in the case of Donne's Gunpowder Day sermon, situated in the discourse about James I's foreign policy regarding Spain as well as the commemoration of his deliverance from the Catholic conspiracy of November 5, 1605.

Another aspect of this project has been to assemble what we do know and to explore this knowledge in progress, to reconsider what was involved in actually staging a Paul's Cross sermon, including such considerations as the need to gain and sustain a congregation's attention, the need to accommodate into the performance the realities of ambient noise, and the need to deal with problems of audibility and crowd response. This process has, over time, opened up new areas of inquiry, raising questions, as we have noted, about how the preacher dealt with the sound of the bell tolling the hour and the quarter-hour, how the preacher sought congregational engagement, how he sustained the congregation's attention over the two-hour duration of his sermon, and how we might come to a fuller sense of his congregation's responses to his preaching. The Virtual Paul's Cross Project thus represents an opportunity to restore direct experience to the repertoire of tools we employ to understand the character of worship and religious life in the early modern period. Although much of what constituted the real event we seek to model has of necessity been excluded from this project—people always ask me about the smell, for example—we believe the Virtual Paul's Cross Project demonstrates how multimedia modeling is beginning to fulfill the digital humanities' promise of providing what David Berry has deemed a "new way of working with representation and mediation," enabling us "to approach culture in a radically new way" (2012, 2). One aspect of this "new way" of working is the restoration of multi-sensory and real-time experience to our understanding of human experience and expression. We can now include in our understanding of the past the word spoken and heard, the word performed, the word as interchange, situated in the specifics of a particular time and place.

Our multimedia model therefore integrates into one experiential presentation all the information we have today and all we might reasonably infer about the place, the space, and the physical circumstances of a Paul's Cross sermon into a single experiential and interactive model, enabling us to reach conclusions based on our experience with that model rather than on our experience with the text alone. The Paul's Cross model has thus come to mediate between the historical record and the conclusions we reach based on our virtual experience, moving us into an area in which our conclusions have no direct historical evidence to support them.

We now approach a paradox of scholarly inquiry. At the very moment when we find ourselves at the boundaries of traditional scholarly methods in the humanities, we, as McCarty once suggested to me in conversation, approach the methodologies of the sciences, where modeling reality is often the preferred mode of learning about reality. The digital humanities are closest to the sciences in practice when we engage in thought experiments through modeling, all the while aware that our models—no matter how accurate and how much historical information they enfold—are always approximations of reality, thus always partial, incomplete, fragmentary. We may understand this somewhat paradoxical situation more fully if we consider some of the ways in which this project is changing our understanding of the subject of our study. When scholars customarily attend to the early modern sermon, they view it as a text, as a composition of words on the page of a printed book or written out on the page of a manuscript. We speak confidently of what the preacher said, and quote from these texts, and yet these are only assertions on our part. Sermons from the past are too often regarded much like theological essays, experienced primarily in private, in the quiet, air-conditioned comfort of our offices or, perhaps, in the hushed surroundings of a research library, as opposed to the public performances for which they were prepared.

Now, however, and as a result of this project's recovery of real-time experience of hearing, our understanding of the texts of these sermons, whether in their printed or manuscript forms, shifts perceptibly. We can begin to regard these texts—even though we are dependent on them for our knowledge of what was said in the delivery of the sermon— as a set of traces of their performance, traces among other traces, pointers to the early modern sermon but not co-equal to it. We can begin to recognize the distance between the actual sermon performances—which is really what interests us—and the texts that survive from those performances. We recognize that those texts are at best memorial reconstructions of the content of oral performances, written out after the fact of their delivery, that what we really wish to study are the actual performances, improvised from notes, delivered in an interactive environment on specific occasions in chill, drafty church buildings, or, as in the case of Paul's Cross sermons, outdoors at all times of the year.

Digital modeling thus challenges us on several fronts. It at once enables us to integrate into a single model a wide variety of historical evidence while it also invites us to consider possibilities for understanding past events that are based on our experience with the model but not supported by traditional forms of scholarly evidence. It distances us from traditional understandings of what we study as scholars in the humanities while it invites us to seek new modes of study that at least raise the possibility of getting closer to those aspects of past events that most concern us. In other words, in the case of the Virtual Paul's Cross Project, we at once realize that the surviving text of a sermon is not the version of the sermon that really interests us, while we also begin to see how we might read that text in ways that open to us its antecedents as a performance.

Figure 2. Paul's Churchyard, looking east, from the west. From the visual model, constructed by Joshua Stephens, rendered by Jordan Gray.

One of the many things the Virtual Paul's Cross Project has taught us is that in spite of our efforts to document each of the many elements that go together to compose our models, we have become increasingly aware of the limits to our knowledge. These limits need to be recognized regardless of whether one is dealing with traditional forms of historical evidence or with conclusions we reach from our experience with the digital models. On the one hand, the structures one sees in the Paul's Cross visual model[3] aspire to the highest possible degree of accuracy. They integrate the historic visual record with both historic[4] and contemporary surveys of the surviving physical traces of these buildings, some of which remain in the ground in the City of London. We have high confidence in the accuracy of this aspect of the model. Yet other measurements—the height of the Paul's Cross preaching station, for example—are lost to us, or come to us with conflicting evidence. We know from a nineteenth-century survey of the foundations of the preaching station that the base of the foundation was 34 feet across, and the base of the structure sheltering the preacher was 17 feet across (Penrose 1883, 390). Yet visual images of the preaching station show it was a relatively slim building, totally different in proportion to the dimensions derived from this survey.

In addition, the detailed visual depictions of St Paul's Cathedral that survive from the early modern period often contain contradictory elements; the more closely one looks at them, the more discrepancies among them one recognizes. For instance, Wenceslaus Hollar's drawing of the north side of the cathedral differs significantly in detail from his engraving of the same view of the cathedral.

Note the differences in the detailing of the north front of the north transept between Figures 3 and 4, as well as the door that suddenly appears on the ground level of the first bay of the Choir as one's eye moves westward from the east end of the building. Presumably, the engraving is based on the drawing, and both come from the hand, and from the eye, of the same artist; we had to decide, as we built the visual model of St. Paul's, which set of details Hollar saw, as well as what explanations there might be for the differences between them.

---

[3] Constructed by Joshua Stephens and rendered by Jordan Gray while they were students in the graduate program of the College of Design at North Carolina State University, under the supervision of associate professor of architecture David Hill.

[4] Interestingly, the dimensions of St Paul's itself—the height of the vaulting, roofs, and interior ceilings—are taken from a survey done by Christopher Wren in the early 1660s as part of a project to repair and remodel the pre-Fire cathedral.

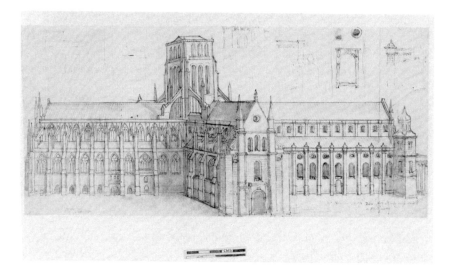

Figure 3. Wenceslaus Hollar, drawing of St. Paul's Cathedral, north side (1650?). Image courtesy Sotheby's, London.

Figure 4. Wenceslaus Hollar, engraving, St. Paul's Cathedral, north side. From William Dugdale, *History of St Paul's Cathedral* (1658). Image courtesy the Wenceslaus Hollar Digital Collection, University of Toronto.

To note one more example, the most significant visual account of a Paul's Cross sermon is that of John Gipkin (1616), yet Gipkin chose in his depiction to truncate dramatically the length of the Choir and, even more dramatically, the length of the nave, where ten bays have shrunk to three.

Figure 5. John Gipkin, painting of Paul's Cross (1616). Image courtesy of the Bridgeman Art Library, New York, and the Society of Antiquaries, London.

Even more challenging is the fact that—thanks to surveys made after the Great Fire in 1666 to secure a record of property lines—we have measurements of the foundations and verbal descriptions of the buildings surrounding the churchyard to the north, but we do not have visual depictions of them (Blayney 1990). Tantalizingly close to these buildings in seventeenth-century London were the buildings depicted in a drawing from 1585, showing structures at the intersection of Paternoster Row and Cheapside. This image brings us so close to the part of London that chiefly concerns us in this Project that the archway shown on the right side of this drawing is Paul's Gate, leading directly into the Cross Yard. But, to our frustration, the houses shown in the drawing face outward from the churchyard.

Figure 6. Drawing, London, intersection of Paternoster Row and Old Change (1585). British Museum BM Crace 1880-11-13-3516. Image courtesy British Museum.

What one sees in this part of our model, therefore, are representative structures based on the appearance of comparable buildings from the early modern period that survive in London or in cathedral towns in other parts of England.

The images of the visual model that one sees on the Paul's Cross website, therefore, represent our interpretation of the evidence we have available. The visual model offers a range of visual approximations, from the high degree of accuracy we believe characterizes our model of the cathedral itself to the slightly more approximate depiction of the Paul's Cross Preaching Station using survey data plus the visual record, plus informed estimation of unknowns such as the height of the structure, to the more representational visual depiction of the mixed-use buildings around the churchyard to the north and east.

Gazing into Imaginary Spaces: Digital Modeling and the Representation of Reality 297

Figure 7. Paul's Churchyard, looking north, from the Sermon House. From the visual model, constructed by Joshua Stephens, rendered by Jordan Gray.

Figure 8. Paul's Churchyard, looking east, from the west. From the visual model, constructed by Joshua Stephens.

In addition, we have sought to incorporate into the appearance of the visual model evidence of other visual aspects of the occasion we are representing. Unlike many digital reconstructions of historic buildings, which present everything in a pristine state,[5] apparently freshly constructed and of the identical vintage, the visual model in the Virtual Paul's Cross Project shows the effects of time, climate, and season. The sermon at the heart of our model was preached on November 5, 1622, outside a building already several hundred years old and in a city where, according to the work of Ken Hiltner (2009, 429), a particularly acidic form of coal was being burnt to provide heat both for comfort and cooking. As a result, we are told, "a seemingly perpetual cloud of sulfurous smoke [hung] over London." To extend the range of historical information we incorporate into this model, we have included traces of the effects of time and weather, and show the relative ages of the buildings we display. The gathering darkness of London in the fall deepens as the days grow shorter; since London in 1622 was still using the Julian calendar in the early 1600s, November 5 of that year was November 10 on the Gregorian calendar. Our visual model adds to the effects of smoke the general gloom of a mid-November day in London, when the weather was, more likely than not, overcast and chilly, the sun rising at 7:20 a.m., setting at 4:12 p.m., and never getting higher in the sky than 20 degrees of elevation, thus casting a long shadow even at noon (www.weatherspark.com).

The result is, we believe, a visually compelling, historically appropriate, and convincing model. Yet that very fact cloaks the visual model in the illusion of completeness and authenticity, obscuring the extent to which much of the model is at best only representationally accurate, and even the most thoroughly documented parts lack historic depth and authentic particularity. Also, we are able to model the visual and acoustic space, but not other elements of the experience. Conveying the smell, for instance—likely to be a mix of horse manure, human sweat, and rotting garbage—has so far eluded us. Nor have we been able to incorporate the likely, although randomly occurring or unpredictable, event, like a dog fight breaking out in one corner of the crowd, or someone collapsing after standing too long in the churchyard. We are modeling elements of reality, not engaging in time travel; we can at best achieve what Samuel Johnson called "the just representation of the general nature" of our subject.[6]

---

[5] See, for example, Rome Reborn (http://romereborn.frischerconsulting.com/), brilliantly visualized but with no signs of climate, weather, or the relative ages of the various structures.

[6] From Johnson's preface to his edition of Shakespeare (London, 1765), viii.

Gazing into Imaginary Spaces: Digital Modeling and the Representation of Reality 299

Figure 9. Paul's Churchyard, looking east. From the visual model, constructed by Joshua Stephens, rendered by Jordan Gray.

Hence, there will always be distance between our very best modeling efforts and the real places and events we seek to know through digital modeling. If we are to make appropriate use of digital modeling, therefore, we need to keep both the opportunities and the limits of the practice before us. We need to be aware that our models—no matter how accurate they were and how much historical information they were able to enfold—are always, and to a greater or lesser degree, approximations of reality, never the events, times, or places being modeled.

The Virtual Paul's Cross Project exemplifies this tension between modeled depiction and experiential reality because the sermon at its heart, and therefore the event of its delivery, although intended for Paul's Cross, was actually "Preached in the Church ... by reason of the weather" (Donne 1959, 235). In this model, in some sense, Donne finally gets to preach the sermon at Paul's Cross he had intended to preach there, but was delayed in doing so for 400 years.

Similar opportunities and limitations apply to the acoustic model developed for the Virtual Paul's Cross Project. Recreating authentically the sound of the human voice after 400 years of silence seems, on the surface, to be an even more daunting task than recreating the visual appearance. We have very little idea how John Donne might have sounded. Yet, as is the case for the rest of the project, our not knowing everything does not mean that we know nothing. Sorting out what we do know, what we can model, is a valuable and instructive task in its own right, offering numerous opportunities for new understandings of the historic event it recreates.

To create the acoustic model, we delivered a simplified version of our visual model, together with a list of the materials out of which the objects would have been constructed in 1622—stone, wood, plaster, brick, dirt, and the bodies and clothing of the congregation—to our acoustic engineers Ben Markham and Matthew Azevedo, who then imported it into an auralization program. The resulting acoustic model now functions like the visual model created by Hill, Stephens, and Gray, allowing us to gather what we know or can surmise, in this case about sounds of preaching, about the sound of spoken English, and about the ambient sounds of urban life in early modern London.

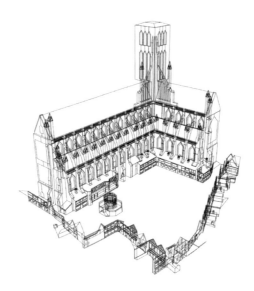

Figure 10. Wireframe image of the acoustic model of Paul's Churchyard, the Cross Yard, based on the visual model constructed by Joshua Stephens.

In fact, the acoustic model may well be the most consistently accurate feature of the whole project. For an accurate model of the acoustic properties of a space, one needs only the dimensions of the space, the arrangement of basic geometric forms in the space, and the materials from which objects in the space are made. Sound, when it encounters an object, is either (and to some degree) absorbed, reflected, or dispersed. The acoustic modeling program incorporates the absorptive, reflective, or dispersive qualities of the forms and materials in the space—in this case, the space of Paul's Churchyard—into the acoustic model it produces. To avoid bringing into the acoustic model traces of our environments, Crystal's recording of Donne's sermon was made in an anechoic recording studio—a space designed to absorb rather than reflect the sounds made inside of it so that it adds no acoustic properties of its own to the recording but yields only pure source sound, without any of the ambient qualities of the room in which the sound was recorded. When one introduces sounds recorded in an anechoic chamber into the model, the output from the model enables us to experience the source sound as though we were hearing it from inside the acoustic space.

The process of creating the sounds one hears from inside the acoustic model of Paul's Churchyard has involved careful review of what we do know about the ambient sounds and human voices made in early modern London. During an event of at least two hours' duration in the open air in the middle of a city of 200,000 people, there definitely would have been ambient noise randomly occurring from a variety of sources and in a variety of intensities. The fullest survey of the ambient soundscape of early modern London is that of Bruce R. Smith (1999, 49–95), which describes the sounds of moving water and the practice of trades and crafts as well as the sounds of nature and other sounds of the practices of daily life. Rather than try to incorporate a full inventory of all possible kinds of ambient noise and to assess the degree of impact for each of these sounds, we have chosen to represent this noise by concentrating on three sources of noise—the birds, the horses, and the dogs—all documented as sound sources in Paul's Churchyard through their appearance in Gipkin's painting, and all occurring randomly during the delivery of Donne's sermon.

These sounds do not represent a literal reconstruction of the ambient noise audible in Paul's Churchyard in 1622, for the sounds one would have heard in Paul's Churchyard on that Tuesday morning in early November must have been many and varied both in sources, volume, and distance from the churchyard. Instead, these sounds serve as representatives of that ambient noise, reminders that when we consider the experience of the Paul's Cross sermon we need to include in our thinking the presence of such sounds as part of that experience.

Figures 11a–11c. John Gipkin, painting of Paul's Cross (1616). Details showing birds (a), horse (b), and dog (c). Image courtesy of the Bridgeman Art Library, New York, and the Society of Antiquaries, London.

Thanks to Tiffany Stern's recent research (2012), we also know that part of the soundscape of Paul's Churchyard was the sound of a bell activated by a clock mechanism that rang on the quarter-hours and hours of each passing day.[7] We have incorporated into the acoustic model the sound of a struck bell ringing out the hours and the quarter-hours. Yet this, too, is at best a representative approximation; Stern points out that there were more than a hundred parish churches in the City of London in the 1620s with clocks that rang out the hours (but not the quarter-hours). She also notes that there were no guarantees that any of these clocks rang the hours in coordination with any other clock, so the time it took for all these clocks to ring the hour must have taken several minutes for quiet to be restored once more.

The most important sound one hears in the Virtual Paul's Cross model is of course the sound of the preacher's voice, the sound of the actor Crystal performing Donne's sermon for November 5, 1622. In some ways, our effort to reproduce the sound of Donne preaching is the most speculative part of this project. After taking stock of what we did know, and applying the principles of approximation we used to develop the visual model, we found that we knew or could surmise a great deal about the content of Donne's sermon and the sound of Donne's voice. Even with the limited knowledge we have, however, we can still use this project to explore questions about style of delivery, the congregation's response to different kinds of passages in the sermon, and the preacher's use of the time of delivery in making his points. Here, the accuracy we are dealing with is clearly at a very general level intended far more to raise possibilities and introduce new questions than it is to provide final answers.

To help Crystal prepare for his performance of Donne's sermon, he was provided with two types of information. One was a general description of the kind of space in which, in the Virtual Paul's Cross Project, he would be heard performing Donne's sermon. The second was a compilation of contemporary accounts of Donne's preaching, available on the Virtual Paul's Cross website (http://vpcp.chass.ncsu.edu/donne-preaching/). These accounts of Donne's style of delivery support a characterization of Donne's preaching style as

---

[7] English cathedrals in the sixteenth century are known to have had such mechanical clocks, often without dials; the ringing of a bell was its sole means of communicating the time. One such faceless clock survives at Salisbury Cathedral. Interestingly, the sounding of the bell in these clocks was not discretionary. It could not be stopped and started again without disrupting its accuracy as a timepiece (Maltin and Dannemann 2013).

multi-vocal, varying in mode of delivery from section to section of the sermon, if not moment to moment of his delivery, depending on the content of each passage and its relationship to the overall structure and argument of the sermon as a whole.

Again, while there is much we don't know, and can never know, about how Donne sounded, that does not mean we know nothing. Donne preached from notes; the texts of his sermons that come down to us are memorial reconstructions after the fact of delivery. As we have already noted, Donne wrote out a memorial reconstruction of his Gunpowder Day sermon for 1622 shortly after delivering it at the request of King James. James hoped that this sermon would defend him against concerns that his efforts to secure a political alliance with the Spanish, and to seal the alliance with the marriage of his son Charles, the Prince of Wales, and the daughter of the King of Spain, a devout Catholic, indicated that James was sympathetic to Catholicism.

We have not been so bold as to try to recapture potential variations in Donne's content; the recording on the Virtual Paul's Cross Project website follows almost precisely the wording in Donne's manuscript. We consider that manuscript to be a trace of the sermon as delivered rather than a documentary presentation of the sermon itself. We take some comfort in the fact that the manuscript text of this sermon, while still an after-the-fact reconstruction, is closer in time than any of the texts of Donne's other sermons, and therefore is likely to have been the text closest to what Donne actually said on November 5, 1622.

The fact that Donne did not read his sermons as one would a formal lecture, but performed them as one would a public oration, or, perhaps better, as one would perform a character or a role in a play, reminds us that Donne's preaching took place within the context of a tradition of practice, informed by manuals of oratory, that defined the preacher's social and professional role in relationship to other roles played by members of his congregation. In addition, since his sermons were performed from notes that guided his delivery rather than from a fully developed manuscript that controlled the wording, he would have followed his pre-arranged structure while improvising the actual words in the process of delivery. Thus, he would have been able to develop, expand, or contract specific sections in response to external elements like the regular sounding of the clock bell or the irregular responses of the congregation to this or that point, a gesture, or a distinctive passage.

Listeners to the sermon performance on the project website have commented especially on the distinctive accent employed by Crystal in his performance, a result of the fact that Crystal is using a script of Donne's sermon that incorporates early modern London pronunciation prepared by the linguist David Crystal. Our decision to use an original pronunciation script was prompted by the fact that Donne was a Londoner by birth. In addition, Crystal delivered his performance after considering two kinds of information about this project: the first having to do with general principles of delivery with special regard for audibility in large reverberant spaces, and the second having to do with what we do know about Donne's preaching style.

Figure 12. John Gipkin, painting of Paul's Cross (1616). Detail. Images courtesy of the Bridgeman Art Library, New York, and the Society of Antiquaries, London.

In regard to the former of these considerations, Crystal was told that evaluating the audibility of a preacher delivering a sermon without amplification in a large open space was a major concern of this project. He thus paid special attention to the pacing of his delivery, knowing that his performance would be presented in a space with significant reverberation, and accompanied by the noise of a large crowd. The sermon was recorded from multiple angles, and Crystal was encouraged to visualize the crowd around him and direct his speech to various people, particularly the nobility who would have been seated in the sheltered seats along the cathedral wall. Crystal delivered the sermon in a manner consistent with a practiced orator delivering a speech to a large outdoor crowd: the voice is strong, the cadence measured, a reasonable and appropriate strategy for improving intelligibility. Had he spoken more quickly, the effect of the reverberation in this space would have run together the sounds of his speech, muddying his articulation and making aural comprehension much more difficult.

This is a case, I believe, in which the recreation of original performance conditions in Paul's Churchyard has given us access to Donne's original style of delivery in a particular space. Listeners' surprise at Crystal's choice of pacing is an artifact of our lack of familiarity with the sound of the unamplified human voice in large open spaces; amplified sound, together with the effects of acoustic engineers' ability to shape sound delivered through speakers, means that we have grown accustomed to hearing almost everyone, in any setting, speaking at a conversational pace. I would expect Donne to use a faster pace when audibility was not so much of a concern, in a smaller venue or before a smaller crowd, such as in a parish church, the Chapel Royal, or the Choir of St. Paul's Cathedral, where he preached most frequently in the 1620s.

In addition, Crystal was asked to remember that a sermon is not a theological essay to be read in one's study or a formal lecture to be attended to in silence, but a performance intended to entertain as well as to inform, to move his hearers emotionally as well as to educate them about the content of their faith. Crystal was given a full account of Donne's own comments on preaching: that, for example, the effective preacher sought to engage his hearers and to bring them—through both cognitive and emotional means—to amend their lives in directions set out in the sermon. Donne once described the performance of an effective preacher in terms of a coordinated effort of body, feeling, and ideas, uniting "matter and manner," the quality of the voice—"pleasant"—and personal manner—"acceptably, seasonably, with a spiritual delight," with "a holy delight," toward the goal of "profit" for his congregation (Donne 1955, 167).

Crystal was also given every available account of how Donne's contemporaries described his preaching style, noting especially their emphasis on Donne's wit, his eloquence, and his capacity to express and arouse feeling, to elevate, captivate, and motivate. Donne's sermons, his contemporaries claimed, were delivered in an empathetic, emotionally expressive, responsive, evocative speaking style, always aspiring to intimacy and engagement, sometimes confrontational, sometimes laudatory, but always lively, personable, and connecting with his congregation. Thus we can now experience Donne's sermon not as a theological essay to be experienced with the eye, which can proceed word by word, or can skip over, move around, go back, pause, and resume its activity, but as a performance that unfolds in real time, word by word, an event of the ear. We therefore consider Donne not only as a learned essayist, but also as a skillful performer, one who relied upon a variety of rhetorical techniques to convey his message to an audience.

The goal of this project has been to make available for study our assumptions about the conditions of sermon delivery and reception, reminding us that these sermons were originally performances during which preacher and congregation interacted to shape their mutual experience of the occasion and of the sermon itself. We are now able to test the consequences of our assumptions as they are realized in the visual model and played out in the acoustic model, always aware that we can revise the model as we develop our understanding, incorporating new research into an unfolding process of development.

One of these assumptions must be the degree of literacy among Donne's listeners. The best estimates I have found give the literacy rate among adult males in 1600 as about 30 per cent, strongly skewed toward the aristocracy, the clergy, the gentry, and members of the mercantile classes.[8] Women's literacy rates were substantially lower. Indeed, given these limitations on literacy, speaking and hearing rather than writing or print publication are perhaps best seen as the defining modes of human communication and meaning-making in the early modern period. The Virtual Paul's Cross Project encourages us to consider the fluidity of texts and the plurality of their uses and meanings, given the wide range of audiences and the variety of their modes of access to both oral and written cultures (Chartier 1988, 182). Thinking in these terms is especially useful for the study of religion in early modern England because of that culture's mixed status as a world still essen-

---

[8] For a thoughtful summary of the evidence about literacy in the early modern period, see Green 2000, 24–27.

tially medieval in its functioning through the communication techniques of an aural culture, yet increasingly being reshaped by the development of the book trade and the spread of literacy. Such a culture is one in which, as Carol Symes notes, "the gesture, the posture, the sound, and the spoken word were the powerful discursive modes through which writing was publicized and … subordinated" (2010, 287).

While it is true that the English Reformation was enabled by the printing press and defined by a collection of books, notably the Book of Common Prayer, the Bible in English, and the two Books of Homilies, these books were primarily intended for oral performance, rather than for private reading. They provided scripts for worship, with scripture readings and homilies nested within the Prayer Book's liturgies, performed on specific occasions in specific spaces and in the presence of specific assemblies of clergy and their congregations (Booty, Wall, and Siegenthaler 1981).

Rethinking the culture of worship in early modern England as a culture of speaking and hearing is especially important for our understanding of preaching, given renewed vigor by the Reformers' emphasis on the Word read and preached and on worship conducted in the vernacular. As Arnold Hunt has reminded us, "'faith is by hearing' … was a precept that Protestant commentators took very literally" (2010, 22). Edward Vaughan, for example, claimed in 1617: "take away the preacher take away the word, take away the word take away hearing, take away hearing take away Faith, take away Faith take away calling upon God, take away calling upon God take away salvation in Christ" (1617, 25–26).

We can learn something of how such a culture functioned if we attend to the detailed description of worship in the Church of England, written by William Harrison not quite two decades after the Elizabethan Settlement of Religion and the adoption of the Book of Common Prayer of 1559. Harrison gives this account of how parishioners learned their parts in the services of the Book of Common Prayer. The "minister," Harrison notes, "saieth his service commonly in the body of the church with his face toward the people … by which means the ignorant do not only learn divers of the Psalms and usual prayers by heart, but also such as can read do pray together with him, so that the whole congregation at one instant pour out their petitions unto the living God for the whole estate of His church in most earnest and fervent manner" ([1587]1968, 36).

In Harrison's account, the ability to read does not determine the extent of congregational participation in worship, for "ignorant" people learn their parts by listening and memorizing while others learn them by reading. The habitual practices of an aural culture thus make it possible for the "whole congregation at one instant"—that is, in one voice, speaking together in unison, both the "ignorant" and "such as can read" —to "pour out their petitions unto the living God for the whole estate of His church in most earnest and fervent manner." Harrison's interest in people hearing clearly so they will understand and perhaps memorize what is being said extends to sung services as well. Harrison praises choirs and their composers, who make sure "that each one present may understand what they sing, every word having but one note" ([1587] 1968, 34).

From Harrison's description, as well as from our knowledge from other sources of literacy rates in early modern England, we recognize that congregations in early modern England consisted of both the "ignorant" (i.e., the class of non-readers) and "such as can read." In fact, most congregations were almost certainly composed of significantly more non-readers than readers. The ability to read in this period correlates highly with social and economic status; hence, some urban congregations were likely to have a higher percentage of literate parishioners than rural congregations. This would be especially true of the congregation gathered to hear sermons at Paul's Cross, since these events were open to the entire population of London and not just to the more thoroughly literate congregations one would expect in the chapel at Lincoln's Inn, the Chapel Royal, or in the Choir of St Paul's Cathedral.

Thus the Virtual Paul's Cross Project leads us to a reconsideration of the kind of congregation Donne faced, and of his relationship with them—a congregation deeply embedded in an aural culture, accustomed to listening attentively and responding actively to what was being said by the preacher. It also leads us to a fundamental reconsideration of the sermon they would have heard on November 5, 1622, especially in relationship to the texts of the sermon that come down to us. For it turns out that this sermon surely existed in at least six discrete versions. We are familiar with the last two versions of this sermon: the manuscript version Donne prepared at the request of King James shortly after its delivery, and the printed version in *Fifty Sermons* (1649), an anthology of sermons prepared by Donne and printed in London after his death. The printed version differs from the manuscript version chiefly in a number of revisions and additions to the text, all of which reflect second thoughts rather than reversions to an earlier state of the text. Yet we know that these full-text versions of Donne's sermon were preceded by four other versions, or stages, of Donne's sermon. While no direct evidence for the content of these versions comes down to us, we can—by working from Walton's account of Donne's methods of preparation for a sermon performance—reconstruct at least the kinds of material they contained. Doing so helps us clarify the relationship between the surviving texts of this sermon and the version of the sermon Donne delivered on November 5, 1622.

The first version of the sermon is the sermon Donne anticipated giving, developed as he studied the text and deposited it in memory—if Walton is right that after Donne chose a text for a sermon, "he ... cast his Sermon into a form, and his Text into divisions; and the next day betook himself to consult the Fathers, and so commit his meditations to his memory" (Walton [1675] 1927, 67). The second is the set of notes Donne wrote out to take with him into the pulpit to support his memory and guide him in the delivery of the sermon. The third version is the actual sermon he delivered, working from notes and memory, yet able to compose the specific words of the sermon in the process of delivery, and thus improvising from his notes to fit the specific terms of the occasion, responding to the ebb and flow of congregational response, shaping his words to fit the time allotted between the cathedral's bell chiming of the quarter-hours, modifying them to fit the interests and level of engagement he developed with those listening and responding to his delivery.

The fourth version is a manuscript version of that sermon which Donne wrote out and then gave to a professional scribe for copying. The fifth version is the scribal copy of Donne's manuscript, which Donne corrected in own hand and sent to King James, where in time it became MS Royal 17.B.XX in the British Library.[9] It remains, uniquely, the version of a sermon by Donne that is as close to the lost third version, the delivered version, as we are ever likely to get.

The sixth version is the version printed in *Fifty Sermons*, a version very close in its readings to the fifth version, with the addition of some expansions of specific points. These later versions must be seen, therefore, not as the sermon Donne delivered but as various stages of memorial reconstruction on Donne's part of the sermon he delivered as he remembered delivering it.

---

[9] This manuscript version was discovered in the British Library by Professor Jeanne Shami and published by Duquesne University Press as *John Donne's 1622 Gunpowder Plot Sermon: A Parallel-Text Edition* in 1996.

Gazing into Imaginary Spaces: Digital Modeling and the Representation of Reality 311

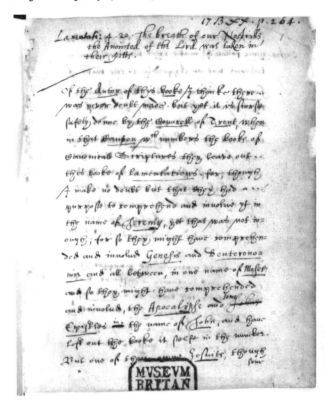

Figure 13. John Donne, *Sermon for Gunpowder Day, November 5, 1622*, Page One. From MS Royal. 17.B. XX. Image courtesy the British Library.

In reading the manuscript text of Donne's sermon along with the printed text from *Fifty Sermons*, however, we must remember that they are best regarded as traces of the sermon that interests us, not the sermon Donne delivered. They are likely to reflect lapses of memory as well as editorial decisions Donne made while reconstructing his sermon. What one remembers saying is not necessarily what one actually said; conscious or unconscious revisions inevitably creep in, reflecting, in addition to what Donne remembered saying, second thoughts about what he wished he had said (clearly the case in the printed version in its expansion of the 1622 manuscript text) or the consequences of a change in audience between the delivered sermon at Paul's Cross and its fully-written-out version prepared at the request of King James.

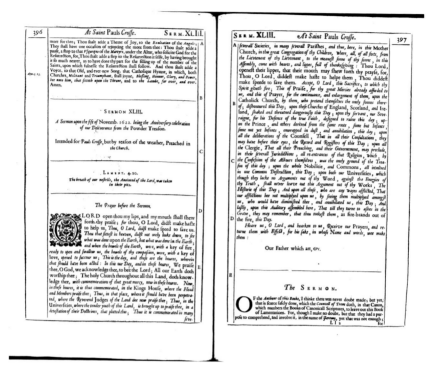

Figure 14. John Donne, *Sermon for Gunpowder Day, November 5, 1622*, Page One. From Fifty Sermons (London, 1649).

Nevertheless, these after-the-fact versions of Donne's sermons may give us clues to the nature and content of prior versions. We may, for example, get some idea of what the notes he took into the pulpit consisted of, and what Donne was able to make of them, if we go to a passage in the sermon about 30 minutes or so into the text, which in the manuscript version reads as follows:

> They would not trust Gods meanes, theire was their first fault; And then though they desird a good thing, and intended to them, yet they fix God his tyme, they would not stay his leasure; and both these, to aske other things then God would giue, or at other tymes then God would giue them is displeasing to him. use his means and stay his leysure. But yet though God were displeasd with them, he executed his owne purpose; he was angry with their manner of asking [for] a King but yet he gaue them a King.

Here Donne is reviewing the history of monarchy in Israel. He wants to defend monarchy, but also make something of the fact that the people of Israel demanded that God provide a king for them before God was ready to do that. He claims that God intended Israel to have a monarchy all along, so that by demanding a king of God before God was ready, the people of Israel were asking for a good thing but simultaneously showed lack of trust in God. Notice the sentence fragment—"use his means and stay his leisure"—in the middle of this passage, part of which repeats the phrase "stay his leisure" from the previous sentence. Note also that this fragment does not add anything to the preceding thought, nor does it provide a transition to the next sentence; in fact, if one reads the passage, leaving this phrase out, the passage makes perfectly good sense. Donne says the people asked too soon for God to name a king, and thus displeased God, but that God did what he wanted anyway, in spite of his displeasure, and gave them a king. I want to suggest that this phrase "use his means and stay his leisure" is in fact a survival of one of Donne's notes to himself, and that the surrounding passage represents Donne's expansion of this note as he performed it in the actual sermon. Hence, Donne would not actually have said "use his means and stay his leisure" because he had already made of that note what he wanted to on that occasion. Happily for us, this note was retained in the written copy of the sermon, created by Donne some time later, although working from the same set of notes.

The existing texts of Donne's sermons represent *ex post facto* reconstructions by Donne of what he said in performance. Even though MS Royal. 17.B. XX represents Donne's memorial reconstruction of the sermon he gave on November 5, 1622, it is nevertheless the product of very different circumstances from those that pertained when Donne preached the sermon on November 5. Gone are the constraints of time limits, the more or less precise sounds of the clock bell ringing out the hours and the quarter-hours, the congregation for whom Donne was performing and whose reactions to the sermon's delivery surely shaped the sermon as it developed. We are fortunate that Donne gives us so many traces of that actual performance. Hence this manuscript version of the sermon is the version that serves as the source for our performance script.

We have now arrived once more at the point of contact with the sciences we mentioned earlier, when we on the one hand are beginning to reach conclusions about the events of November 5, 1622, based on our experience with the models we have created; we have now moved into an area—unlike our choice of the sounds of horses, birds, and dogs as our examples of ambient noise, chosen because they are documented in Gipkin's painting—in which our conclusions have no direct historical evidence to support them. While we are still working with a multimedia model that integrates all the information we have and all we might reasonably infer about the place, the space, and the physical circumstances of a Paul's Cross sermon, including the light and sound, together with the words of Donne's text, the manner of his preaching style, and the behavior of his congregation, into a single experiential and interactive model, we are nonetheless reaching conclusions based on our experience with that model. The model has thus come to mediate for us between the historical record and the conclusions we reach based on our experience.

As uneasy as our approximations in the Virtual Paul's Cross Project make us regarding our conclusions, they also enable us to see and hear the most accurate recreation of Donne's sermon possible with our current knowledge and the technology available. Keeping in mind the limitations of the model— it models but does not reproduce reality—and updating it as new research and technology become available makes the Virtual Paul's Cross a useful tool for exploring our assumptions about an important historical event. Similar virtual models could support other research projects by providing a near-real experience based on existing knowledge, and allow exploration of an approximated but still representative environment.

There are benefits. The Virtual Paul's Cross Project provides us the opportunity to experience the phenomenon of the Paul's Cross sermon, an occasion of outdoor public preaching at Paul's Cross in Paul's Churchyard in early modern London, as an event unfolding in real time, in a space modeled after the space in which it was originally delivered, in the presence of a large public gathering of people, and with at least the suggestion of the conditions and circumstances in which they gathered for its original delivery. As such, it enables us to identify and evaluate our understanding of the Paul's Cross sermon as a social phenomenon, as an interactive occasion, as a public event.

Our process of developing the performance of Donne's sermon has helped us reconfigure our understanding of the early modern sermon, shifting our thinking from regarding the surviving texts of Donne's sermons as the proper objects of our study to regarding them as traces, at best memorial reconstructions, of the actual, delivered text (Wall 2014a). At the same time, we have learned more about the limits of our knowledge, or, perhaps better, learned the various forms in which knowledge can come, learned the risk as well as the value of conceptualizing through approximation. That in itself represents a fundamental shift in our thinking about the early modern sermon, some compensation, perhaps, for all the aspects of Donne's preaching that are still beyond our reach.

WORKS CITED

Berry, David M. 2012. "Introduction: Understanding the Digital Humanities." In *Understanding Digital Humanities*, edited by David M. Berry, 1–20. New York: Palgrave Macmillan.

Blayney, Peter. 1990. *The Bookshops in Paul's Cross Churchyard.* London: Bibliographical Society.

Booty, John E., John N. Wall, and David Siegenthaler. 1981. *The Godly Kingdom of Tudor England: Great Books of the English Reformation.* Wilton, CT: Morehouse-Barlow.

Chartier, Roger. 1988. *The Cultural Uses of Print in Early Modern France*, translated by Lydia G. Cochrane. Princeton, NJ: Princeton University Press.

Donne, John. 1649. *Fifty Sermons.* London: Printed by Ja. Flesher for M.F.J. Marriot and R. Royston.

─────── 1955. "A Lent-Sermon Preached at White-hall, February 12, 1618." In *The Sermons of John Donne*, ed. and introd. George R. Potter and Evelyn M. Simpson. 10 vols. 2: 164–78. Sermon no. 7. Berkeley and Los Angeles: University of California Press.

─────── 1959. "The Anniversary Celebration of our Deliverance from the Powder Treason … preached in the Church, November 5, 1622, on Lamentations 4.20." In *The Sermons of John Donne*, ed. and introd. George R. Potter and Evelyn M. Simpson. 10 vols. 4: 235–63. Sermon no. 9. Berkeley and Los Angeles: University of California Press.

Green, Ian. 2000. *Print and Protestantism in Early Modern England.* Oxford: Oxford University Press.

Harrison, William. (1587) 1968. *The Description of England.* Edited by Georges Edelen. New York: Dover Books.

Hiltner, Ken. 2009. "Renaissance Literature and Our Contemporary Attitude toward Global Warming." *Interdisciplinary Studies in Literature and the Environment* 16: 429–41.

Hunt, Arnold. 2010. *The Art of Hearing: English Preachers and their Audiences, 1590-1640.* Cambridge: Cambridge University Press.

Luhmann, Niklas. 1990. *Essays on Self-Reference*. New York: Columbia University Press.

Maltin, Michael, and Christian Dannemann. 2013. "Dating the Salisbury Cathedral Clock." *ClockNet UK*. http://clocknet.org.uk/wiki/doku.php?id=dating_salisbury.

Morrissey, Mary. 2011. *Politics and the Paul's Cross Sermons, 1558-1642*. Oxford: Oxford University Press.

Penrose, Francis. 1883. "On the Recent Discoveries of Portions of Old St Paul's Cathedral." *Archaeologica* 47: 381-92.

Potter, George R., and Evelyn M. Simpson, eds. 1953-62. *The Sermons of John Donne*. 10 vols. Berkeley: University of California Press.

Shami, Jeanne, ed. 1996. *John Donne's 1622 Gunpowder Plot Sermon: A Parallel-Text Edition*. Pittsburgh: Duquesne University Press.

Smith, Bruce R. 1999. *The Acoustic World of Early Modern England: Attending to the 0-Factor*. Chicago: University of Chicago Press.

Stern, Tiffany. 2012. "'Observe the Sawcinesse of the Jackes': Clock Jacks and the Complexity of Time in Early Modern England." Paper presented at the Convention of the Modern Language Association, Seattle, January 5-8.

Symes, Carol. 2010. "Out in the Open, in Arras: Sightlines, Soundscapes, and the Shaping of a Medieval Public Sphere." In *Cities, Texts, and Social Networks, 400-1500: Experiences and Perceptions of Medieval Urban Space*, edited by Caroline J. Goodson, Anne E. Lester, and Carol Symes, 279-302. Aldershot, UK: Ashgate.

Vaughan, Edward. 1617. *A plaine and perfect method, for the easie vnderstanding of the whole Bible containing seauen obseruations, dialoguewise, betweene the parishioner, and the pastor*. London: Snodham. STC 24600.

Wall, John N. 2012. "Recovering Lost Acoustic Spaces: St. Paul's Cathedral and Paul's Churchyard in 1622." In *Digital Studies/Le Champ Numérique* (SDH-SEMI 2012 Conference Proceedings). http://www.digitalstudies.org/ojs/index.php/digital_studies/article/view/251/310.

———. 2014a. "Transforming the Object of our Study: The Early Modern Sermon and the Virtual Paul's Cross Project." *Journal of*

Digital Humanities 3: 2014. http://journalofdigitalhumanities.org/3-1/transforming-the-object-of-our-study-by-john-n-wall/.

_____ 2014b. "Virtual Paul's Cross: The Experience of Public Preaching after the Reformation." In *Paul's Cross and the Culture of Persuasion in England, 1520–1640*, edited by Torrance Kirby and P.G. Stanwood, 61–92. Leiden: E. J. Brill.

Walton, Izaak. (1675) 1927. *The Lives of John Donne, Sir Henry Wotton, Richard Hooker, George Herbert, and Robert Sanderson.* World's Classics series. London: H. Milford, Oxford University Press.

Cambridge Revisited?: Simulation, Methodology, and Phenomenology in the Study of Theatre History

Jennifer Roberts-Smith
University of Waterloo
j33robertssmith@uwaterloo.ca

Shawn DeSouza-Coelho
University of Waterloo
shawnathanddc@gmail.com

Paul J. Stoesser
University of Toronto
stoesser@chass.utoronto.ca

The Simulated Environment for Theatre (SET) Research Team
set.humviz.org

Introduction: Discontent

The starting point for this essay is the design of the Simulated Environment for Theatre, or SET, a digital system for visualizing theatrical text and performance created by a multi-institutional team over six years. Over the course of its development, the research team came to understand SET to be a potential tool for archival theatre history research. Most recently, building on the seminal archival research that Alan Nelson presented in *Early Cambridge Theatres* (1994), we reconstructed the temporary stage that was set up and taken down on what may have been an annual or even a seasonal basis at Queens' College Cambridge from the 1540s to perhaps 1640 (Nelson 1989, 2.1110ff). The Queens' Cambridge project was conducted in parallel with the larger SET project's preparations for a third version of the software system, and our discussion here arises from the interplay between those two activities—archival research and system design. Our focus, however, is neither primarily on the design of the SET system nor on the Cambridge stage. Rather, we use SET and Cambridge as the contexts for a theoretical exploration of the affordances of digital tools—simulations in particular—for the study of theatre history.

The SET team's theoretical perspective has been grounded since the project's inception in the concept of *theatricality*. Until we began work on Queens' Cambridge, we were especially engaged with what we think of as a *methodological* understanding of theatricality: theatricality, that is, as something that is *done*. But our recent work on Cambridge has necessarily refocused our attention on the *phenomenological* senses of theatricality; that is, on theatricality as something that is *experienced* and *perceived*. This shift in our thinking has encouraged us to look at the design of the SET system differently, and to reconsider the claims we have made for it in past publications as a "Simulation" not just "for Theatre" but of *theatricality*.

Our conclusion is that our past attempts to simulate *methodological theatricality* were both limited in their success and also limited in their relevance to theatre historical research; we predict similar limitations to any attempt we might make to simulate *phenomenological theatricality*. In short: we question whether any version of the SET system that we can imagine, current or future, could ever convincingly imitate either. In response, we propose that we might usefully limit the expectations we have of virtual simulation as a research method, and distinguish among those categories of insight that it can make available. As a first gesture toward that corrective, we offer a summary review and critique of the three principles of *methodologically theatrical simulation* that we have proposed in the past (function, agency, and social context) and a more extensive discussion of three further elements that we propose might pertain to *phenomenologically theatrical simulation* (logistics, semiotics, and reception) in order to interrogate the extent to which each of these might (or might not) help us understand past performance.

1 Methodological Theatricality: Function, Agency, and Social Context

When the SET team has used the term *theatricality* in the past, we have tended to think in the terms William B. Worthen sets out in his introduction to the 2010 monograph *Drama: Between Poetry and Performance*, in which he describes theatrical text as technological. By technological, he means that theatrical text can be co-opted in a potentially infinite range of social contexts for purposes unanticipated by its creator(s). He clarifies this characterization by contrasting a technology with a tool, using the example of a screwdriver to demonstrate: if you use the screwdriver to turn a screw (the purpose for which it was designed to function in a context predicted by its designer), it is a tool; if you use it as a murder weapon (a purpose it can fulfill, but for which it was not designed, and which occurs in a context presumably unanticipated by its designer), it is a technology (21-22; and see Roberts-Smith et al. 2013). That distinction—the difference between tool and technology—is, if we extrapolate from Worthen, at base about three things: the function of an object, the agency of its user, and the context in which it is used: in other words, what can be done with it, by whom, when, and where. Theatricality arises from the occasional, purposeful uses, rather than the stable characteristics, of material things; it is in that sense a *method*.

Worthen is talking specifically about theatrical text. In our work, we have extended his characterization to other material elements of the theatre, arguing, for example, that a stage property is a technology, an actor's body is a technology, and so on (Roberts-Smith et al. 2014). We have also argued that in the context of theatre historical research, documentary and material witnesses to past performances—here we mean records other than the texts of plays, such as posters, set designs, production photographs, reviews, archaeological remains, stage properties, and so on—are technological, because they are the instruments by which we speculatively or imaginatively reconstruct past performances. The SET project's theatre historians treat the archive in the same way that Worthen's theatre artists treat texts, and in so doing, our historians make non-textual records of production into theatrical technologies in the context of theatre historical research (Roberts-Smith et al. 2014). To put this another way, we have understood theatre history to be a discipline that is *methodologically theatrical*.

Our effort to design a simulation system for theatre historical research emerged through several iterations as an effort to simulate the elements we began to understand as foundational to *methodological theatricality*; that is, as an effort to simulate the function of theatrical records, the agency of creators and researchers working with them, and the social contexts in which theatre is made and studied.

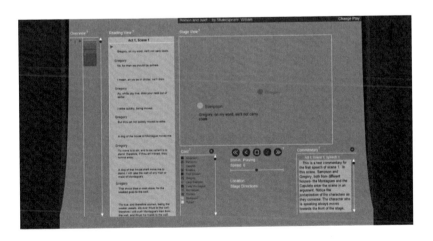

Figure 1. *Watching the Script*, the predecessor to SET.

The predecessor to SET, a text visualization system called *Watching the Script*, made a rudimentary gesture toward acknowledging that the intended *function* of theatrical text is its articulation by actors moving around in space. It parsed texts into speeches, and associated them with movable dots color-coded to represent speakers (see Figure 1). It would be more accurate to describe *Watching the Script* as a bookish simulation than a theatrical one, since it essentially reorganized and re-emphasized information that would be available to us in a print copy of the same text. The first design decision we made for Version 1 of the SET system (see Figure 2) was to extend the implication that theatrical text is intended to be spoken in space to the full acknowledgment that an intended theatrical performance is time-based, occurs in three dimensions, and is presented by speaking bodies (see Figure 2).

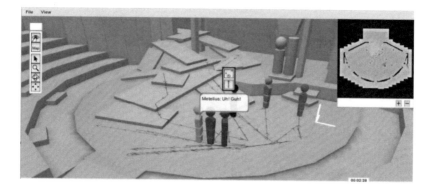

Figure 2. The Stage View concept for SET, showing a simulation of the 2008 performance of Shakespeare's *Julius Caesar* on a set designed by William Chesney, in the University of Waterloo's Theatre of the Arts.

The Stage View of the SET 1.0 interface might be described as a more *theatrical* simulation than Watching the Script had been (at least insofar as it offered a fuller representation of the intended *function* of a theatrical text). But because the system accommodated only one performance simulation at a time, its Stage View implied that its simulation was *the* performance rather than *a* performance. There could be nothing unexpected about the use of a text because there could be only one use of it; it was consequently still a tool in Worthen's terms, rather than a technology. We had the capacity to simulate a performance in a theatre, but not to simulate methodological theatricality.

When the design team began to think more systematically about Worthen's concept of theatricality, we abandoned the attempt to simulate theatrical text or performance, and attempted instead to design a system that would enable the movement between text and multiple potential performances (Roberts-Smith et al. 2013).

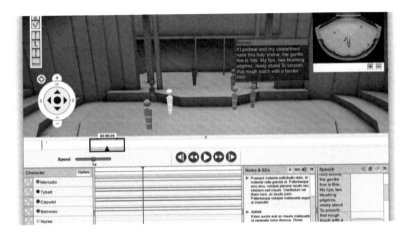

Figure 3. SET Version 1, showing an imaginary performance of Shakespeare's *Romeo and Juliet* in the University of Waterloo's Theatre of the Arts.

Since users could move avatars around inside the SET interface (and replay, export, and revise their blocking patterns), they had some agency to determine the way(s) in which the text was instantiated as a performance. The system permitted and predicted multiple, future, unknown instantiations of the text in simulated representations of different simulated performance contexts. We argued that any instantiation made by a SET 2.0 user was technological, and hence theatrical. Nonetheless, users' agency to co-opt texts for their own purposes was limited, since both the avatars and the stage models available had been designed by the SET team and pre-loaded into the simulation. Perhaps more importantly, the social contexts in which users might instantiate their performance ideas were restricted to other SET system sessions.

In the next major phase of development, we abandoned the idea that SET was a simulation of the theatrical creation process and began instead to think of Version 2 as an environment that could simulate the research methods of theatre historians. We began using SET 2.0 to reconstruct past performances, at first simulating performances that members of the research team had witnessed (see Roberts-Smith et al. 2012 and Kovacs et al. 2015), then simulating performances that we could access only by means of surviving theatre historical documents. Our experience led to a proposed system design that would present users with a blank screen inviting them to pose research questions (see Figure 4a), then to select and import reproductions of relevant archival materials (see Figure 4b), and then interpret those materials by creating their own timed and dimensional models of the performance elements they represented (see Figure 4c, and see Roberts-Smith, Harvey et al. 2013).

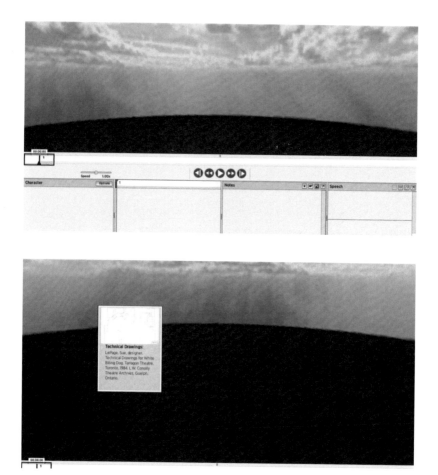

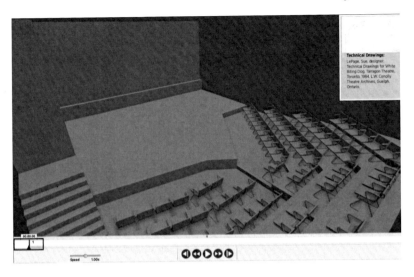

Figures 4a–4c. SET Version 2.0, showing the process of modeling the set, designed by Sue LePage, which was used in the Tarragon Theatre's 1984 production of Judith Thompson's *White Biting Dog*.

Although our ambition for an in-house user-controlled construction kit was not fulfilled at this stage, an important extension of SET's functionality was the addition of its annotation feature (see Figure 5). Annotation allowed users to import digitized records and made transparent the epistemological agency of the researcher in the research process.

We also added video to the range of media that could be accommodated in annotations, so that researchers could compare alternative performances of the same text (see Figure 6). The video feature diminished the status of the performance simulated in the central stage view from *the* performance to *a* performance, providing an analogy for the functional potentiality of all the texts and records that had been used as sources for the Stage View's visualization.

326   Jennifer Roberts-Smith, Shawn DeSouza-Coelho, and Paul J. Stoesser

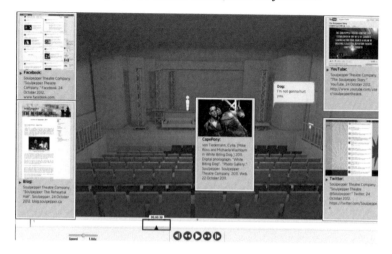

Figure 5. Our reconstructive visualization of Soulpepper Theatre's 2011 production of Judith Thompson's *White Biting Dog*, designed by Christina Poddubiuk, annotated with excerpts from publicity for the production.

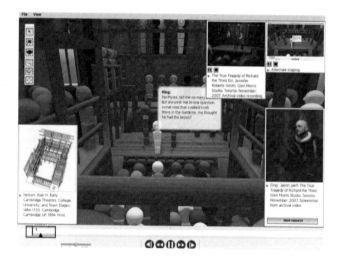

Figure 6. A hypothetical performance of the Queen's Men's *True Tragedy of Richard the Third* (performed ca. 1583; published 1594) on our reconstruction of the stage used at Queens' College, Cambridge. The annotations in the upper right corner show two alternative stagings, one in the form of archival footage of a live reconstructive performance on a set designed by Paul Stoesser, and the other in the form of a screen-capture video of another virtual performance in the SET environment.

Since each performance simulation (in the Stage View or in video annotations) might have been generated by a different researcher or team of researchers, SET 2.5 also characterized the theatre historian as *a* researcher rather than *the* researcher, and by visualizing the differing perspectives of *some* researchers, it arguably encouraged the emergence of a community of interpretive agents that might itself become a social context uncontrolled by SET's creators for the theatrical interpretation of records of past performance (Roberts-Smith et al. 2014).

Because it explicitly invites multiple users to make and compare multiple, differing, valid interpretations of records of past performance, SET 2.5 might be said to encourage an epistemological mode of what we have been calling *methodological theatricality*, in the sense that it gives agency to researchers to interpret performance records for purposes not fully predictable by their creators or by the creators of the SET system. Nevertheless, the extent to which SET 2.0 fully *simulates* methodological theatricality is doubtful, because of the ways in which it limits users' activities. By "limits," we do not mean merely limiting the functions that are still missing from the system, but the fact that it is a programmed software system at all. We could design a Version 3, for example, that solved one of the problems we identified in Version 2 in the form of an in-world construction kit for making custom avatars and performance environments (see DeSouza-Coelho et al. 2012). We could even design a system with a workflow in which users were required to make interpretive decisions before they could proceed, guaranteeing that they would engage with theatrical records in a methodologically theatrical way (see Roberts-Smith, Harvey et al. 2013). But of course, the more SET 3.0 guided users, the less agency they would have. To put this another way: because the SET system is a programmed electronic simulation, it cannot ultimately be used for any purpose for which it was not intended; its functions can only be co-opted for the purposes afforded by the system. Insofar as it is an electronic simulation, the SET system itself is anti-theatrical (see also Roberts-Smith, DeSouza-Coelho, and Malone, forthcoming 2016).

2 Phenomenological Theatricality

When we completed the design of SET 2.5, we were also conscious that rather than creating a tool that might be of use to theatre historians in their research, we had focused our attention on creating a simulation (however imperfect) of historiographical methods. Our system was a successful research prototype (see Ruecker 2015) in that its development had provided a fruitful context for developing a theory of *methodological theatricality*; but we were not sure that the system had affordances of use to anyone who wanted to learn about past performances. Consequently, our most recent phase of development—a review of the affordances of Version 2.5 and the development of design specifications for Version 3—has been conducted in parallel with our first attempt to distinguish the actual outcomes of a theatre historical research project conducted using the SET system from the system's existing potential and hypothesized future affordances. That is: we began to ask not just what we *might* do with SET, but what we *could actually* learn by using it. Since the goal of all theatre historical research might be characterized as an attempt to better understand what the experience of a past performance might have been for the artists and audiences who participated in it, we have come to understand this development phase as an attempt to define SET's affordances for simulating *theatrical phenomenologies* rather than *methodologies*.

Our test case has been an attempt to understand the phenomenology of the temporary stage that, as Alan Nelson has demonstrated, seems to have been used for performances in Queens' College Cambridge's Old Hall for almost a century starting in the mid-1540s. It is documented in an inventory of parts, with assembly instructions, dated 1640 (Nelson 1989, 1.688–93), which Nelson used to create reconstructive drawings of the stage structure (Nelson 1994, 21, Figure 10). We have extended Nelson's work, using the same archival inventory, to create a virtual, three-dimensional scale model of the stage, placing our model inside a scale model of Queens' Old Hall, populating it with avatars representing both actors and audience, and using the model to interpret some of the remaining archival records referring to performances held there over the decades in which it was used. In particular, we have focused on the records related to what appears to have been the major period of the stage's initial construction between 1547 and 1549 (although later alterations were made; see Nelson 1989, 2.1118–23). In order to position our stage model in the context of its venue, we looked to some archaeological records that were not available to Nelson at the time he was writing; but the principal methodological difference between our work and his was that whereas Nelson was working from static two- and three-dimensional representations, which could only be disseminated in print, we have been working with time-based three-dimensional representations, which we are able to disseminate electronically.

Our research process has led us to critique our past claims for the SET system in terms of a different, although in some ways parallel, theoretical frame. Whereas in the past we were interested in the function of theatrical things, the agency of theatre makers, and the contexts for theatrical production—all elements that we might describe as belonging to the process of generating theatre, to *methodological theatricality*—we are now more interested in elements that might belong more fully to the process of experiencing theatre, namely materiality, semiotics, and reception. By materiality, we mean the people and things that make meaning: which material objects are in play in time and space, what they are made of, where they come from and where they go both before and after a performance, who has to deal with them, how much they cost, and so on. By semiotics, we mean the ways in which those things are meaningful: if we examine people and objects—material things—what we can learn from their structure and their arrangement in relation to other material things. By reception, we mean how individual human beings experience those meanings as unique, historically and socially positioned, emotionally affected and intellectually engaged members of a community. These are of course far from original theoretical approaches to the study of theatre; we might cite the entire documentary theatre history movement led by the Records of Early English Drama project on historical materiality; theorists from Patrice Pavis to Ric Knowles on the relationship between materiality and semiotics; iconoclasts from Richard Schechner to Peggy Phelan on reception. But these are perhaps less familiar approaches to the study of tools in the digital humanities, and we have found them useful means of shaping our thinking about how our tool might most usefully be characterized.

3 Materiality

Three-dimensional, time-based, digital representation was particularly helpful in revising Nelson's hypothesis about the process that was used to construct one end of the stage structure, called the Great Gallery, which was erected at the high end (the east end) of Queens' Cambridge Old Hall. The Great Gallery consisted of priority seating behind and above the stage, so that dignitaries attending performances would have the best opportunity to see and be seen.

Figure 7. The Great Gallery at the high end (the east end) of Queens' Cambridge Old Hall, seen from the screens gallery at the low end.

According to the surviving inventory, the Great Gallery was the first portion of the Cambridge stage structure to be built, so it needed to provide stability for the rest of the stage scaffolding, which wrapped all the way around the room. Among the first pieces to be set into place was a "girt," a huge beam that stretched from one side of the room to the other.[1] Based on archival evidence alone, Alan Nelson proposed that the girt must have been anchored in a hole in the interior wall of the Hall (Nelson 1994, 18–21). To test the practical implications of his hypothesis, we created a video reconstruction of the process of mounting a beam of those dimensions and weight within the confines of the dimensions of Queens' Old Hall, and extrapolated from our reconstruction the demands on human laborers of proceeding as Nelson proposed (see Figure 8).

---

[1] See *OED*, "girt," n. 2.a. The term is used in the 1640 inventory, and is interchangeable with "girder."

Cambridge Revisited? 331

Figure 8. Alan Nelson's proposal for the installation of the first supporting girt in the construction process of the Queens' Cambridge stage. Our video visualization is available at http://ems.itercommunity.org.

Our conclusion is that it is very unlikely the girt was supported by holes in the hall walls; instead, a fastening system that included the use of wedges on a slightly shorter girt is much more likely (see Figure 9).

Figure 9. Our alternative proposal for the installation of the first supporting girt in the construction process of the Queens' Cambridge stage. Our video visualization is available at http://ems.itercommunity.org.

As a research process, our video simulation offered us an efficient way to test and illustrate our new hypotheses about the construction of the Queens' College stage as we formulated them. We benefited by our freedom from the physical constraints of materially reconstructing the stage, such as, for example, the costs that a material reconstruction would incur, not to mention the backs of the laborers who would have to raise the beam; these would ordinarily make research of this scale and nature impossible. The extremely quick rate at which we were able to accomplish revisions allowed almost immediate comparison of alternatives with negligible consequences in terms of the time and cost it took to make changes to our model. Consequently, this particular experiment confirmed our sense that a useful function for Version 3 of the SET system would be an in-world construction kit that allowed users to animate objects. It might provide a fluid and efficient means of investigating some kinds of theatrical materiality.

Nonetheless, the materially-oriented piece of our theatre-historical research also drew our attention to SET's pretensions to imitative *simulation*, pretensions that are expressed as visualizations—in this case, visualizations of material objects: pieces of wood. As simulations, our reconstructive videos lack significant sensory elements. We can see an abstraction of the dimensional challenges of moving that large girt around within the limits of Old Hall's walls, and we can calculate the force required to mount it, but we cannot actually feel the girt's weight, smell the scent of its oak fabric, or touch its texture. Was it planed and sanded at first? Rubbed with oil? Did it develop smooth grooves over time or did it dry out and splinter? What sound did the men who lifted it make as it slid successfully into place? Or didn't, and slipped to the ground? Was it chipped and dented from the dropping and banging of each installation? None of these material phenomena are available to us in our exploration of the experience of the artists—in this case the carpenters and their crew—who made the Queens' Cambridge stage.

Because it is a virtual system, SET's ability to explore and communicate the sensory nature of theatrical materiality is at best limited and at worst nonexistent. We might argue that our visualization of the construction process for the Cambridge stage engages at least one of our senses, but we might also notice that vision is not the only, and perhaps not the most important, sense engaged by the stage's original builders. What about proprioception (the kinesthetic sense), for example? Or nociception (the sense of pain)? Most importantly for our argument, because SET limits our experience of theatrical materiality to virtual reality, it limits the range of potential functional appropriations of its simulations as performances; it cannot be used to perform smells or textures or movement or exertions. In other words: when we simulate a material thing used in the theatre, we reduce it from technology to tool, its avatar limited to the sensory channels that its platform can engage. The materiality of the theatre is multi- and inter-medial, whereas a simulation is platform-specific. Simulation is therefore a barrier, rather than an aid, to *theatrical materiality*.

4 Semiotics

The SET system's navigation, in its Version 2.5 iteration, allows users to navigate its virtual spaces, viewing them from a variety of in-world perspectives. In our work on the Queens' stage, we used this function to develop a hypothesis about the semiotic arrangement of theatrical space during performance. It helped us understand the ways in which the stage both expressed and reconfigured the social hierarchies of the College. The stage's Great Gallery, as we have mentioned, seated college and visiting dignitaries behind the stage (see Figure 7); on the two additional tiers of audience space above the dignitaries' heads, there was standing room only (Nelson 1994, 36). Each of these tiers had a different perspective on stage action during a performance. The dignitaries' was both the most intimate (see Figure 10a) and the most public, since the dignitaries themselves were so visible to other audience members (see Figure 10b). The Upper Great Gallery's view was the most awkward, both for sight lines, since audience members were looking down on the tops of the performers' heads and did not likely have access to the desirable view of the dignitaries at all; and also for comfort, since audience members would have had to strain forward, perhaps precariously, to see over the Upper Gallery's rail (see Figure 10c).

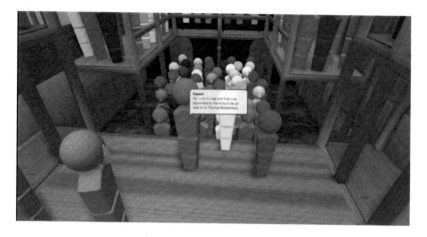

Figure 10a. View from the Dignitaries' Gallery.

Figure 10b. View of the Dignitaries' Gallery.

Figure 10c. View from the Upper Great Gallery.

The Great Gallery's semiotics expressed very clearly the social hierarchy of the College by means of its occupants' proximity to the stage and visibility from elsewhere in the room.

Our simulation also revealed that the structure of the Great Gallery overlays important permanent features of the Hall, which were themselves symbols of the social arrangement of the Cambridge community; these included the dais at the high end of Old Hall, the Hall's windows (which cast light on the high table that would on ordinary days likely have occupied the dais), its fireplace, and its very expensive wainscoting. In that context, the process of erecting the stage (thinking back to our video simulation in Figure 9), which involved the first face of the stage being hoisted from the ground into place by laborers, might be described as a kind of theatrical process of co-opting the hall for a purpose for which it was not originally intended in the particular social context of the College's performance festivals. Because it is officially sanctioned, the process in some ways reinforces the Hall's role as a semiotic expression of the social organization of the College, and its outcome in some ways preserves its non-theatrical semiotics, since it leaves the College dignitaries in their customary position at the high end of the hall. In other ways, the stage construction upsets the Hall's usual semiotic functions, for example, by making it impossible to light a fire to warm the dignitaries (or displacing responsibility for their comfort from the exclusivity of the fire to the commonality of the body heat generated by other audience members), and blocking the view of the magnificent wainscoting backdrop and the slant of the light from the ornamental windows against which they would normally be admired (see Figure 11).

Figure 11. The fireplace in the west hall of Queen's College Cambridge Old Hall, blocked by the structure of the temporary stage.

Any individual performance on the stage had a similar capacity to reorganize the semiotic hierarchy of space in the hall. For example, there is a reference in an archival record of the stage's construction to a heavens, which it seems likely had a lift that could raise performers up from stage level into the air above the stage, parallel with the upper tiers of its galleries (Nelson 1989, 2.1125). In the act of raising performers out of the stage's usual performance space, the region of embodied and visual proximity to the dignitaries and of the dignitaries' privilege, the dignitaries' status as the focal point of the performance is disrupted. In our view, the performance's focal point moves up the height of the room, with the performer being cranked upwards by the lift, and settles on the low-status audience members who are stuck in the least comfortable, most crowded, dangerous, standing-room-only viewing positions in the Upper Great Gallery instead (see Figure 12).

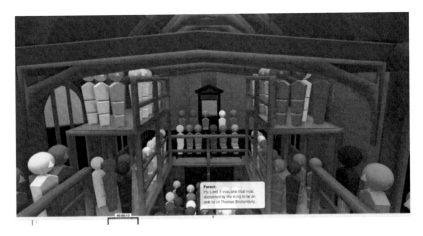

Figure 12. Inverted Spatial Semiotics: View from the "lift"?

In this example, the semiotics of the stage itself are inverted by the functioning of the lift (that is, in its use for a particular purpose in a particular context), and although its mechanical capacity was undoubtedly intended by its creator(s), the particularities of all the performative contexts in which it would be employed (the plays, the occasions, the specific dignitaries in attendance) were unlikely to have been. In other words, we propose that the Queens' Cambridge lift was a theatrical technology that had the capacity—albeit playfully and temporarily—to reconfigure the social hierarchy of the College community. If the next iteration of SET could combine the visualization of the lift's mechanics (using the construction kit function we proposed in section 2.1 above) with the possibility of observing its mechanics from multiple audience perspectives (as Version 2.5 already does), it might arguably have the ability to simulate the conditions under which a *theatrical semiotics* occurred in the past. It would not, however, imitate *theatrical semiotics*, because the system is neither capable of convincingly imitating the material elements that make meaning, nor capable of imitating the processes by which the materiality of performance is perceived and interpreted.

5 Reception

The limitation of SET's ability to simulate *theatrical reception* lies in the way it locates interpretive agency. The agency to interpret theatrical semiotics during a live performance is distributed, communicative, and inescapably cumulative over time: at Queens' Cambridge in 1547, for example, we need a group of laborers to erect the stage and install the lift, and we need performers, dignitaries, and students to instantiate and then reverse the spatial hierarchy the stage creates in Old Hall. Interpretive agency in the SET system, by contrast, is located in and limited to a single user at a time. SET makes interpretive agency exclusive, introspective, and independent of continuous time; the experience of it is arguably opposite to the experience of *theatrical reception*. Even the perspective of the user, who possesses the ability to alternate between the first-person perspective of any actor or audience member (see Figures 13a and 13b), and a third-person perspective in which visual manipulation along all three axes is possible (see Figure 13c), is, phenomenologically, an anti-theatrical affordance of the system, since live performance can only be experienced by means of the spatially situated body of an individual participant.

Figure 13a. The perspective of an actor in SET's simulation of a performance on the Queens' College, Cambridge stage.

Figure 13b. The perspective of an audience member standing at floor level in SET's simulation of a performance on the Queens' College, Cambridge stage.

Figure 13c. The SET user's third person perspective, manipulable along all three spatial axes.

As we have demonstrated above, SET excludes certain kinds of theatrical or theatre historical evidence because it cannot accommodate them. Some of these we have characterized as static (the weight, smell, or texture of wood, for example), and some as procedural (the actions performed by participatory agents that result, for example, in the meaning generated by the raising of a performer on a lift). We can make an animated simulation of the lift, but because we exclude certain features of the materiality of the experience of the lift, and because we (as system designers) control the entire experience of the lift for our users, our animation is a fixed semiotic system. Any computational system, by virtue of its selection and arrangement of the processes it makes available to its users, is a closed system; the phenomenology of the theatre, on the other hand—because agency in the theatre is distributed and communicative—is the experience of an open system. The outcome of this line of argument is, we think, the conclusion that there can be no *theatrical phenomenology* in a computational system at all.

6 Conclusion: Why Use New Technologies for the Study of Renaissance Theatre?

If there can be no theatricality – methodological or phenomenological – in a closed system, then we and perhaps others in the field need to be cautious about the implicit claims that our virtual simulations of performance have made—merely by virtue of their being simulations—to be reproductions rather than representations of performance. Some simulations have provided considerably more detail and complexity in their virtual analogues for embodied historical phenomenologies than the SET system has: *Virtual Vaudeville*, for example, used motion capture technology to create a much more convincing avatar of its principal performer than SET's abstract avatars, and even programmed an entire audience of anthropomorphic avatars responding to its performance. To the extent that *Virtual Vaudeville* claims—implicitly, by means of its technological ambition (although its creators never make this explicit claim)—to reproduce the past performance it represents, it is making as false a claim as any the SET team has made explicitly for our system. To put this another way: any virtual simulation of live performance (past or present) that does not make its status as a simulation emphatically explicit risks irresponsibly eliding the difference between reproduction and representation.

On the other hand, if we accept and acknowledge simulation as by definition absent of *theatricality*, we might also accept it as a useful conceptual framework within which to characterize what our system and others like it can do and what they cannot do. If SET is a simulation, we need not expect it to deliver theatricality, but can instead appreciate its ability—because it is free of the phenomenological constraints of time, space, and specific, situated individual perspectives—to make the evidence that underlies our visualized historical arguments explicit. Indeed, it might be quite freeing to think of all theatre historical research as a form of simulation, since it hypothesizes and models scenarios that account for the fragmentary surviving documentary record of past performances, but never actually reproduces historical performance events. Jeremy Lopez, for example, has made a strong case against the use of live performance to communicate such models, as the "original practices" movement claims to do. Lopez argues that the affectivity of original practices performances prevent (perhaps deliberately) evidentiary transparency (2007); to extend his argument: phenomenology may, in some contexts at least, undermine scholarship.

In the past, we conceived of our tool, the Simulated Environment for Theatre, as having the potential to make two principal contributions to knowledge in the fields of new technologies and Renaissance studies. The first was that SET might become a useful research tool for other scholars in other projects. If we are able to complete Version 3.0, SET might achieve this goal. We think we have enough experience with SET and with other visualization environments as tools for theatre historical research that we can say with some confidence that SET 3.0 would be useful. The affordances we hope to offer would make our work on the Queens' Cambridge stage, for example, even more efficient, transparent, and vividly illustrated than it has been to date, and we would be capable of more, and more ambitious, iterative revisions to our hypothesized performance scenario in a smaller space of time.

The second contribution we imagined for SET was that it might open up theoretical discourses or design rhetorics that contributed productively to the creation or critique of other, different tools or systems; to our understanding of historiographical methodologies; or to our understanding of the theatre as an art form. Speaking only for our own team, SET has done this. We have ourselves designed spin-off systems, including *Staging Shakespeare*, an educational game for high school students (see Roberts-Smith, DeSouza-Ceolho and Malone, forthcoming 2016); we have published on SET as a historiographical tool (Roberts-Smith et al. 2012; Kovacs et al. 2015; Roberts-Smith, Harvey, et al. 2013); and we have, between the first contribution to a conference proceedings we ever drafted (Roberts-Smith et al. 2009) and the present article, significantly expanded and enriched our understanding, as a research team, of the thing we have aimed to represent, namely: theatricality. In the case of this project on the Queens' Cambridge stage in particular, SET has operated in a sense like a two-way theatrical time machine: we have looked forward into the theatrical future by using it to plan future historically communicative productions using a reconstructed Cambridge stage (*pace* Lopez); we have looked backward into the theatrical past by using it to hypothesize productions at Queens' that are lost to history; and we have done those things simultaneously in the same virtual space. We have occupied a phenomenology of situated scholarly simulation that we find quite interesting, and we are looking forward to living in it for a while.

As a methodology, however, the process of designing, building, applying, and interpreting SET has also suggested a third contribution to knowledge that we did not foresee. In our work on the Cambridge stage, as in all our earlier work together, SET has taken us year by year into our shared and individual intellectual futures. That is to say: the system itself is not a future, nor is any simulation we might create within it; but the process of developing the tool has been intellectually productive in itself. That process is a kind of processual epistemology, by which we mean the knowledge is in the doing, rather than subsequent to, an outcome of, or an ancillary to the doing; the process is our product and our argument. The knowledge we are still developing is not unlike the participatory, performative epistemology of the theatre, the kind of knowledge we can only have by continuing to be there doing things together. At this moment, we think the projects of theatre-making, Renaissance theatre history-making, and digital tool-making may belong to the same category of epistemological activity, one that is related to but also different from the activities our field has referred to as "critical making" and digital "protoyping." Critical making is normally deeply invested in the semiotic and rhetorical presences of the materiality of the physical objects it generates (see for example Ratto 2011), and humanities prototyping is similarly invested in the semiotics of the digital objects we make (see Galey and Ruecker 2010). SET-making, by contrast, has, in retrospect, been invested in the semiotic and rhetorical expressiveness of its participants in collaborative action: what we do together, how what we do changes us, how our changing together is what we do next. For us now, the production of SET has become one of those processes of knowledge production whose distinguishing features, best practices, and evaluative measures—as the performance as research movement's early history illustrates—have traditionally been difficult to translate into words (see Piccini and Kershaw 2004). As paradoxical (or perhaps as obvious) as this may seem, the intersections between new technologies and Renaissance studies, in the context of early English theatre history, have turned our attention to the ways in which we as a group of present individuals generate knowledge together and articulate it together in the things we do together.

WORKS CITED

DeSouza-Coelho, Shawn, Jennifer Roberts-Smith, and the SET project. 2012. "Manipulating Time and Space in Virtual Worlds: Design Directions for the *Simulated Environment for Theatre*." Society for Digital Humanities/Société pour l'étude des médias interactifs, Congress, May 26–28.

DeSouza-Coelho, Shawn, and the SET project. 2012. "Version 3—Implementation Recommendations." Unpublished white paper.

Galey, Alan, and Stan Ruecker. 2010. "How a Prototype Argues." *Literary and Linguistic Computing* 25 (4): 405–24.

Kovacs, Alexandra (Sasha), Jennifer Roberts-Smith, Shawn DeSouza-Coelho, Teresa Dobson, Sandra Gabriele, Omar Rodriguez-Arenas, Stan Ruecker, and Stéfan Sinclair. 2015. "An Interactive, Materialist-Semiotic Archive: Visualizing the Canadian Theatrical Canon in the Simulated Environment for Theatre (SET)." In *Place and Space: Cultural Mapping and the Digital Sphere*, edited by Ruth Panofsky and Kathleen Kellett, 51–88. Edmonton: University of Alberta Press.

Lopez, Jeremy. 2007. "Imagining the Actor's Body on the Early Modern Stage." *Medieval and Renaissance Drama in England* 20: 187–203.

Nelson, Alan H., ed. 1989. *Records of Early English Drama: Cambridge*. 2 vols. Toronto: University of Toronto Press.

———. 1994. *Early Cambridge Theatres: College, University, and Town Stages, 1464–1720*. Cambridge: Cambridge University Press.

Piccini, Angela, and Baz Kershaw. 2004. "Practice as Research in Performance: From Epistemology to Evaluation." *Digital Creativity* 15 (2): 86–92.

Ratto, Matt. 2011. "Critical Making: Conceptual and Material Studies in Technology and Social Life." *The Information Society* 27 (4): 252–60.

Roberts-Smith, Jennifer, Sandra Gabriele, Stan Ruecker, and Stéfan Sinclair with Matt Bouchard, Shawn DeSouza-Coelho, Diane Jakacki, Annemarie Akong, David Lam, and Omar Rodriguez. "The Text and the Line of Action: Re-conceiving Watching the Script." *New Knowledge Environments: Research Foundations for Understanding Books and Reading in the Digital Age* 1 (2009). Papers drawn from the INKE 2009: Birds of a Feather conference. 23–24 October 2009. http://journals.uvic.

ca/index.php/INKE. Rpt. in *Scholarly and Research Communication* 3.3 (2012) http://src-online.ca/index.php/src/article/view/96.

Roberts-Smith, Jennifer, Shawn DeSouza-Coelho, Teresa Dobson, Sandra Gabriele, Omar Rodriguez-Arenas, Stan Ruecker, Stéfan Sinclair, Annemarie Akong, Matt Bouchard, Diane Jakacki, David Lam, and Lesley Northam. 2013. "Visualizing Theatrical Text: From *Watching the Script* to the *Simulated Environment for Theatre* (SET)." *Digital Humanities Quarterly* 7 (3).

Roberts-Smith, Jennifer, Shawn DeSouza-Coelho, Teresa Dobson, Sandra Gabriele, Omar Rodriguez-Arenas, Stan Ruecker, Stéfan Sinclair, Alexandra (Sasha) Kovacs, and Daniel So. 2014. "SET Free: Breaking the Rules in a Processual, User-Generated, Digital Performance Edition of *Richard the Third*." In *The Shakespearean International Yearbook* 14, 69–99.

Roberts-Smith, Jennifer, Shawn DeSouza-Coelho, Teresa Dobson, Sandra Gabriele, Stan Ruecker, Stéfan Sinclair, Alexandra (Sasha) Kovacs, and Daniel So. 2012. "Visualizing Theatre Historiography: Judith Thompson's *White Biting Dog* (1984 and 2011) in the *Simulated Environment for Theatre* (SET)." *Digital Studies/Le champ numérique* 3 (2).

Roberts-Smith, Jennifer, Kathryn Harvey, Shawn DeSouza-Coelho, Teresa Dobson, Sandra Gabriele, Alexandra (Sasha) Kovacs, Omar Rodriguez-Arenas, Stan Ruecker, and Stéfan Sinclair. 2013. "Is There an Archivist in the Sim?: Teaching Literacy through Experience in a Post-positivist, Mixed-media Virtual Theatre Archive." *Canadian Theatre Review* 156: 40–45.

Roberts-Smith, Jennifer, Shawn DeSouza-Coelho, and Toby Malone. Forthcoming, 2016. "Staging Shakespeare in Social Games: Towards a Theory of Theatrical Game Design." In "Social Media Shakespeare," special issue, *Borrowers and Lenders: The Journal of Shakespeare Appropriation*, edited by Maurizio Calbi and Stephen O'Neill.

Ruecker, Stan. 2015. "The Multiple Uses of Prototypes in the Digital Humanities and Design." Sustaining Partnerships to Transform Scholarly Production: *An INKE-hosted Partner Gathering*. Whistler, BC, 27 January. Conference paper.

Worthen, William B. 2010. *Drama: Between Poetry and Performance.* Chichester, UK: Wiley-Blackwell.

Staying Relevant: Marketing Shakespearean Performance through Social Media

Geoffrey Way
Arizona State University
gway1@asu.edu

Given the widespread prominence of social media among their audiences, the marketing departments of Shakespearean theatres and festivals are continually finding themselves in new territory. Considering the main goal of marketing departments is to establish and maintain their institutional relevance with audiences, social media can be a boon, facilitating direct interaction with their audiences directly through sites like Facebook, Twitter, and Pinterest. However, even though social media provide the means to generate such audience–institution interactions, the fact is that engaging theatre audiences online presents a new series of challenges these organizations must navigate to be successful. Effective digital marketing requires approaches that create meaningful interactions with audiences instead of delivering content. For the Oregon Shakespeare Festival (OSF; Ashland, OR), the Royal Shakespeare Company (RSC; Stratford-upon-Avon, England), Shakespeare's Globe (London), and the Stratford Festival (Stratford, ON), this means focusing less on their websites, which are designed to sell tickets and deliver content. Their websites are also host to a large amount of information on the history of these institutions and their work, but usually fail in offering audiences interactive experiences. When they do try to create interactive spaces, usually in the form of blogs, these experiences are often buried within the site's navigation and not readily accessible. However, these spaces fail at attracting audiences because audiences visit these websites to gain information, not to seek out interactive experiences.

If Shakespearean performance institutions hope to engage their audiences online, they have to design new ways of doing so. The website is seen more often as a necessity than a powerful tool for interacting with audiences. Anita Gaffney, executive director of the Stratford Festival, states that digital marketing is "table stakes. It's what you have to do to survive. It's not what you do to exponentially grow your business. If we didn't have a social media presence, or if we didn't have a website, we'd be extinct" (2013). Mallory

Pierce, director of marketing and communications for the Oregon Shakespeare Festival, holds a similar view: "If you don't have a website, you don't really exist in a marketing sense" (2013). Both Gaffney and Pierce highlight the need for digital marketing in terms of survival and relevance; through digital media, theatres and festivals are able to remain relevant and compete with other institutions and media online. Since these organizations are location-based and require their audiences to visit the physical site to maintain financial stability, digital media may not at first glance seem to be beneficial for institutions to invest their time and effort in.

However, Shakespearean theatres and festivals are faced with the realities of the current economy, and as travel prices continue to rise, it becomes more difficult for audiences to visit these sites, especially on an annual basis. This may result in years between visits for regular audience members, and may give pause to those thinking about visiting a site for the first time. It is in these gaps before or between visits that social media become powerful marketing tools. Two particular aspects of social media, presence and persistence, speak to their effectiveness in digital marketing. Social media allow institutions to create online presences and facilitate experiences for audiences from a remote setting.[1] They are also digitally persistent: active and accessible for audiences at their leisure (Smith 2004). To utilize social media effectively means more than just creating an accessible online space. As Ryan Nelson argues, "when an organisation's online presence may not just be an audience's first port of call, but their only point of contact, the very concept of the theatre audience needs to be re-imagined" (2014). To address this challenge, Shakespearean theatres and festivals are relying on both the newness of social media and the nostalgia of their audiences to encourage their audiences to *interact* and *engage* with their digital presences in meaningful ways.

From Web 1.0 to 2.0: Performing Newness

As they work to compete with other institutions and media, Shakespearean theatres and troupes are faced with the reality that to thrive, they have to ensure their financial viability. Marketing departments support this goal through the promotion of ticket sales, but they are equally important in establishing the cultural relevance of their institutions to guarantee their financial stability. Through justifying their cultural relevance, theatres and festivals are able to generate donations and government funding to support

---

[1] My definition of presence is based on Marvin Minsky's telepresence, the idea that future robotic technologies and remote control tools can "feel and work so much like our own hands that we won't notice any significant difference" (1980).

their work, which has become a necessity in the wake of smaller organizations such as Shakespeare Santa Cruz and Georgia Shakespeare closing due to a lack of financial support, although the former has found new life as Santa Cruz Shakespeare (Baine 2013). Along with the need to justify their relevance, theatres and festivals are focusing on their cultural impact, a concept that has become more integral to determining government funding for cultural institutions (Belfiore and Bennett 2007). The valuation of cultural relevance represents a shift away from the model of the early to mid-1990s when, as Michael Bristol argues, "Commercial profit rather than a wish to guarantee the durable public value of Shakespeare [was] the motive that best accounts for the diverse enterprises of book publishers, theatre managers, film-makers and television producers" (1996, x). There are multiple motives that influence the enterprises of Shakespearean theatres and festivals, from economic to cultural to social, and while commercial profit is integral to their continued existence, the work they do very much serves to "guarantee the durable public value of Shakespeare." This new model emphasizes cultural impact not as a model for financial profit, but as one for financial survival.

What has driven this shift is the fact that Shakespearean performance has become even more of a niche market in recent years; although Shakespeare's cultural value may still be strong, theatres and festivals have to compete for their audiences against a multitude of other entertainment and media options. Initially, when these organizations moved online and created websites, they were competing in an information economy. However, over time there has been a move online from an information economy to one of attention (Lanham 2006). Now institutions are focused on attracting audiences to their work by gaining their attention, but more static offerings like websites cannot compete for an audience's attention as they once did. Thus, these organizations have made the move into social media to interact and engage with their audiences. They hope to remain relevant through the adoption of new technologies, as Diana Henderson argues: "As digitalization has again made the medium the message in a particularly urgent way, there is a perceived imperative to use new media whenever possible, a perceived imperative current not only among practitioners in new media and in marketing, but in traditional fields" (2002, 123). To adapt and compete in the attention economy, theatres and festivals are now faced with courting the attention of their audiences, instead of relying on the innate cultural value of their work to maintain that attention.

One of the major aspects of this shift has been the move from a model of stickiness to a model of spreadability. Stickiness is exemplified by Web 1.0

technologies designed to operate in an information economy, "centralizing the audience's presence in a particular online location to generate advertising revenue or sales" (Jenkins, Ford, and Green 2013, 4). The stickiness model generally scaffolded traditional marketing strategies onto digital technologies, and its success relied on users actively visiting websites on their own. Although users had the ability to take the content shared on institutional websites and post it elsewhere, sticky content was designed to be accessed on the websites themselves rather than shared among users' networks. As this approach has become less effective, theatres and troupes have had to adjust their strategies, which means shifting away from passively delivering content to audiences and embracing technologies that encourage audience-institution interactions, such as social media (Chan-Olmsted 2002). The continued rise of social media has brought with it a new model: spreadability. Coined by Henry Jenkins, Sam Ford, and Joshua Green, spreadability "refers to the potential—both technical and cultural—for audiences to share content for their own purposes, sometimes with the permission of rights holders, sometimes against their wishes" (2013, 3). Social media such as Facebook, Twitter, and Pinterest are built on the model of spreadability, and theatres and festivals are adapting to this new model by creating content that users can share quickly and easily with their various online networks.

The challenge for Shakespearean performance institutions is that they cannot simply enter these online spaces and immediately engage their audiences. They have to embrace the reality that, as Paul Edmondson and A. J. Leon argue, "Changing a culture for Shakespeare online involved to a great extent knowing how the web works" (2014, 200). This has led to these organizations hiring individuals and teams that come not from Shakespeare or theatre backgrounds, but digital marketing ones. These individuals/teams, with titles such as digital officer, digital communications associate, digital marketing officer, and digital media producer, often have to implement strategies that rely on their knowledge of social media, but they may be reticent to engage with online audiences in depth when it comes to theatre or Shakespeare.[2] Through the creation of these positions, institutions are moving beyond what Donald Hedrick calls entertainment logic, which pushes the "economic principle of

---

[2] Sian-Estelle Petty, digital officer at Shakespeare's Globe, highlights the need for collaboration when reaching out to other Shakespeare institutions: "It's something I'd like to do a lot more of, and part of it is that I don't have a Shakespeare academic background, so I'm a bit scared to get involved a lot of times. So we have our research team, several of them on Twitter, and they tend to lead a little bit more on things like that" (2013).

return as opposed to strict *value*, to its extreme or limiting case: the ability to make a profit off of *nothing*" (2010, 205). They are instead actively investing in building and maintaining strong presences on social media sites to interact with and engage their audiences in these spaces.

Digital marketing represents a new set of stakes for institutions, emphasizing a focus on engagement over sales. As Kathleen McLuskie argues, the more audiences engage with an organization and its work, the more they become assets, as "Levels of engagement, that can be quantified, become the direct justification and product for funding and investment, giving a reality to the aspirations for democracy and access to the arts" (2011, 10). By focusing on engaging their audiences online, Shakespearean theatres and festivals have new ways to justify the cultural value of their work. However, using social media to engage audiences online, especially in an attention economy, is more easily said than done. Marketing teams have to understand the shifting affordances and limitations of each social media site they establish a presence on, for while those affordances "do not dictate participants' behavior" on these sites, "they do configure the environment in a way that shapes participants' engagement" (boyd 2010, 39–40). They cannot just hope to embrace the positive rhetoric of digital media; marketing teams and institutions have to account for both technological affordances and user practices on social media sites if they hope to facilitate meaningful forms of audience engagement (Rumbold 2010). This means embracing the real-time, persistent nature of social media, and working with the reality that Facebook and Twitter users are more focused on the present moment than on older posts or content.

In using social media as a marketing tool, Shakespearean theatres and festivals are participating in what W. B. Worthen terms Shakespeare 3.0, in which content is released from a single platform but accessed through any number of other platforms or media (2008, 60). Social media provide the means to distribute content to their audiences broadly, which can in turn create opportunities for online audiences to engage with these institutions. However, each theatre and festival has to address the fact that it is location-based and relies on audiences physically visiting its site. As they use social media to market themselves online, institutions have to consider how to navigate between encouraging online interactions and redirecting "audiences to identifiable, palpable sites of the real" (Huang and Ross 2009, 9). Any digital marketing and outreach theoretically serves to reinforce the cultural relevance of these institutions and persuade their audiences to invest in them through purchasing tickets and physically attending performances. However, marketing through social media allows performance institutions to craft their digital presences to engage audiences and attract their attention to the cultural impact of their current work without always redirecting the audiences' attention back to the physical sites of the theater.

One way that theatres and troupes can engage their audiences online is through working to create online spaces that may develop into what James Paul Gee calls affinity spaces, where "people 'bond' first and foremost to an endeavor or interest and secondarily, if at all, to each other" (2007, 98). Social media sites can provide the groundwork for building such spaces, as institutions are able to create spaces where their online audiences can congregate around their work. To do so, theatres and festivals cannot operate their marketing through social media as they would with the website. A major aspect of affinity spaces is that they allow for users and audiences to interact—with the content, with the institutions, and with each other—in meaningful ways. This means that audiences play a critical role in the creation of affinity spaces, a role that relies on their interaction and engagement. Organizations that depend solely on the newness of social media to attract audiences to their outreach will find little success, because newness alone offers nothing new for audiences. The newness of social media and other digital technologies is not what competes for audiences' attention; it is what those new technologies are able to facilitate that is of value: meaningful interaction.

An emphasis on interaction in digital marketing is not a new concept, but the interactions that social media can facilitate between institutions and their audiences are. In theorizing interactivity, Marie-Laure Ryan discusses two types of interaction: exploratory and ontological (2001). Exploratory interaction allows users to navigate different sections of content, but does not allow for any deeper type of user engagement. An example of this is the Stratford Festival's Behind the Scenes app, created for the iPad and released in late 2012. The app provides a multimedia look behind the scenes of the Festival's 2011 season, with rehearsal images, video interviews, audio recordings, and computer-generated set images, all available in the app for users to explore. Although Behind the Scenes offers an interesting look into how the festival works, the content on the app remains the same as it was when the app was originally released. Not only will the novelty of the app wear off the older it gets, but there is little interaction that the app affords users beyond navigating the different types of content. So even though the app displays content with a high production value while trying to embrace the newness of the iPad as a platform for engaging audiences, it ultimately serves as just another method of content delivery instead of a deep, interactive experience for users.

Where social media can help institutions is in their ability to facilitate ontological interactions between them and their audiences. Ontological interaction, as Ryan discusses, is interaction that allows users to generate meaning and value in the interactive experience, and not just navigate content. Social media sites like Facebook, Twitter, and Pinterest are fertile ground for such interactions, as they have low barriers to participation for users and many members of the institutions' audiences are already active on these sites. When they enter into these online spaces to engage their audiences, though, theatres and festivals also have to provide their audiences with a reason to engage with them. To accomplish this goal, they cannot rely on a one-size-fits-all approach for their social media outreach. Institutions have to know who they are trying to target online if they want to compete for their audience's attention, which means recognizing that while social media provide "the opportunity to interact with large and diverse audiences—dozens, hundreds, thousands, and sometimes even millions of people," there is no guarantee as to the actual size of their social media audiences (Litt 2012, 332). There is potential for institutions to use social media to tap into large online audiences, but they have to know their target audiences if they hope to attract and keep the attention of those audiences through social media. While institutions have access to some demographic information, they target broader audiences that generally range in age from 20 to 55, representing the most likely groups that will use social media to engage with theatres and festivals online.

Social media provide Shakespearean performance institutions access to new audiences, but these audiences are not always unique. For example, the Globe averages more than 700,000 visitors to the physical site in a year; in 2011, they had more than 2 million visitors on their website, and in 2012, were mentioned by users on Twitter more than 100,000 times (Nelson 2014; Estelle-Petty 2013). While it would be great for the Globe if a seventh of their on-site visitors tweeted about their visits, or that each of the 2 million hits represented a new user interacting with the Globe's website, how much overlap occurs is unknowable. Audience overlap occurs across the various social media sites on which theatres and festivals are engaging their audiences as well. Between visitors to the physical sites and their digital audiences, organizations are reaching out to their audiences in more fragmented ways. Katherine Rowe has discussed the trade-offs between extensibility and audience fragmentation in new media studies, and theatres and festivals have to consider how their online audiences are already composed of different groups with varied interests when designing their digital outreach (2010,

65). Since one of their main goals is to cultivate and maintain their cultural relevance with online audiences through digital marketing, their outreach must offer something different from what they offer at their physical sites while accounting for the affordances and limitations of the social media sites on which they cultivate presences. If they target their audiences too broadly by delivering the same content across all platforms, or too specifically by designing a unique experience on one site that is completely inaccessible to their audiences on any other sites, they run the risk of losing their audiences' attention. If they hope to build affinity spaces around their work on social media sites, theatres and troupes have to avoid overextending, and, in the process of trying to offer something unique to everyone, failing to provide anything accessible or meaningful for anyone.

One of the most important things for which theatres and festivals can account is the portions of their online audiences that do not overlap with their physical ones. As Ryan Nelson points out, "a significant portion of the 'Globe audience' is comprised of people who only encounter its online presence" (2014). Building affinity spaces around their work on social media sites offers institutions a means to tap into audiences that have never visited their physical sites. If they are able to do so successfully, the reward is a means of cultivating their cultural value with their online audiences, many of whom may not have had points of contact with the institution before. By using social media to create ontological interactions for their audiences, organizations may be able to create such affinity spaces, uniting their online audiences around their shared interest in these organizations' work. The opportunity to build and maintain connections with their online audiences is integral for institutions working to maintain their cultural value and relevance. As Sylvia Morris argues, social media can help change organizational cultures and assist "in promoting internationalism and reducing fragmentation" (2014, 181). While theatres and festivals employ the newness of digital media to accomplish these ends (in terms of their recency and popularity), I would argue that the most successful digital marketing techniques are the ones that utilize not only the newness of digital media, but also the audience's nostalgia, to achieve their goals.

Selling Nostalgia

As they integrate social media into their marketing strategies, theatres and festivals are working to reassert Shakespeare and Shakespearean performance over social media as worthy of the audience's attention. Institutions can tap into the potential offered by social media to facilitate ontological interactions with their audiences and engage them online. There is a real benefit for organizations that can accomplish this, as Morris argues: "Personal blogs, Facebook pages and Twitter accounts require a more friendly communication style. Routinely acknowledging responses, for instance, and taking the time to follow up a comment can reap rewards both in enlarging readership and providing new subjects for research" (2014, 183).

To court their audiences' attention, the OSF, RSC, Globe, and Stratford have all turned to a common approach to facilitate interaction with their audiences: nostalgia. Defined by Susan Bennett, "nostalgia is constituted as a longing for certain qualities and attributes in lived experience that we have apparently lost, at the same time as it indicates our inability to produce parallel qualities and attributes which would satisfy the particularities of lived experience in the present" (1996, 5). For theatres and festivals, nostalgia is a powerful tool to engage their audiences through social media by drawing attention to the history of the institutions and their work. In doing so, they link their current relevance to their past work and their heritage and tradition as Shakespearean performance institutions. It is through using the past that theatres and festivals strive to remain relevant in the present while working to secure their futures.

A common example of this on Twitter (and to a lesser extent on Facebook) is each institution's participation in Throwback Thursday, a weekly practice in which users share older images with their networks every Thursday, usually accompanied by a brief explanation of the content and the hashtags #tbt or #throwbackthursday. The practice is itself nostalgic in nature, as users re-experience events by sharing older images with their networks. For theatres and festivals, Throwback Thursday is a user practice they can easily reproduce on their own accounts to perform their relevance by sharing images of their past work to draw attention to their current productions or events. The Stratford Festival, in particular, shares numerous Throwback Thursday posts on their accounts. Many of their updates showcase images from previous performances of current or upcoming productions, or of cast and crew members that were part of the Festival in either the past or the present. For instance, several Throwback Thursday posts from 2014 are images of Colm Feore's work with the Festival throughout his career, which celebrated his past work and contribution to the Festival while simultaneously advertising his current performances at the time in *King Lear* and *The Beaux' Stratagem* (see Figure 1). Others showed side-by-side images of past and current performances, highlighting current productions and rooting them in the longer history of the Festival and its work (see Figure 2). However, it is debatable whether institutions gain from participating in Throwback Thursday. Sharing older images with their online audiences every Thursday offers the potential for synergistic moments, but most Throwback Thursday posts tend to promote little audience interaction. Although they attempt to pull on the nostalgia of past productions to instill value and relevance in current ones, such outreach tends to target a smaller portion of their online audience: those who have some investment in these past productions or are able to see their current ones. Such outreach excludes a larger portion of the audience and falls short of engaging them in part because of the narrow focus, and in part because it is too invested in nostalgia for the institution's work. Although organizations commonly use social media for this type of outreach, posts like this do not ultimately encourage much audience interaction.

Figure 1. Throwback Thursday post featuring Colm Feore on the Stratford Festival's Twitter Feed (http://www.twitter.com/stratfest).

Staying Relevant: Marketing Shakespearean Performance through Social Media   355

Figure 2. Throwback Thursday post juxtaposing *Hay Fever* performance images from 1977 and 2014 on the Stratford Festival's Twitter feed (http://www.twitter.com/stratfest).

Much as they rely on their performance histories in practices like Throwback Thursday, all four institutions also use the national histories integral to their identities to attempt to engage their audiences online. Such content helps each institution to reassert its national identity for its primary audience, even as it attempts to reach out to an international audience online. For example, the OSF shared a recent post on its Facebook page regarding Ashland's annual Martin Luther King, Jr. Holiday Celebration (see Figure 3). While the update may seem to have little to do with Shakespeare directly, the OSF used the post to appeal to their American audience by honoring Dr. King's legacy. Although not necessarily related to the majority of the Festival's work, the event does speak to the OSF's use of the "cultural richness of the United States" as a source of inspiration, as outlined in their

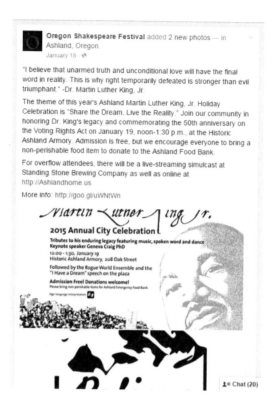

Figure 3. Post advertising Ashland, Oregon's Martin Luther King, Jr. Holiday Celebration on the Oregon Shakespeare Festival's Facebook page (https://www.facebook.com/shakespearefestival).

institutional mission statement.[3] Similar practices can be found in the social media outreach of other institutions, such as the Globe's promotion of their upcoming performance of *King John* in their 2015 season to coincide with the 800th anniversary of the signing of the Magna Carta (see Figure 4). Just as they use their own traditions and histories as means of adding value to their institutional brands, these theatres and festivals also rely on the nostalgia

---

[3] The OSF's full mission statement is: "Inspired by Shakespeare's work and the cultural richness of the United States, we reveal our collective humanity through illuminating interpretations of new and classic plays, deepened by the kaleidoscope of rotating repertory." More about the OSF's mission can be found at https://www.osfashland.org/about/what-is-osf.aspx.

Staying Relevant: Marketing Shakespearean Performance through Social Media   357

linked to their national histories and traditions to appeal to their audiences and keep them engaged with their work. Although these appeals to national identity often tend to focus on information or content delivery rather than deeper levels of audience interaction, they do serve as a means for organizations to locate the cultural value of their work in their institutional brands and identities.

Figure 4. Post advertising *King John* as a celebration of the 800th anniversary of the Magna Carta's signing on the Globe's Facebook page (https://www.facebook.com/ShakespearesGlobe).

As they use nostalgia to appeal to their audiences through social media, theatres and festivals have to address a key issue regarding how their audiences interact with their digital outreach. To move onto social media and allow audiences to participate in the process of creating meaning around an institution and its work requires embracing spreadability and relinquishing some control over content and audience interactions. Such a move presents a risk for institutions, as they are providing their audiences with the opportunity to co-create the narrative of these institutions' relevance and cultural value. The results of allowing audiences into the process may have the potential to backfire if the audiences go against the institutional narrative, but as Morris argues above, there are benefits to letting audiences in and engaging with them. In discussing the use of nostalgia in storytelling at the Riverside Museum in Glasgow, Kirsty Devine concludes:

> We need to question what and why we are being told something and equally what has been missed out. In this way, we avoid selective editing out of negative or unpalatable subject matter. That said, the role of nostalgia does not mean a complete fabrication of previous events or indeed the smoothing out of the rough edges of history. Combined with other primary source material, it helps us to understand how people make sense of both their individual and collective pasts and what they judge as important to them. (2013, 7)

Theatres and festivals want to maintain control of their content so they can shape the narratives around their culture value and smooth out the rough edges, especially if a particular performance or experience is or was not well received by their audiences. The stickiness model gave institutions the power to shape the content so it was received in a particular way, but spreadability allows for content to move out of institutional control and be used in ways that may run counter to their intentions or desires.

Thus, to open up the meaning-making process and allow audiences to engage with their content in meaningful ways through social media requires a good deal of trust. If theatres and festivals are serious about attracting their audiences' attention in order to create affinity spaces around their work, they can use social media to bring the audience into the meaning-making process. Although this move presents risk, it allows institutions to add key perspectives to the narratives of their cultural value, that of their audiences. By shifting their approach to interacting with their audiences through social media, institutions can work to compete against other institutions and media for their audiences' attention. This approach can be seen in some of the updates from the RSC's Facebook page. Many of the updates focus on sharing content, such as Throwback Thursday posts, or are informative in nature, whether it is about the RSC's website being down or providing details on an upcoming production. These posts tend to receive some feedback from the

RSC's Facebook audience in the form of likes or brief comments, but for the most part they do little to engage audiences. Some posts, however, begin to go a bit further, asking questions of the audience about current and upcoming performances and events, such as, "Which cinema will you be watching from on 3 September?" or, with regard to their 2014 production of *The Roaring Girl*, "Did you see the show? Tell us your favourite Moll moment?" (see Figure 5). These types of posts tend to generate a bit more interaction from the audiences in the form of comments because they are slightly more open-ended, but they still do not always present moments for users to engage with the RSC's content in an extensive way because they tend to target narrower segments of their online audience.

Figure 5. Post calling for audience feedback on the 2014 production of *The Roaring Girl* on the RSC's Facebook page (https://www.facebook.com/thersc).

Throughout the posts made by the RSC on their Facebook page in 2014, there is one that stands out among the others. In early June 2014, the RSC posted a rather simple status update asking, "What are your favourite Stratford summertime memories?" accompanied by the hashtag #RSCSummer (see Figure 6). The post received 483 likes, 116 user comments, and was shared by users 52 times. Although other posts have generated similar numbers of likes and shares, no other post has generated anywhere close to the number of user comments that this post received. There are a few reasons that this particular post was able to generate such a large amount of user interactions. First, instead of focusing on advertising something specific, it focuses more broadly on the audiences' experiences with the RSC and Stratford-upon-Avon. It asks a simple question that allows for a multitude of responses; by structuring the post in this way, the RSC demonstrates that it is more interested in what its audiences have to say than generating positive feedback about a specific performance or production. It also offers a relatively low barrier for audiences to participate, as users do not need to have a deep level of engagement with Shakespeare or his works, nor do they need to have attended a current RSC production to be able to participate in the conversation. This post succeeds by allowing for an ontological interaction between the institution and the audience, as the multitude of audience responses work to co-create meaning in the interaction through the individual experiences of users. Simply put, it was easy for users to engage with this post because the question it posed was both personal and open-ended.

Where the #RSCSummer post truly succeeded was in its ability to tap into the audiences' nostalgia. The post modeled the type of meaningful interactions that can occur in an affinity space by generating a rather large set of individual responses that reflected back positively on the RSC and its cultural value. I do not want to argue that this one post provides an easily replicated model for theatres and festivals to implement by just asking general, open-ended questions to the audience, but that it utilizes the audience's nostalgia as a means of engaging them in an ontological interaction focused on their experiences with Stratford-upon-Avon and the RSC. The response around this post exemplifies the potential social media offer to institutions, allowing them to establish affinity spaces around their work and cultural value. The post also highlights the social nature of constructing meaning through an appeal to nostalgia; as Bennett argues, "how we construct and engage memories cannot be seen as an individualized act but, instead, something prepared by the dissemination of a collective history and lodged in the physical selves of its subjects" (1996, 9). In responding to this post, users came together and revealed a collective history through their individual experiences; by sharing their stories, they created narratives around their experiences with the RSC and Stratford-upon-Avon, narratives that spoke to the impact and value of the institution among its online audience.

Figure 6. Post asking individual users to post their favorite memories of Stratford-upon-Avon in the summer on the RSC's Facebook page (https://www.facebook.com/thersc).

Another recent example of this approach, albeit on a much larger scale, was #LoveTheatre Day on November 19, 2014, organized by the Guardian Culture Professionals Network, Twitter UK, and the CultureThemes blog. In this project, "arts organisations, practitioners, artists, actors and audiences, but in particular, those working on the front and back-lines of our beloved theatres" were asked to participate through Twitter by sharing why they loved theatre.[4] The main hashtag was #LoveTheatre, but throughout

---

[4] For more, see Matthew Caines, "Introducing #LoveTheatre Day," *The Guardian*,

the day three other hashtags, each meant to focus on a different theme, were used: #BackStage, #AskATheatre, and #Showtime. Created to bring exposure to the work theatres do as they faced deep funding cuts, it allowed individuals from all areas of theatre to come together and discuss the reasons for their love and value of theatres and their work. Although it originated in the United Kingdom, institutions from around the world participated, and the OSF, RSC, Globe, and Stratford all took part in the event online throughout the day (see Figures 7–10). #LoveTheatre Day opened up the dialogue between institutions and audiences, with participants from both sides coming together to discuss why they loved theatre. On a broader scale, institutions and audiences came together to explain the value of theatre in their lives, co-creating a narrative regarding the value of theatre in general. The hashtag also led to discussions about the work and value of individual theatres and festivals, and the Twitter feeds for each institution were devoted to sharing posts and responding to users throughout the day. Events such as #LoveTheatre Day allow theatres and festivals to tap into their audiences' past experiences and use those experiences as proof of the cultural impact of their work. As they do so, they open up the narratives of their cultural value to include both practitioners and audiences in the process.

Figure 7. #LoveTheatre post shared on the OSF's Twitter feed (https://twitter.com/osfashland).

November 5, 2014.

Staying Relevant: Marketing Shakespearean Performance through Social Media 363

Figure 8. #LoveTheatre post shared on the RSC's Twitter feed (https://twitter.com/thersc).

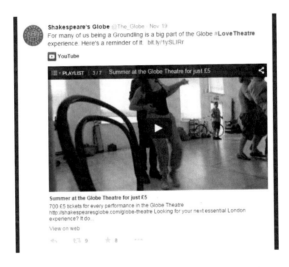

Figure 9. #LoveTheatre post shared on the Globe's Twitter feed (https://twitter.com/the_globe).

Figure 10. #LoveTheatre post shared on Stratford's Twitter feed (https://twitter.com/stratfest).

Through turning to nostalgia and the past to engage their audiences via social media, theatres and festivals are able to show how their audiences' past experiences emphasize their cultural value and relevance in the present. However, since theatres and festivals are, as Douglas Lanier states, "devoted to the performance of [Shakespeare's] works in the present," (2002, 145) how can they utilize nostalgia in more targeted ways to engage their audiences in their current work? These four organizations all strive to fashion themselves as institutions that perform work their audiences can relate to in order to remain relevant, but how can they use social media to convey that relevance, especially for their online audiences who may not be able to see those productions live? While one way is to appeal to audiences through tapping into the nostalgia related to their broader experiences and histories, another way to utilize nostalgia is by focusing more on the recent past. Whereas the #RSCSummer post allowed audiences to talk about their experiences with the RSC and Stratford-upon-Avon over the years, the Globe used nostalgia in a different and more focused way in the Bankside Bottoms event that accompanied their 2013 production of *A Midsummer Night's Dream*. As they celebrated the opening of *Midsummer* that year, the Globe posted a large blue banner on Bankside, which included Bottom's ears from his transformation into an ass. Anyone who visited the Globe in person during the event could have their picture taken and tweet it to the Globe using the hashtag #BanksideBottoms; the Globe would then share their pictures via Twitter and

Pinterest (see Figure 11).[5] More than 170 people participated in the event, and over two years later their Bankside Bottoms are still available to view on Pinterest. Like the #RSCSummer post, Bankside Bottoms had a low barrier for participation, although it required users to visit the Globe in person to participate in the event.

Figure 11. Pinterest board for Bankside Bottoms on the Globe's Pinterest Account (https://www.pinterest.com/the_globe/bankside-bottoms).

Unlike the response generated by the single RSC post, the participation in Bankside Bottoms spanned from May to September 2013. Even though Bankside Bottoms engaged audiences over a longer period of time than the #RSCSummer post did, it was a more focused experience that pulled on short-term nostalgia to engage audiences. Through Bankside Bottoms, the Globe took advantage of the immediacy of social media to allow members of its audience to participate and co-create meaning in the event. The Globe, like the RSC, had to trust its audience to participate in Bankside Bottoms in ways that would not run counter to the Globe's mission or reputation, especially considering the freedom users had in taking and sharing their individual images. At first glance, Bankside Bottoms may not seem nostalgic, given that it was centered on users actively participating in the event during the 2013 production and audiences could view these photos before attending a production of the play, or without ever visiting the Globe itself. However, when looking at the event

---

[5] The Globe's board can be viewed at http://www.pinterest.com/theglobe/bankside-bottoms.

after the fact, whether considering the experience of an individual user or the event as a whole, there is a use of nostalgia, although that nostalgia is for the recent past rather than for a time long gone. Nostalgia for the recent past can play as important a role as nostalgia for the more distant past, especially for institutions concerned about their relevance with audiences in the present. As Pat Gill argues, "The recent past, one that seems much simpler and slower and still (almost) graspable, becomes a reassuring construct that allows for a stabilizing self-definition in the present" (1997, 164). Whereas the #RSCSummer post focuses on the RSC's relevance with its audiences by asking a broad question that could be answered through any number of individual experiences, Bankside Bottoms facilitated the participation of the Globe's audience by centering on a particular experience. For online audiences, once an image was posted to the Globe's Twitter and Pinterest accounts, it became an immediate reminder of their experiences at the Globe, and through their participation they shaped the Bankside Bottoms event and its success online whether they participated directly in the event or not.

As individual users shared their Bankside Bottoms with their Twitter followers, they extended the event's reach to their own networks, providing the Globe with a greater amount of exposure than it would have received if the images had been shared only via its institutional accounts. The event also presented the Globe with an opportunity to use its audience's experiences and images to shape a narrative of cultural value, one that focuses on its relevance in the present by utilizing the recent past. The recent audience experience with Bankside Bottoms provided a means for the Globe to stabilize and reinforce its relevance in the present through the participation of the audience. Instead of using nostalgia to focus on the theatre's history and tradition, the Globe used the recent past as an indication of its institutional value through engagement, as well as innovation. Bennett discusses the drive behind "staging old texts to explore the possibilities of performance in the present," noting how such stagings "rely on willing audiences who recognize and are nostalgic for the classical text but who are attracted to the event for its innovation with and renovation of that text" (1996, 20). Bennett's line of thinking here can also be applied to the social media outreach of theatres and festivals and their usage of nostalgia, recent or not. As institutions continually work to build narratives of cultural value that rely on their relevance and impact, they need to continually provide an innovative theatrical experience. The relevance of Shakespearean performance institutions relies as much on their work with Shakespeare as on their ability to re-present the performance experience to their online audiences using innovative methods.

Future Directions

While using nostalgia to construct their relevance through longer traditions and histories is an important part of the cultural value of the Oregon Shakespeare Festival, the Royal Shakespeare Company, Shakespeare's Globe, and the Stratford Festival, they also have to utilize the recent past as a means of justifying their continued relevance with audiences. As evidenced by the examples discussed in this chapter, nostalgia can be an effective tool for engaging audiences through social media, particularly to encourage ontological interactions between institutions and audiences. By turning to the past, organizations can engage their audiences in the present, and must do so if they hope to compete for their audience's attention through social media. Theatres and festivals have to embrace the real-time nature of social media and adjust to the "fundamental shift from the static archive toward 'flow' and the 'river'" (Lovink 2011, 11) as they develop digital outreach strategies to engage their audiences. If cultural institutions utilize the newness of social media to engage their online audiences in tandem with an appeal to their audience's nostalgia, instead of privileging one over the other, they can configure themselves as institutions that provide interactive, engaging experiences focusing on their audiences, and not just the institutions themselves.

#Throwback Thursday, the Martin Luther King, Jr. and Magna Carta posts, #RSCSummer, #LoveTheatre Day, and Bankside Bottoms are all examples of social media outreach that show some of the ways in which Shakespearean performance institutions are reimagining their approaches to digital marketing. Rather than relying on the stickiness model and delivering content, theatres and festivals are finding success by tapping into the potential of spreadability through social media. While these various examples offer potential models for future outreach, however, digital outreach that engages audiences in ontological interactions is still the exception, not the rule. Theatres and festivals need to continue to work on new approaches to engaging their audiences through social media, and through other digital technologies, especially if they want to remain competitive in the attention economy. Future outreach can focus less on individual productions or performances and more on embracing the active role of audiences in the entire experience, allowing them to take control and co-create meaning in the narratives of cultural value integral to these institutions' continued relevance. Although institutions may not always be able to control where and how their content is interacted with or spread by the audience, they can realize and take advantage of the fact that as content is circulated among social media networks, value is added to that content in the process (Lash and Lury 2007, 4–5).

By shifting their approaches to digital outreach, Shakespearean performance institutions can work to remain competitive in the attention economy, and in doing so build their individual brands around the experiences they offer their audiences, digital or otherwise. As they work to build their individual brands around Shakespeare and the performance experience, their work and digital outreach adds to the overall "*impression* of a Shakespeare brand" (Rumbold 2011, 26). Through their different approaches to digital outreach, the OSF, RSC, Globe, and Stratford all participate in the larger process of strengthening Shakespeare's value with their audiences through their work. For institutions competing for their audiences' attention in order to remain culturally relevant, they can all stand to benefit from the outreach of the others in revaluing Shakespeare. The newness of digital media and the nostalgia of audiences offer two tools for gaining that attention in order to revalue Shakespeare and Shakespearean performance. However, it is up to the institutions themselves to embrace these new methods of outreach, and trust in their audiences to be active participants in the process of making these institutions, and Shakespeare, continually relevant.

Although social media are only one aspect of the network of marketing practices that theaters and festivals use to shape the cultural value and impact of their work, the use of sites such as Facebook, Twitter, and Pinterest to facilitate institution–audience interactions is quickly becoming a standard and widespread practice. The attention economy of social media offers fertile ground for the scholarly community to understand how Shakespeare's cultural value is being established and maintained by theaters and festivals as they compete for the attention of their online audiences. Not only do social media make audience–institution interactions readily and continuously accessible for study, they can also offer insight into the ways in which the cultural impact of Shakespeare and Shakespearean performance is shaped through institutions' various online presences. As social media allow audiences to engage with institutional content in more active ways, greater attention needs to be paid to the shifting dynamics of audiences' online participation. Institutions can reach larger audiences than they have been able to in the past, but audiences are also able to make their voices heard online and more easily exert pressure on institutions and their work. Even though the newer technologies and strategies utilized by Shakespearean performance institutions encourage audiences to participate and engage with institutions' work, how successful these approaches are and how they contribute to the larger narratives regarding the cultural value of Shakespeare and Shakespearean performance warrants greater consideration in the future.

WORKS CITED

Baine, Wallace. 2013. "UCSC Arts Dean Defends Closing Shakespeare Santa Cruz." *Santa Cruz Sentinel*, August 29. http://www.santacruzsentinel.com/santacruz/ci_23977879/ucsc-arts-dean-defends-closing-shakespeare-santa-cruz.

Belfiore, Eleonora, and Oliver Bennett. 2007. "Rethinking the Social Impacts of the Arts." *International Journal of Cultural Policy* 13 (2): 135–51. doi:10.1080/ 10286630701342741.

Bennett, Susan. 1996. *Performing Nostalgia: Shifting Shakespeare and the Contemporary Past.* New York: Routledge.

boyd, danah. 2010. "Social Network Sites as Networked Publics: Affordances, Dynamics, and Implications." In *Networked Self: Identity, Community, and Culture on Social Network Sites*, edited by Zizi Papacharissi, 39–58. New York: Routledge.

Bristol, Michael D. 1996. *Big-time Shakespeare.* London: Routledge.

Caines, Matthew. 2014. "Introducing #LoveTheatre Day." *The Guardian*, November 5. http://www.theguardian.com/culture-professionals-network/2014/nov/05/love-theatre-day-twitter-uk.

Chan-Olmsted, Sylvia M. 2002. "Branding and Internet Marketing in the Age of Digital Media." *Journal of Broadcasting and Electronic Media* 46 (4): 641–45.

Devine, Kirsty. 2013. "Removing the Rough Edges?: Nostalgia and Storytelling at a UK Museum." *Consumption Markets & Culture* 17 (2): 208–14. doi:10.1080/10253866.2013.776311.

Edmondson, Paul, and A. J. Leon. 2014. "Changing a Culture with the Shakespeare Birthplace Trust: Championing Freedom and Democracy." In *Shakespeare and the Digital World: Redefining Scholarship and Practice*, edited by Christie Carson and Peter Kirwan, 193–201. Cambridge: Cambridge University Press.

Gaffney, Anita. 2013. Interview by Geoffrey Way. Phone interview. May 7.

Gee, James Paul. 2007. *Good Video Games + Good Learning: Collected Essays on Video Games, Leaning and Literacy.* New York: Peter Lang.

Gill, Pat. 1997. "Technostalgia: Making the Future Past Perfect." *Camera Obscura: A Journal of Feminism, Culture, and Media Studies* 40–41: 163–80.

Hedrick, Donald. 2010. "Forget Film: Speculations on Shakespearean Entertainment Value." In *The English Renaissance in Popular Culture: An Age for All Time*, edited by Greg Colón Semenza, 199–216. New York: Palgrave Macmillan.

Henderson, Diana. 2002. "Shakespeare: The Theme Park." In *Shakespeare After Mass Media*, edited by Richard Burt, 107–26. New York: Palgrave.

Huang, Alexander C.Y., and Charles S. Ross. 2009. *Shakespeare in Hollywood, Asia, and Cyberspace*. West Lafayette, IN: Purdue University Press.

Jenkins, Henry, Sam Ford, and Joshua Green. 2013. *Spreadable Media: Creating Value and Meaning in a Networked Culture*. New York: New York University Presos.

Lanham, Richard A. 2006. *The Economics of Attention: Style and Substance in the Age of Information*. Chicago: University of Chicago Press.

Lanier, Douglas. 2002. *Shakespeare and Modern Popular Culture*. Oxford: Oxford University Press.

Lash, Scott, and Celia Lury. 2007. *Global Culture Industry: The Mediation of Things*. Cambridge: Polity Press.

Litt, Eden. 2012. "Knock, Knock. Who's There? The Imagined Audience." *Journal of Broadcasting and Electronic Media* 56 (3): 330–45. doi:10.1080/08838151.2012.705195.

Lovink, Geert. 2011. *Networks Without a Cause: A Critique of Social Media*. Cambridge: Polity Press.

McLuskie, Kathleen. 2011. "The Commercial Bard: Business Models for the Twenty-First Century." Special issue, *Shakespeare Survey* 64: 1–12.

Minsky, Marvin. 1980. "Telepresence." *OMNI Magazine,* June. http://web.media.mit.edu/~minsky/papers/Telepresence.html.

Morris, Sylvia. 2014. "Gamekeeper or Poacher?: Personal Blogging/Public Sharing." In *Shakespeare and the Digital World: Redefining Scholarship and Practice*, edited by Christie Carson and Peter Kirwan, 176–86. Cambridge: Cambridge University Press.

Nelson, Ryan. 2014. "Developing a Digital Strategy: Engaging Audiences at Shakespeare's Globe." In *Shakespeare and the Digital World: Redefining Scholarship and Practice*, edited by Christie Carson and Peter Kirwan, 202–11. Cambridge: Cambridge University Press.

Petty, Sian-Estelle. 2013. Interview by Geoffrey Way. Phone interview. May 16.

Pierce, Mallory. 2013. Interview by Geoffrey Way. Personal interview. April 26. Oregon Shakespeare Festival, Ashland.

Rowe, Katherine. 2010. "Crowd-Sourcing Shakespeare: Screen Work and Screen Play in Second Life in After Shakespeare on Film." Special issue, *Shakespeare Studies* 38: 58–67.

Rumbold, Kate. 2010. "From 'Access' to 'Creativity': Shakespeare Institutions, New Media, and the Language of Cultural Value." Special issue, *Shakespeare Quarterly* 61 (3): 313–36.

——— 2011. "Brand Shakespeare?" *Shakespeare Survey* 64: 25–37.

Ryan, Marie-Laure. 2001. "Beyond Myth and Metaphor: The Case of Narrative in Digital Media." *The International Journal of Computer Game Research* 1 (1): http://www.gamestudies.org/0101/ryan/.

Smith, Abby. 2004. "Preservation." In *A Companion to Digital Humanities*, edited by Susan Schriebman, Ray Siemens, and John Unsworth, 576–91. Malden, MA: Blackwell.

Worthen, W. B. 2008. "Shakespeare 3.0: Or Text Versus Performance, the Remix." In *Alternative Shakespeares 3*, edited by Diana E. Henderson, 54–77. New York: Routledge.

Contributors

Anupam Basu is an Assistant Professor of English at Washington University in Saint Louis, where he is currently developing a web portal that seeks to make the EEBO-TCP corpus tractable for large-scale computational analysis. He is also working on a monograph on the representation of crime and social change in early modern England.

Craig A. Berry holds a Ph.D. in English from Northwestern University and has two forthcoming articles on Chaucer and Spenser. He has been working as a professional programmer since the 1980s, with intermittent forays into the digital humanities since the 1990s. He is currently Digital Projects Editor at *The Spenser Review* and serves on the Executive Committee of the International Spenser Society.

Mattie Burkert is an Assistant Professor in the Department of English at Utah State University. She received her Ph.D. from the University of Wisconsin-Madison, where her doctoral research was supported by a Mellon-ACLS Dissertation Completion Fellowship. She is currently at work on a book project that examines the relationship between public finance and the London theater in the decades following the 1688 Revolution Settlement. This project draws on insights from a prototype database of late-seventeenth-century performance records, which she is expanding across the long eighteenth century. Her work on George Eliot has appeared in *Romanticism and Victorianism on the Net*, and she has an essay on Colley Cibber forthcoming in *Modern Philology*.

Philip R. Burns is Senior Developer in Academic and Research Technologies at Northwestern University. His academic and professional background lies primarily in mathematical and statistical computing applied across a broad range of disciplines. His involvement in humanities projects reaches back to the early 1970s.

Timothy W. Cole is Mathematics Librarian and Coordinator for Library Applications in the Center for Informatics Research in Science and Scholarship at the University of Illinois at Urbana-Champaign. Prior appointments at Illinois include: Interim Head of Library Digital Services and Development, Systems Librarian for Digital Projects, and Assistant Engineering Librarian for Information Services. His research interests include metadata, Linked Open

Data, digital library interoperability, annotation systems, and XML-based library applications. He is Principal Investigator for *Exploring the Benefits for Users of Linked Open Data for Digitized Special Collections*, a research project funded by the Andrew W. Mellon Foundation. He is Co-PI for *Emblematica Online*.

Shawn DeSouza-Coelho recently completed his M.A. (English, Experimental Digital Media) at the University of Waterloo, exploring the intersections between narrative theory and the act of play in video games. He is a writer, theatre theorist/practitioner, and professional magician based in Toronto, Ontario. His book, *Metamagic: An Introduction*, released in 2013, explores the art of magic as a medium for discourse. Shawn is currently working on his next book: the authorized biography of retired stage manager Nora Polley.

Maciej Eder is Associate Professor at the Institute of Polish Studies at the Pedagogical University of Kraków, Poland, and at the Institute of Polish Language at the Polish Academy of Sciences. He is interested in European literature of the Renaissance and the Baroque, classical heritage in early modern literature, and scholarly editing (his most recent book is a critical edition of sixteenth-century Polish translations of the *Dialogue of Salomon and Marcolf*). A couple of years ago, while doing research on anonymous ancient texts, Eder discovered the fascinating world of computer-based stylometry and non-traditional authorship attribution. His work is now focused on a thorough re-examination of current attribution methods and their application to non-English languages, such as Latin and Ancient Greek.

Laura Estill is an Associate Professor of English at Texas A&M University, where she edits the *World Shakespeare Bibliography* (worldshakesbib.org) and *DEx: A Database of Dramatic Extracts*. Her specialties are early modern drama, print and manuscript culture, and digital humanities. She is the author of *Dramatic Extracts in Seventeenth-Century English Manuscripts: Watching, Reading, Changing Plays* (2015). Her articles and chapters have appeared in *Shakespeare Quarterly*, *Huntington Library Quarterly*, *Early Theatre*, *The Oxford Handbook of Shakespeare*, and *Shakespeare and Textual Studies*, among others. She has thoroughly enjoyed collaborating with Diane, Michael, and everyone who has worked hard to create this collection.

Myung-Ja ("MJ") K. Han is a Metadata Librarian/Associate Professor at the University of Illinois at Urbana-Champaign. Her main responsibilities consist of developing application profiles for digital collections and evaluating and enhancing cataloging and metadata workflows. Her research interests include interoperability of metadata, issues of bibliographic control in the digital library environment, and semantic web and linked data. She is Co-PI for *Emblematica Online*.

Jonathan Hope is Professor of Literary Linguistics at Strathclyde University in Glasgow. He has published widely on Shakespeare's language and the history of the English language. His most recent book, *Shakespeare and Language: Reason, Eloquence and Artifice in the Renaissance* (2010), seeks to reconstruct the linguistic world of Shakespeare's England and measure its distance from our own. With Michael Witmore (Folger Shakespeare Library), he is part of a major digital humanities project, funded by the Mellon Foundation, to develop tools and procedures for the linguistic analysis of texts across the period 1450–1800. Early work from this project is blogged at winedarksea.org.

Diane K. Jakacki is the Digital Scholarship Coordinator and Affiliated Faculty in Comparative Humanities at Bucknell University. Her research specialties include digital humanities methodologies as applied to early modern British literature and drama, and the ways in which pedagogy can be transformed by means of digital interventions. She is currently editing *Henry VIII* for the *Internet Shakespeare Editions*, and serves as the ISE Technical Editor; she is also editing *The Shoemaker's Holiday* for Broadview Press. Articles and essays have appeared in *Early Theatre*, *Medieval and Renaissance Drama*, and *Digital Scholarship in the Humanities*, among others. She is grateful to Laura and Michael for one of the most rewarding collaborative experiences of her career and looks forward to many more!

Janelle Jenstad is Associate Professor of English at the University of Victoria. She directs *The Map of Early Modern London* (MoEML) and serves as Associate Coordinating Editor of the *Internet Shakespeare Editions*, for which she is editing *The Merchant of Venice*. With Jennifer Roberts-Smith, she co-edited *Shakespeare's Language in Digital Media* (forthcoming from Taylor & Francis/Routledge). Her essays and book chapters have appeared in *Shakespeare Bulletin, Elizabethan Theatre, EMLS, JMEMS*, among others. For a full list, see janellejenstad.com.

Martin Mueller, Professor Emeritus of English and Classics at Northwestern University, is the author of a monograph on the *Iliad* and of *Children of Oedipus, and Other Essays on the Imitation of Greek Tragedy, 1550–1800*. He is also co-editor of the Chicago Homer and editor of WordHoard, both Web-based applications that focus on the analysis of deeply tagged texts.

Fabrizio Nevola is Chair and Professor of Art History and Visual Culture at the University of Exeter, specializing in the urban and architectural history of early modern Italy. He was involved in the exhibition *Renaissance Siena: Art for a City* (National Gallery, London, October 2007–January 2008), and is the author of *Siena: Constructing the Renaissance City* (2007). His edited volumes include *Locating Communities in the Early Modern Italian City* (2010), *Tales of the City: Outsiders' Descriptions of Cities in the Early Modern Period* (2012), and *Experiences of the Street in Early Modern Italy* (2013). He is PI for the digital humanities project *Hidden Florence*, published as an app on AppStore and GooglePlay in 2014, and accompanying website hiddenflorence.org.

Rebecca Niles is Digital Texts Editor and Interface Architect for *Folger Digital Texts*, and manages the Folger Shakespeare Library Editions. Rebecca has been involved in the development of various applications of the Folger Editions content, including the Folger Luminary Shakespeare apps for iPad. Rebecca holds Masters degrees from the University of Toronto in English literature (with a focus on textual scholarship), and Information Studies (concentrating on the development and use of digital tools and environments for expressing and studying pre-digital texts). She was also a researcher in the Implementing New Knowledge Environments (INKE) project (inke.ca).

Michael Poston is Digital Texts Editor and Encoding Architect for *Folger Digital Texts*. He is currently assisting the Folger Shakespeare Library's *Early Modern Manuscripts Online* (EMMO) and *Digital Anthology of Early Modern English Drama* transcription efforts. He has designed and developed numerous digital projects for the Folger, including scholarly resources such as the *Union First-Line Manuscript Index*, PLRE.Folger (*Private Libraries in Renaissance England*), a paleography transcription and collation site called Dromio, a tool for teaching the history of the book called Impos[i]tor, and digital displays for Folger Exhibitions.

Jennifer Roberts-Smith is an Associate Professor of Theatre and Performance at the University of Waterloo. Her research and creative practice focus on Elizabethan performance techniques, inter-medial theatre research methods, and experience design in the digital humanities. She is Principal Investigator of two multi-institutional research projects, *The Simulated Environment for Theatre (SET)* and *The Stratford Festival Online: Games and Virtual Learning Environments for Education and Audience Engagement*. She also serves as co-leader of the Interface cluster in the Implementing New Knowledge Environments (INKE) research project (inke.ca).

Paul J. Stoesser teaches Canadian theatre history and praxis-based scenography and theatre production in the graduate program of the University of

Toronto's Centre for Drama, Theatre and Performance Studies, where he is also Technical Director. His most recent collaboration with Roberts-Smith, DeSouza-Coelho et al. appears in *Shakespeare International Yearbook 2014*. Dr. Stoesser's teaching and research examines opsis in conjunction with the history and development of theatre production technology, especially regarding modern applications of Renaissance production techniques.

Jesús Tronch is Senior Lecturer at the University of Valencia. His research focuses on the transmission and editing of texts (especially of early modern drama) and the reception and translation of Shakespeare in Spain. He has published *A Synoptic 'Hamlet'* (2002), co-edited English-Spanish editions of *The Tempest* (1994) and *Antony and Cleopatra* (2001), and, with Clara Calvo, *The Spanish Tragedy* for Arden Early Modern Drama (2013). He is currently editing *Timon of Athens* for *Internet Shakespeare Editions,* and collaborating in an open-access, hypertextual, and multilingual collection of early modern European theatre within the ARTELOPE project at the University of Valencia.

Michael Ullyot is an Associate Professor of English at the University of Calgary, specializing in early modern literature and the digital humanities. He has published articles on anecdotes, abridgements, and Edmund Spenser. Current projects include a monograph on the rhetoric of exemplarity and a program that detects rhetorical figures of repetition and variation in literary texts. Co-editing this volume with Laura Estill and Diane Jakacki has been a distinct pleasure.

Mara R. Wade is a Professor in the Department of Germanic Languages and Literatures at the University of Illinois at Urbana-Champaign. Her research focuses on early modern studies, with special emphasis on court studies and emblematics. She is PI for *Emblematica Online* (funded by the National Endowment for the Humanities). Her most recent edited volumes include *Emblem Digitization: Conducting Digital Research with Renaissance Texts and Images, Early Modern Literary Studies,* Special Issue 20 (2012); with Sara C. Smart, *The Palatine Wedding of 1613: Protestant Alliance and Court Festival* (Wiesbaden, 2013); and *Gender Matters: Discourses of Violence in Early Modern Literature and the Arts* (Amsterdam, 2014). She is the editor of *Emblematica: An Interdisciplinary Journal for Emblem Studies.*

John N. Wall is a Professor of English at North Carolina State University. His fields of inquiry in research and scholarship include early modern English literature, religion and literature, and the digital humanities. He, along with David Hill and Yun Jing of NC State University, is Principal Investigator of the Virtual Paul's Cross Project and the Virtual St. Paul's Cathedral Project, both supported by Digital Humanities grants from the National Endowment for the Humanities. He is the author of *Transformations of the Word: Spenser, Herbert, Vaughan* (U of Georgia Press 1988), co-author, with John Booty and David Siegenthaler, of *The Godly Kingdom of Tudor England* (Morehouse 1981), and the editor of *George Herbert: The Country Parson and The Temple* (Paulist Press 1981). He is also the author of numerous articles and reviews which have appeared in *John Donne Journal, Studies in Philology, Anglican and Episcopal History, Renaissance Papers,* the *Journal of Digital Humanities,* and *Digital Studies/ Le Champ Numérique,* among others.

Geoffrey Way is a Ph.D. Candidate in Literature and an Instructor of English at Arizona State University. He has recently completed his dissertation, "Digital Shakespeares and the Performance of Relevance," which explores the myriad ways that Shakespearean theaters and festivals incorporate digital media into their marketing and performance practices.

Michael Witmore is Director of the Folger Shakespeare Library, Washington, DC. His publications include *Landscapes of the Passing Strange: Reflections from Shakespeare* (2010, with Rosamond Purcell); *Shakespearean Metaphysics* (2008); *Pretty Creatures: Children and Fiction in the English Renaissance* (2007); and *Culture of Accidents: Unexpected Knowledges in Early Modern England* (2001). With Jonathan Hope (Strathclyde University, Glasgow), he is part of a major digital humanities project, funded by the Mellon Foundation, to develop tools and procedures for the linguistic analysis of texts across the period 1450–1800. Early work from this project is blogged at winedarksea.org.